Technological Eco-Innovations for the Quality Control and the Decontamination of Polluted Waters and Soils

Technological Eco-Innovations for the Quality Control and the Decontamination of Polluted Waters and Soils

Special Issue Editors

Massimo Zacchini
Paras Ranjan Pujari

MDPI • Basel • Beijing • Wuhan • Barcelona • Belgrade • Manchester • Tokyo • Cluj • Tianjin

Special Issue Editors

Massimo Zacchini
Research Institute on Terrestrial
Ecosystems, National Research
Council of Italy
Italy

Paras Ranjan Pujari
National Environmental
Engineering Research Institute,
Council of Scientific and
Industrial Research
India

Editorial Office
MDPI
St. Alban-Anlage 66
4052 Basel, Switzerland

This is a reprint of articles from the Special Issue published online in the open access journal *Water* (ISSN 2073-4441) (available at: https://www.mdpi.com/journal/water/special_issues/Polluted_Waters_Soils).

For citation purposes, cite each article independently as indicated on the article page online and as indicated below:

LastName, A.A.; LastName, B.B.; LastName, C.C. Article Title. *Journal Name* **Year**, *Article Number*, Page Range.

ISBN 978-3-03928-464-1 (Pbk)
ISBN 978-3-03928-465-8 (PDF)

© 2020 by the authors. Articles in this book are Open Access and distributed under the Creative Commons Attribution (CC BY) license, which allows users to download, copy and build upon published articles, as long as the author and publisher are properly credited, which ensures maximum dissemination and a wider impact of our publications.

The book as a whole is distributed by MDPI under the terms and conditions of the Creative Commons license CC BY-NC-ND.

Contents

About the Special Issue Editors vii

Preface to "Technological Eco-Innovations for the Quality Control and the Decontamination of Polluted Waters and Soils" ix

Fabrizio Pietrini, Laura Passatore, Valerio Patti, Fedra Francocci, Alessandro Giovannozzi and Massimo Zacchini
Morpho-Physiological and Metal Accumulation Responses of Hemp Plants (*Cannabis Sativa* L.) Grown on Soil from an Agro-Industrial Contaminated Area
Reprinted from: *Water* **2019**, *11*, 808, doi:10.3390/w11040808 1

Gordana Medunić, Prakash Kumar Singh, Asha Lata Singh, Ankita Rai, Shweta Rai, Manoj Kumar Jaiswal, Zoran Obrenović, Zoran Petković and Magdalena Janeš
Use of Bacteria and Synthetic Zeolites in Remediation of Soil and Water Polluted with Superhigh-Organic-Sulfur Raša Coal (Raša Bay, North Adriatic, Croatia)
Reprinted from: *Water* **2019**, *11*, 1419, doi:10.3390/w11071419 11

Elżbieta Mierzejewska, Agnieszka Baran, Maciej Tankiewicz and Magdalena Urbaniak
Removal and Ecotoxicity of 2,4-D and MCPA in Microbial Cultures Enriched with Structurally-Similar Plant Secondary Metabolites
Reprinted from: *Water* **2019**, *11*, 1451, doi:10.3390/w11071451 25

Francesca Bosco, Annalisa Casale, Fulvia Chiampo and Alberto Godio
Removal of Diesel Oil in Soil Microcosms and Implication for Geophysical Monitoring
Reprinted from: *Water* **2019**, *11*, 1661, doi:10.3390/w11081661 41

Michael Santangeli, Concetta Capo, Simone Beninati, Fabrizio Pietrini and Cinzia Forni
Gradual Exposure to Salinity Improves Tolerance to Salt Stress in Rapeseed (*Brassica napus* L.)
Reprinted from: *Water* **2019**, *11*, 1667, doi:10.3390/w11081667 59

Inga Tabagari, Maritsa Kurashvili, Tamar Varazi, George Adamia, George Gigolashvili, Marina Pruidze, Liana Chokheli, Gia Khatisashvili and Peter von Fragstein und Niemsdorff
Application of *Arthrospira (Spirulina) platensis* against Chemical Pollution of Water
Reprinted from: *Water* **2019**, *11*, 1759, doi:10.3390/w11091759 81

Anna Wyrwicka, Magdalena Urbaniak, Grzegorz Siebielec, Sylwia Siebielec, Joanna Chojak-Koźniewska, Mirosław Przybylski, Aleksandra Witusińska and Petra Susan Kidd
The Influence of Bottom Sediments and Inoculation with Rhizobacterial Inoculants on the Physiological State of Plants Used in Urban Plantings
Reprinted from: *Water* **2019**, *11*, 1792, doi:10.3390/w11091792 89

Fabrizio Pietrini, Monica Carnevale, Claudio Beni, Massimo Zacchini, Francesco Gallucci and Enrico Santangelo
Effect of Different Copper Levels on Growth and Morpho-Physiological Parameters in Giant Reed (*Arundo donax* L.) in Semi-Hydroponic Mesocosm Experiment
Reprinted from: *Water* **2019**, *11*, 1837, doi:10.3390/w11091837 111

Barbara Casentini, Marco Lazzazzara, Stefano Amalfitano, Rosamaria Salvatori, Daniela Guglietta, Daniele Passeri, Girolamo Belardi and Francesca Trapasso
Mining Rock Wastes for Water Treatment: Potential Reuse of Fe- and Mn-Rich Materials for Arsenic Removal
Reprinted from: *Water* **2019**, *11*, 1897, doi:10.3390/w11091897 131

David Rossi, Anna Barra Caracciolo, Paola Grenni, Flavia Cattena, Martina Di Lenola, Luisa Patrolecco, Nicoletta Ademollo, Ruggiero Ciannarella, Giuseppe Mascolo and Stefano Ghergo
Groundwater Autochthonous Microbial Communities as Tracers of Anthropogenic Pressure Impacts: Example from a Municipal Waste Treatment Plant (Latium, Italy)
Reprinted from: *Water* **2019**, *11*, 1933, doi:10.3390/w11091933 . . . . **149**

Magdalena Urbaniak, Anna Wyrwicka, Grzegorz Siebielec, Sylwia Siebielec, Petra Kidd and Marek Zieliński
The Application of Different Biological Remediation Strategies to PCDDs/PCDFs Contaminated Urban Sediments
Reprinted from: *Water* **2019**, *11*, 1962, doi:10.3390/w11101962 . . . . **169**

Isabel Nogues, Paola Grenni, Martina Di Lenola, Laura Passatore, Ettore Guerriero, Paolo Benedetti, Angelo Massacci, Jasmin Rauseo and Anna Barra Caracciolo
Microcosm Experiment to Assess the Capacity of a Poplar Clone to Grow in a PCB-Contaminated Soil
Reprinted from: *Water* **2019**, *11*, 2220, doi:10.3390/w11112220 . . . . **185**

Rajshree Patil, Saurabh Levin, Samuel Rajkumar and Tahmina Ajmal
Design of a Smart System for Rapid Bacterial Test
Reprinted from: *Water* **2020**, *12*, 15, doi:10.3390/w12010015 . . . . **203**

About the Special Issue Editors

Massimo Zacchini, Ph.D., is currently working at the Research Institute on Terrestrial Ecosystems (IRET) of CNR. His main research interests regard the physiological and biochemical characterisation of plants useful for the decontamination of polluted soils and water (phytoremediation) and the ecotoxicological risk assessment of chemicals in plants. He is author and co-author of more than 60 scientific papers mainly published on International ISI journals. He has been involved in many research projects funded by UE, Italian and Spanish Ministry of Science and Research, and CNR Joint projects with foreign research Institutions. He organised International conferences and workshops dealing with innovative technologies for the decontamination of soil and water, also acting as invited speaker and chairperson in many International conferences on bioremediation and environmental pollution. He carried out academic activity as lecturer and member of the evaluation panels of European and national Ph.D., and tutoring activity for Ph.D. students, Erasmus+ students and European and national research fellows. He is member of the Editorial board of International scientific journals and referee for many International ISI journals dealing with phytoremediation, plant physiology, plant biochemistry and environmental science.

Paras R Pujari, Ph.D., is currently working as Principal Scientist, Critical Zone Research Group, in the CSIR-National Environmental Engineering Research Institute at Nagpur in India. His main research interests are in the Near Surface characterization by Geophysical Tools, Impact assessment of anthropogenic pollution on soil and groundwater, groundwater management. He is author and co-author of more than 35 research papers mainly published in SCI Journals. He has been involved in many research projects funded by the Government, EU, UNISEF, DST of Government of India, DBT Government of India. He was the Indian coordinator of the EU co-supported project TECO. He has organized national and International conferences. He is also Associate Professor in the AcSIR and is presently guiding Ph.D students. He is also on the Board of Studies and Research Advisory Board of many Universities in India. He is reviewer for many International journals dealing with Earth and Environmental Science.

Preface to "Technological Eco-Innovations for the Quality Control and the Decontamination of Polluted Waters and Soils"

The impact of industrial development on the environment is often severe and many times dramatically destructive. In these last years, the technologies based on a better understanding of natural processes and the discovery of new materials and analytical techniques have been further developed, offering a wide array of innovative solutions to recover critical situations due to the over-exploitation of the natural resources. Such technologies represent extraordinary tools for addressing the environmental challenges we are facing worldwide, in order proposing sustainable approaches to the issues related to the water and soil quality and the decontamination of polluted environmental matrices. In this regard, the Special Issue "Technological Eco-Innovations for the Quality Control and the Decontamination of Polluted Waters and Soils" offers an updated overview of the eco-innovative solutions, carried out both at field and laboratory scale, targeting different aspects of the environmental contamination such as the chemical nature of the pollutants, the characteristics of the substrates and the effects on living organisms. A particular emphasis is given to reports focused on the use of biological organisms to monitor the environmental quality and to decontaminate polluted water and soils. The Special Issue "Technological Eco-Innovations for the Quality Control and the Decontamination of Polluted Waters and Soils" is carried out within the dissemination activities of the European Union-India Research and Innovation Partnership "TECO" project, co-funded by the European Union.

Massimo Zacchini, Paras Ranjan Pujari
Special Issue Editors

Article

Morpho-Physiological and Metal Accumulation Responses of Hemp Plants (*Cannabis Sativa* L.) Grown on Soil from an Agro-Industrial Contaminated Area

Fabrizio Pietrini, Laura Passatore, Valerio Patti, Fedra Francocci, Alessandro Giovannozzi and Massimo Zacchini *

Research Institute on Terrestrial Ecosystems, National Research Council of Italy, Section of Montelibretti, Via Salaria km 29.300, 00015 Monterotondo, Italy; fabrizio.pietrini@cnr.it (F.P.); laura.passatore@iret.cnr.it (L.P.); ittapoirelav@gmail.com (V.P.); fedrafrancocci@gmail.com (F.F.); a.giovannozzi@tiscali.it (A.G.)
* Correspondence: massimo.zacchini@cnr.it; Tel.: +39-0690672537

Received: 1 April 2019; Accepted: 15 April 2019; Published: 18 April 2019

Abstract: Hemp is a promising plant for phytomanagement. The possibility to couple soil restoration to industrial crop cultivation makes this plant attractive for the management of contaminated sites. In this trial, *Cannabis sativa* L. plants were grown in a greenhouse on soils from two sites of "Valle del Sacco" (Lazio Region, Italy), a wide area contaminated by agro-industrial activities. One site was representative of moderate and diffuse metal(loid) multi-contamination, above the Italian concentration limit for agriculture (MC—moderately contaminated). The second site showed a metal(loid) content below the aforementioned limit, as a typical background level of the district (C—control). After 90 days, biometric and physiological parameters revealed satisfactory growth in both soil types. MC-grown plants showed a slight, but significant reduction in leaf area, root, and leaf biomass compared with C-grown plants. Chlorophyll content and chlorophyll fluorescence parameters, namely the quantum yield of primary photochemistry (F_v/F_m) and the Performance Index (PI_{ABS}), confirmed the good physiological status of plants in both soils. Metal(loid) analyses revealed that As, V, and Pb accumulated only in the roots with significant differences in MC- and C-grown plants, while Zn was found in all organs. Overall, preliminary results showed a satisfactorily growth coupled with the restriction of toxic metal translocation in MC-grown hemp plants, opening perspectives for the phytomanagement of moderately contaminated areas.

Keywords: chlorophyll fluorescence; industrial crop; metal tolerance; phytomanagement; phytoremediation; soil pollution

---

1. Introduction

In the agricultural soil, the overuse of fertilisers and waste release from industrial processes has resulted in a large number of contaminated sites over Europe requesting to be reclaimed, contributing to exacerbated land degradation problems [1]. A recent survey reported that in the European Union, approximately 340,000 contaminated sites are present, most of them polluted by metal(loid)s [2]. The awareness about the harmful effects produced by such contaminants on human health, through plant cultivation and animal breeding, is forcing the characterisation of different environmental restoration technologies, among which eco-sustainable ones are largely studied and successfully applied, especially in sites characterised by moderate and diffuse contamination [3,4]. In this regard, the phytomanagement is recognised as an effective approach to carry out a risk management strategy [5,6], being constituted by an array of gentle remediation options (GROs) technologies that can be applied as a part of integrated site risk management solutions. In this sense, the phytomanagement approach represents an application of the phytoremediation biotechnology, exploiting in a broader way the ecological benefits offered by plants. Specifically, phytomanagement relies on the choice of suitable plant species for the purpose of the management of the contaminated sites [7]. Among polluted sites, agricultural areas represent a particular concern for safe food production [8], forcing local administrations to restrict their exploitation with relevant loss of income for farmers. Therefore, particular attention is currently being paid to investigating the possibility of cultivating non-food crops on contaminated lands that, besides the capability to remediate soils, could satisfactorily grow and produce biomass and other bio-products for multiple profitable uses, avoiding the metal transfer to the food chain [9].

Because of its biological characteristics, such as rapid growth, high biomass production, wide root system, high genetic variability, remarkable ability to adapt to different environmental conditions, and low susceptibility to disease and pests [10,11], hemp (*Cannabis sativa* L.) is a plant species of notable interest for phytomanagement. In fact, because of the multiple non-food uses [12], it can offer a good opportunity to integrate soil recovery with the cultivation of a commercially exploitable resource. Specifically, hemp plants involved in phytomanagement strategies can profitably produce fibres that could be commercialised as insulating or composite material, cellulose materials from stem (suitable for packaging industry), and seeds representing a source of oil for biofuel production. The hemp potential for phytotechnologies has been poorly explored so far; most studies have focused on tolerance and accumulation of metals [13–17] and few reports deal with plant behaviour versus organic contaminants such as chrysene and benzopyrene [18] and radio-compounds [19]. Therefore, a better characterisation of the growth and physiological responses of hemp plants in metal-contaminated soil under controlled conditions is requested. In this context, the application of the chlorophyll *a* fluorescence analysis, which represents a rapid and non-destructive technique to analyze the changes at the physiological and biochemical level in the photosynthetic apparatus under stress conditions, and the evaluation of the chlorophyll content, also performed with non destructive methods, are particularly useful. Specifically, OJIP fluorescence transient analysis, known as the JIP test [20], has been developed for the quantification of several phenomenological and biophysical expressions together with the energy flux parameters of photosystem II (PS II), and may be used to assess metal stress on plants in vivo [21]. This approach could be of relevance for a successful screening of plant materials for phytoremediation. In the present study, an ex situ pot experiment was performed with soil sampled from the "Valle del Sacco" area. It is a contaminated site recognised as a National Interest Site by the Italian Ministry of Environment (L. 248/2005) for its huge extension (7235 ha) and diffused contamination by metals and organics that prevents any food crop cultivation and grazing, which formerly characterised the area. This preliminary experiment, performed in a greenhouse, aimed to investigate the growth potential of hemp plants on moderately metal-contaminated soils, evaluating the presence of the pollutants in the biomass for possible exploitation as a bio-resource.

2. Materials and Methods

2.1. Plant Material and Growth Conditions

Seeds of *Cannabis sativa* L. (cv. Codimono, an Italian monoecious variety cultivated for fiber production) were placed in plates filled with moistened agriperlite in darkness at room temperature (R.T.) until germination occurred. Then, germinated seeds were transferred in 11 L pots filled with soil collected from the topsoil (up to 25 cm after removing 5 cm top layer) of the two sampling points (41°44.245′ N, 13°00.732′ E; 41°44.263′ N, 13°00.875′ E) within the area of "Valle del Sacco" polluted site (Municipality of Colleferro, Rome, Italy), characterised by different metal concentration levels.

The first soil was sampled as representative of the moderate and diffuse multi-contamination by metal(loid)s, specifically Pb, V, As, and Zn (as highlighted by a survey previously carried out within the activities of the Commissioner Officer for the "Valle del Sacco" National Interest Site, ref.

07384-022R01E03/2006 by Studio Geotecnico Italiano SRL), slightly above the Italian concentration limit for green areas, except for Zn (namely, 20 ppm for As, 100 ppm for Pb, 90 ppm for V, 150 ppm for Zn). The second one was chosen for its metal(loid) content below the aforementioned limit, as a typical background level of the district. The two soils were referred as moderately contaminated (MC) and control (C) soil, respectively.

Agronomic characteristics were analysed by a certified environmental analysis laboratory following the guidelines of the Italian Ministry of Agriculture and Forestry (DM 13/09/99) [22] as an Italian official reference for soil chemical analyses. The total nitrogen content was evaluated by dry combustion in an elemental analyser (Met. XIV.1, DM 13/09/99); the determination of available phosphorous was performed following the Olsen method, based on bicarbonate extraction and photometric detection (Met. XV.3, DM 13/09/99); potassium and magnesium content was determined by atomic-absorption spectrophotometry (Met. XIII.5, DM 13/09/99). Inorganic and organic contamination of soil samples was assessed by a certified environmental analysis laboratory based on a previous survey carried out within the activities of the Commissioner Officer for the "Valle del Sacco" National Interest Site (see above). Organic contaminant levels were far below the Italian legal limit (as for DL n.152/06) [23], in most cases even under the detection limits, and thus organic compounds were not considered in this study (data not shown). The experiment was set up in a completely randomized design with four replicates (pots) for each of the four variants (C-soil with plant, C-soil without plant, MC-soil with plant, MC-soil without plant). Pots were placed during summertime (June–August) in a greenhouse under natural photoperiod (about 14 h), with mean (night–day) temperatures of 21.1–28.8 °C and relative humidity of 50%–60%; four pots per treatment (C and MC) were prepared. Pots were sealed at the bottom to avoid metal leaching. Plants were irrigated daily by supplying the water loss for evapotranspiration to maintain 50% of the water holding capacity, evaluated before starting the experiment for each kind of soil. The water supply for each treatment was determined daily by weighing each pot using an electronic balance (SB 32,001 Delta Range; Mettler-Toledo Inc., Columbus, OH). Total cumulative transpiration was calculated by subtracting, for the whole cultivation period, the daily amount of water loss in C- and MC-soil without plant from that in C- and MC-soil with plant, respectively. No fertilisation and pest management was carried out. At the end of the growth period (90 days), during seed production, after measuring leaf area (LI 3100, LI-COR Inc., Lincoln, NE, USA) and other biometric and physiological parameters, plant organs (root, stem, leaves, and inflorescence) were harvested and stored in an oven at 60 °C until a constant weight was reached.

2.2. Chlorophyll Content Determination and Chlorophyll Fluorescence Analysis

At the end of the experiment, total leaf chlorophyll content was estimated by a SPAD-502 Chlorophyll meter (Minolta Inc., Osaka, Japan), as reported by Pietrini et al. [24]. The measurements were taken from at least two fully developed leaves per plant. Four SPAD readings were taken from the widest portion of the leaf lamina, while avoiding major veins. The four SPAD readings were averaged to represent the SPAD value of each leaf. SPAD values were converted to chlorophyll content (μg cm^{-2}) using the equation by Cerovic et al. [25]; that is, chlorophyll content = (99 × SPAD value)/(144 − SPAD value).

On the same leaves chosen for SPAD readings, the chlorophyll fluorescence transient (OJIP transients) was measured using a plant efficiency analyser (PEA, Hansatech Instruments Ltd., King's Lynn, UK). The measurements were performed on leaves that were previously adapted to the dark for 60 minutes for the complete oxidation of the photosynthetic electron transport system, and the fluorescence intensity was measured for 1 s after the application of a saturating light pulse of 3000 μmol m^{-2} s^{-1}. For each experimental treatment, at least eight measurements were performed. The quantum yield of primary photochemistry (F_v/F_m) and performance index (PI_{ABS}) of photosystem II (PSII) were determined as described by Strasser et al. [20].

2.3. Metal(loid) Determinations

Soils and oven dried plant samples were finely ground (Tecator Cemotec 1090 Sample Mill; Tecator, Hoganas, Sweden), weighed, and mineralized. Mineralization was performed by microwave assisted acid digestion (U.S. EPA. 2007) by a certified environmental analysis laboratory. Specifically, the method EPA 3051A/2007 was used. Briefly, samples were dissolved in concentrated nitric acid and concentrated hydrochloric acid using microwave heating. The samples and acids were placed in fluorocarbon polymer (PFA) microwave vessels. The vessels were sealed and heated in the microwave unit (175 °C for 15 min). After cooling, the vessel contents were filtered, centrifuged, and then diluted to volume and analyzed. The total metal(loid) concentration was measured by ICP-MS in plant samples, following the UNI EN ISO 17294-2:2016 method, and by ICP-OES in soil samples, following the EPA 6010D/2014. Method detection limits (for both soil and plant samples) were as follows (mg/kg d.w.): As 0.5, Pb 1.0, V 0.5, Zn 1.0. As the QC procedure, a daily five-point calibration with multi elemental standard followed by a secondary standard calibration, analyses of a certified reference material (CRM) added to samples, and a matrix spike and matrix spike duplicate (MS/MSD) were performed. The bioconcentration factor (BCF) refers to the ratio between the metal(loid) concentration in plant organs and in the soil, as reported in Iori et al. [26].

2.4. Statistical Analysis

The data reported in the tables refer to four replicates (four pots per thesis). Normally, distributed data were processed by one-way analysis of variance (ANOVA), using the SPSS (Chicago, IL, USA) software tool, with the statistical significance of mean data assessed by t-test ($p \leq 0.05$).

3. Results and Discussion

As reported in Table 1, the analyses of agronomic parameters revealed a higher content of organic matter (SOM), total N, available K, and exchangeable Mg in C-soil. Both samples showed sandy loam ground, slightly alkaline pH, and a similar cationic exchange capacity (CEC). Metal (loid) concentrations of As, V, and Pb were below the threshold fixed by the Italian Law (for sites devoted to green areas as for Decree Law n.152/06 [23]) in C-soil and beyond the legal limit in the soil representing a moderate contamination (MC).

Table 1. Physico–chemical parameters and metal(loid) content of soil sampled in two different sites (C—control and MC—moderate contamination) in the "Valle del Sacco" area (Lazio Region, Central Italy).

Soil Type	Texture	pH	SOM (%)	N Total (‰)	P_2O_5 (ppm)	K_2O (ppm)	Mg (ppm)	CEC (meq/100g)	As (ppm)	Pb (ppm)	V (ppm)	Zn (ppm)
C	Sandy loam	7.93	3.7	2.56	78	779	101	16.9	17.3	77	76.5	67.4
MC	Sandy loam	8.07	2.2	1.43	87	667	77	15.7	22.6	115	106.7	92.8

Abbreviations: SOM, soil organic matter; CEC, cation exchange capacity.

In Table 2, some biometric parameters regarding hemp plants grown in pots filled with soils of the two sampled points are shown. Plant height and stem diameter were not affected by the different substrate composition, while leaf area was higher in plants grown in C-soil when compared with those grown in MC-soil. Anyway, the growth of plants (i.e., plant height, stem diameter, above ground, and root dry mass) can be considered as satisfactory with regards to data reported in field trials on different cultivars of fiber hemp [10,27], despite the notable differences in the growth conditions.

Table 2. Biometric parameters and total transpiration measured in hemp plants (*Cannabis sativa* L., cv. Codimono) after 90 days of growth in a greenhouse in pots filled with soil sampled in two different sites (C and MC) in the "Valle del Sacco" area (Lazio Region, Central Italy). In each column, mean values (n = 4 ± SE) are reported. Different letters indicate significant different values (t-test, $p \leq 0.05$).

Soil Type	Plant Height (cm)	Leaf Area (cm^2)	Stem Diameter (cm)	Total Transpiration L/plant
C	158.8 (5.1) a	2507 (249) a	0.88 (0.02) a	16.58 (0.2) a
MC	154.3 (20.7) a	1657 (88) b	0.70 (0.05) a	14.44 (0.38) b

Biomass production data are reported in Table 3. The results revealed that the growth of hemp plants in MC-soil was reduced compared with that observed for plants cultivated in C-soil. In particular, the main differences were found at the root and leaf level. This latter result was consistent with previous reported data on leaf area in Table 2. Interestingly, stem and inflorescence biomass was not affected by the different soil characteristics. The calculation of the total amount of water loss for transpiration by plants revealed a higher total transpiration over the cultivation period in C-soil grown plants.

Taken together, results on growth performances highlighted a good adaptation of hemp plants to the soil sampled in the contaminated area. In this regard, it can be hypothesised that growth was supported mainly by the high SOM and nutrient content, especially with regards to C-soil grown plants. Moreover, it can be underlined that plant growth was not negatively affected by the presence of metal(loid)s in the soil, confirming previous data reported by several authors [14–16]. Notably, the presence of vanadium in the soil was also not deleterious for hemp plants, while in the literature, such metal is reported to be extremely toxic for plant growth at concentrations even lower than those tested in the present work [28,29].

Table 3. Dry biomass of different organs of hemp plants (*Cannabis sativa* L., cv. Codimono) obtained after 90 days of growth in a greenhouse in pots filled with soil sampled in two different sites (C and MC) in the "Valle del Sacco" area (Lazio Region, Central Italy). In each column, mean values (n = 4 ± SE) are reported. Different letters indicate significantly different values (t-test, $p \leq 0.05$).

Soil Type	Biomass (g dw plant $^{-1}$)			
	Root	Stem	Leaf	Inflorescence
C	3.73 (0.27) a	12.44 (1.42) a	9.33 (0.70) a	6.06 (1.19) a
MC	2.16 (0.41) b	8.14 (1.98) a	6.24 (0.39) b	3.79 (0.18) a

To investigate the responses of hemp plants grown in the two different soils at the photosynthetic level, measurements of leaf chlorophyll content and chlorophyll fluorescence were performed (Table 4).

Leaf chlorophyll content is one of the most important factors in determining photosynthetic potential and primary production [30]. Our results showed that a higher leaf total chlorophyll content was found in plants grown in C-soil compared with those grown in MC-soil. These data are in accordance with the higher leaf biomass and area found in plants grown in C-soil (Tables 2 and 3). The analysis of chlorophyll fluorescence was focused on two parameters—the quantum yield of primary photochemistry (F_v/F_m) and performance index (PI_{ABS}) of PSII. Among the chlorophyll fluorescence parameters, F_v/F_m is recognized as a good indicator for photo-inhibitory or photo-oxidative effects on PSII [31]. However, the most widely used parameter from the chlorophyll fluorescence OJIP transient is the PI_{ABS}, which provides quantitative information about the general state of plants and their vitality. PI_{ABS} is the product of three independent characteristics—the concentration of reaction centers per chlorophyll, a parameter related to primary photochemistry, and a parameter related to electron transport [20]. PI_{ABS} reflects the functionality of both PSI and II and produces quantitative information of the plant performance, especially under stress conditions [20,32]. Contrarily to chlorophyll content, the results showed that no differences between plants cultivated in both types of soil were observed concerning the F_V/F_m and PI_{ABS} parameters. Therefore, these data confirmed the good physiological status of hemp plants grown in both soil conditions, as also visually observed for the lack of damage symptoms such as chlorosis or necrosis. Our results are in line with those reported by other authors [33,34], who found a reduction of chlorophyll fluorescence parameters such as F_V/F_m, only under elevated metal concentrations, highlighting the ability of hemp plants to tolerate metal stress and to grow in contaminated soil.

Table 4. Leaf chlorophyll content and chlorophyll fluorescence parameter analysis (the quantum yield of primary photochemistry (F_V/F_m) and performance index (PI_{ABS}) of PS II) in hemp plants (*Cannabis sativa* L., cv. Codimono) after 90 days of growth in a greenhouse in pots filled with soil sampled in two different sites (C and MC) in the "Valle del Sacco" area (Lazio Region, Central Italy). In each column, mean values (n = 4 ± SE) are reported. Different letters indicate significantly different values (t-test, $p \leq 0.05$).

Soil Type	Chl Content (µg cm^{-2})	Chl Fluorescence Parameters (r.u.)	
		PI_{ABS}	F_V/F_m
C	60.4 (3.3) a	18.7 (1.7) a	0.793 (0.003) a
MC	44.4 (0.9) b	17.0 (1.6) a	0.791 (0.006) a

In Table 5, the concentration of the analysed metal(loid)s in the hemp plant organs is reported. Notably, no detection of the most toxic metal(loid) elements, namely, As, V, and Pb, was observed in the above ground organs of plants (stem, leaves, and inflorescence). Conversely, as expected for an essential micronutrient, Zn was found in all organs, especially in the inflorescence, followed by leaves and stem. Contrarily to above ground organs, the presence of metal(loid)s was detected at the root level. In this case, differences in plants were found for As, Pb, and V, with their concentrations being higher in plants cultivated in C-soil compared with MC-soil.

Table 5. Metal(loid) content in the organs of hemp plants (*Cannabis sativa* L., cv. Codimono) after 90 days of growth in a greenhouse in pots filled with soil sampled in two different sites (C and MC) in the "Valle del Sacco" area (Lazio Region, Central Italy). In each column, mean values (n = 4 ± SE) are reported. Different letters indicate significantly different values (t-test, $p \leq 0.05$); nd = not detected.

Plant Organs	Soil Type	Metal(loid) Concentration (ppm)			
		As	Pb	V	Zn
Root	C	1.9 (0.2) a	12.3 (2.7) a	11.4 (1.8) a	13.8 (1.7) a
	MC	0.8 (0.1) b	3.6 (0.2) b	4.1 (0.1) b	14.1 (3.9) a
Stem	C	nd	nd	nd	9.1 (2.2) a
	MC	nd	nd	nd	9.5 (1.1) a
Leaves	C	nd	nd	nd	21.1 (3.5) a
	MC	nd	nd	nd	35.2 (6.2) a
Inflorescence	C	nd	nd	nd	59.1 (1.7) a
	MC	nd	nd	nd	62.7 (5.5) a

The higher capability of C-soil-grown plants to accumulate the metal (loid) elements was in accordance with the higher growth performance highlighted by biomass production, leaf area, chlorophyll content, and leaf transpiration (Tables 2–4), associated with a better nutritional status (Table 1). Moreover, the hypotheses of a more recalcitrant soil pool in MC-soil or a more active excluder mechanism in MC-soil-grown plants cannot be ruled out.

The low ability of hemp plants to accumulate toxic metals from soils is consistent with previously reported works showing the preferential metal allocation in the root apparatus [14,17,35], even if a low metal translocation to shoots was reported in these investigations. Accordingly, a large variability for this trait, because of genetic and environmental factors, was reviewed [36]. In this regard, because of the remarkably higher Cd accumulation in the roots compared with the shoots, hemp was defined as a Cd-excluder plant [16]. Preliminary observations of this work suggest that hemp plants can restrict the uptake of other metal(loid) elements, namely, As, Pb, and V. Further investigations are needed to better evaluate this aspect.

The calculation of the bioconcentration factor (BCF) (Table 6) revealed that this index was higher in the roots of C-soil grown plants than MC-soil grown plants for all the metal(loid)s analysed, except for Zn, as a result of the different concentrations found in the plants and soils types, respectively. Anyway, the values of BCF found for hemp plants grown in both types of soils are to be considered very low following the literature on the matter [37]. According to the present data, low BCF for roots in hemp plants treated with Pb was also found [38–40]. A similar low BCF for As was also reported in hemp plants collected in a survey on a metal-contaminated site [40].

Table 6. Bioconcentration factor (BCF) calculated for roots (n = 4) of hemp plants (*Cannabis sativa* L., cv. Codimono) after 90 days of growth in a greenhouse in pots filled with soil sampled in two different sites (C and MC) in the "Valle del Sacco" area (Lazio Region, Central Italy). In each column, mean values (n = 4 ± SE) are reported. Different letters indicate significantly different values (t-test, $p \leq 0.05$).

Soil Type	Bioconcentration Factor			
	As	Pb	V	Zn
C	0.11 (0.008) a	0.16 (0.02) a	0.14 (0.01) a	0.21 (0.01) a
MC	0.03 (0.001) b	0.03 (0.005) b	0.03 (0.007) b	0.15 (0.02) a

Overall, results on metal(loid) accumulation evidenced that under our experimental conditions, hemp plants were able to exclude toxic metals—namely, As, Pb, and V—from entering the vascular tissues and being transported in the above ground organs. In fact, even though a high transpiration associated to a considerable biomass production was observed, metal loading by the rooting system was reduced and translocation to aerial parts was negligible. This aspect opens up intriguing questions about the possible toxic metal excluder mechanisms activated by hemp plants to cope with the metal presence in the soil to be addressed in future investigations. Contrarily, Zn was taken up and transported in stem, leaf, and inflorescence, where it is requested for physiological functions associated to its role as a micronutrient. Regarding the metal absorption capability of the hemp plants assayed in this trial, it should be taken into due account that the sub-alkaline condition of both sampled soils (Table 1) could have negatively affected the mobility of metals in the circulating soil solution, which is generally favored by a sub-acidic soil pH [41].

4. Conclusions

The results of this trial, obtained in a greenhouse pot experiment, put in evidence that hemp plants cv. Codimono can satisfactorily grow in soil moderately contaminated by metal(loid)s, producing biomass and inflorescence with seeds. Some growth parameters were slightly reduced in MC-soil grown plants compared with those grown in C-soil. Notably, the lack of toxic metal(loid)s in the above ground organs represents a valuable indication for the profitable cultivation of hemp plants in moderately metal contaminated sites. The slight alkaline soil pH reported in this trial is a factor to bear in mind to consider the uptake of metal(loid)s by plants, affecting the metal mobility and bioavailability. Taken together, the results of the present investigation, though preliminary, indicated that hemp plant cultivation could be exploited within a phytomanagement strategy on moderately contaminated areas, where the restriction for crop cultivation/grazing produces a loss of income for farmers. In such cases, the production of fibres, energy from biomass, and oil from seeds could be profitably achieved.

Author Contributions: M.Z. and F.P. conceived, designed, and supervised the experiments; F.F. managed the relations with the public and private bodies and organised the soil sampling procedures; L.P. supervised and performed the experimental work on plant growth; V.P. and A.G. performed the experimental work; all authors performed the biometric parameters analyses; F.P. performed chlorophyll content and chlorophyll fluorescence parameters analyses; M.Z. and F.P. performed the data elaboration; M.Z. wrote the paper with the contribution of F.P., L.P., and F.F.

Funding: This research did not receive any specific grant from funding agencies in the public, commercial or not-for-profit sectors.

Acknowledgments: Authors wish to thank Rachele Invernizzi (South Hemp Techno) for the seeds supply, the Municipality of Colleferro (Rome) and the Associazione Radicati for their support, Marco Giorgetti and Alberto Rinalduzzi for their technical assistance.

Conflicts of Interest: The authors declare no conflict of interest.

References

1. Panagos, P.; Borrelli, P. All that soil erosion: The global task to conserve our soil resources. In *Soil Erosion in Europe: Current Status, Challenges and Future Developments*; Soil Environment Center of the Korea: Seoul, Korea, 2017; pp. 20–21.
2. Panagos, P.; Van Liedekerke, M.; Yigini, Y.; Montanarella, L. Contaminated sites in Europe: Review of the current situation based on data collected through a European network. *J. Environ. Pub. Health* **2013**, *2013*, 158764. [CrossRef] [PubMed]
3. Vangronsveld, J.; Herzig, R.; Weyens, N.; Boulet, J.; Adriaensen, K.; Ruttens, A.; Thewys, T.; Vassilev, A.; Meers, E.; Nehnevajova, E.; et al. Phytoremediation of contaminated soils and groundwater: Lessons from the field. *Environ. Sci. Poll. Res.* **2009**, *16*, 765–794. [CrossRef]
4. Bianconi, D.; De Paolis, M.R.; Agnello, A.C.; Lippi, D.; Pietrini, F.; Zacchini, M.; Polcaro, C.; Donati, E.; Paris, P.; Spina, S.; et al. Field-scale rhyzoremediation of a contaminated soil with hexachlorocyclohexane (HCH) isomers: The potential of poplars for environmental restoration and economical sustainability. In *Handbook of Phytoremediation*; Golubev, I.A., Ed.; Nova Science Publishers, Inc.: New York, NY, USA, 2011; Chapter 31; pp. 1–12.
5. Evangelou, M.W.; Papazoglou, E.G.; Robinson, B.H.; Schulin, R. Phytomanagement: Phytoremediation and the production of biomass for economic revenue on contaminated land. In *Phytoremediation: Management of Environmental Contaminants*; Ansari, A.A., Gill, S.S., Gill, R., Lanza, G.R., Newman, L., Eds.; Springer International Publishing: Cham, Switzerland, 2015; Volume I, pp. 115–132.
6. Cundy, A.B.; Bardos, R.P.; Puschenreiter, M.; Mench, M.; Bert, V.; Friesl-Hanl, W.; Müller, I.; Li, X.N.; Weyens, N.; Witters, N.; et al. Brownfields to green fields: Realising wider benefits from practical contaminant phytomanagement strategies. *J. Environ. Manag.* **2016**, *184*, 67–77. [CrossRef] [PubMed]
7. Domínguez, M.T.; Maranón, T.; Murillo, J.M.; Schulin, R.; Robinson, B.H. Trace element accumulation in woody plants of the Guadiamar Valley, SW Spain: A large-scale phytomanagement case study. *Environ. Pollut.* **2008**, *152*, 50–59. [CrossRef] [PubMed]
8. Komínková, D.; Fabbricino, M.; Gurung, B.; Race, M.; Tritto, C.; Ponzo, A. Sequential application of soil washing and phytoremediation in the land of fires. *J. Environ. Manag.* **2018**, *206*, 1081–1089. [CrossRef]
9. De Medici, D.; Komínková, D.; Race, M.; Fabbricino, M.; Součková, L. Evaluation of the potential for cesium transfer from contaminated soil to the food chain as a consequence of uptake by edible vegetables. *Ecotoxicol. Environ. Saf.* **2019**, *171*, 558–563. [CrossRef] [PubMed]
10. Amaducci, S.; Zatta, A.; Pelatti, F.; Venturi, G. Influence of agronomic factors on yield and quality of hemp (*Cannabis sativa* L.) fibre and implication for an innovative production system. *Field Crops Res.* **2008**, *107*, 161–169. [CrossRef]
11. Prade, T.; Svensson, S.E.; Andersson, A.; Mattsson, J.E. Biomass and energy yield of industrial hemp grown for biogas and solid fuel. *Biomass Bioenergy* **2011**, *3*, 3040–3049. [CrossRef]
12. Salentijn, E.M.; Zhang, Q.; Amaducci, S.; Yang, M.; Trindade, L.M. New developments in fiber hemp (*Cannabis sativa* L.) breeding. *Ind. Crops Prod.* **2015**, *68*, 32–41. [CrossRef]
13. Linger, P.; Müssig, J.; Fischer, H.; Kobert, J. Industrial hemp (*Cannabis sativa* L.) growing on heavy metal contaminated soil: Fibre quality and phytoremediation potential. *Ind. Crop. Prod.* **2002**, *16*, 33–42. [CrossRef]

14. Citterio, S.; Santagostino, A.; Fumagalli, P.; Prato, N.; Ranalli, P.; Sgorbati, S. Heavy metal tolerance and accumulation of Cd, Cr and Ni by *Cannabis sativa* L. *Plant Soil* **2003**, *256*, 243–252. [CrossRef]
15. Arru, L.; Rognoni, S.; Baroncini, M.; Bonatti, P.M.; Perata, P. Copper localization in *Cannabis sativa* L. grown in a copper-rich solution. *Euphytica* **2004**, *140*, 33–38. [CrossRef]
16. Shi, G.; Liu, C.; Cui, M.; Ma, Y.; Cai, Q. Cadmium tolerance and bioaccumulation of 18 hemp accessions. *Appl. Biochem. Biotech.* **2012**, *168*, 163–173. [CrossRef] [PubMed]
17. Ahmad, R.; Tehsin, Z.; Malik, S.T.; Asad, S.A.; Shahzad, M.; Bilal, M.; Shah, M.M.; Khan, S.A. Phytoremediation potential of hemp (*Cannabis sativa* L.): Identification and characterization of heavy metals responsive genes. *Clean Soil Air Water* **2016**, *44*, 195–201. [CrossRef]
18. Campbell, S.; Paquin, D.; Awaya, J.D.; Li, Q.X. Remediation of benzo [a] pyrene and chrysene-contaminated soil with industrial hemp (*Cannabis sativa*). *Int. J. Phytorem.* **2002**, *4*, 157–168. [CrossRef]
19. Vandenhove, H.; Van Hees, M. Fibre crops as alternative land use for radioactively contaminated arable land. *J. Environ. Rad.* **2005**, *81*, 131–141. [CrossRef]
20. Strasser, R.; Tsimilli-Michael, M.; Srivastava, A. Analysis of the chlorophyll a fluorescence transient. In *Chlorophyll a Fluorescence*; Advances in Photosynthesis and Respiration; Papageorgiou, G., Govindje, E., Eds.; Springer: Cham, The Netherlands, 2004; pp. 321–362.
21. Singh, S.; Prasad, S.M. IAA alleviates Cd toxicity on growth, photosynthesis and oxidative damages in eggplant seedlings. *Plant Growth Regul.* **2015**, *77*, 87–98. [CrossRef]
22. Ministro per le Politiche Agricole. Approvazione dei "Metodi ufficiali di analisi chimica del suolo". Available online: http://ctntes.arpa.piemonte.it/Bonifiche/Documenti/Norme/13_Set_99.pdf (accessed on 17 April 2019).
23. Decreto Legislativo. Norme in Materia Ambientale. Available online: http://www.conou.it/wp-content/uploads/2015/11/Dlgs-152_2006-Norme-in-materia-ambientale.pdf (accessed on 17 April 2019).
24. Pietrini, F.; Iori, V.; Bianconi, D.; Mughini, G.; Massacci, A.; Zacchini, M. Assessment of physiological and biochemical responses, metal tolerance and accumulation in two eucalypt hybrid clones for phytoremediation of cadmium-contaminated waters. *J. Environ. Manag.* **2015**, *162*, 221–231. [CrossRef]
25. Cerovic, Z.G.; Masdoumier, G.; Ghozlen, N.B.; Latouche, G. A new optical leaf-clip meter for simultaneous non-destructive assessment of leaf chlorophyll and epidermal flavonoids. *Physiol. Plant* **2012**, *146*, 251–260. [CrossRef]
26. Iori, V.; Gaudet, M.; Fabbrini, F.; Pietrini, F.; Beritognolo, I.; Zaina, G.; Massacci, A.; Scarascia Mugnozza, G.; Zacchini, M.; Sabatti, M. Physiology and genetic architecture of traits associated with cadmium tolerance and accumulation in *Populus nigra* L. *Trees* **2016**, *30*, 125–139. [CrossRef]
27. Amaducci, S.; Errani, M.; Venturi, G. Plant population effects on fibre hemp morphology and production. *J. Ind. Hemp.* **2002**, *7*, 33–60. [CrossRef]
28. Chongkid, B.; Vachirapattama, N.; Jirakiattikul, Y. Effects of vanadium on rice growth and vanadium accumulation in rice tissues. *Kasetsart J. (Nat. Sci.)* **2007**, *41*, 28–33.
29. Vachirapatama, N.; Jirakiattiku, Y.; Dicinoski, G.W.; Townsend, A.T.; Haddad, P.R. Effect of vanadium on plant growth and its accumulation in plant tissues. Songklanakarin. *J. Sci. Technol.* **2011**, *33*, 255–261.
30. Dai, Y.J.; Shen, Z.G.; Liu, Y.; Wang, L.L.; Hannaway, D.; Lu, H.F. Effects of shade treatments on the photosynthetic capacity, chlorophyll fluorescence, and chlorophyll content of *Tetrastigma hemsleyanum* Diels et Gilg. *Environ. Exp. Bot.* **2009**, *65*, 177–182. [CrossRef]
31. Maxwell, K.; Johnson, G.N. Chlorophyll fluorescence—A practical guide. *J. Exp. Bot.* **2000**, *51*, 659–668. [CrossRef]
32. Iori, V.; Pietrini, F.; Bianconi, D.; Mughini, G.; Massacci, A.; Zacchini, M. Analysis of biometric, physiological, and biochemical traits to evaluate the cadmium phytoremediation ability of eucalypt plants under hydroponics. *iFor. Biogeosci. For.* **2017**, *10*, 416–421. [CrossRef]
33. Linger, P.; Ostwald, A.; Haensler, J. *Cannabis sativa* L. growing on heavy metal contaminated soil: Growth, cadmium uptake and photosynthesis. *Biol. Plant.* **2005**, *49*, 567–576. [CrossRef]
34. Shi, G.R.; Cai, Q.S.; Liu, Q.Q.; Wu, L. Salicylic acid-mediated alleviation of cadmium toxicity in hemp plants in relation to cadmium uptake, photosynthesis, and antioxidant enzymes. *Acta Physiol. Plant.* **2009**, *31*, 969–977. [CrossRef]
35. Angelova, V.; Ivanova, R.; Delibaltova, V.; Ivanov, K. Bio-accumulation and distribution of heavy metals in fibre crops (flax, cotton and hemp). *Ind. Crops Prod.* **2004**, *19*, 197–205. [CrossRef]

36. Griga, M.; Bjelková, M. Flax (*Linum usitatissimum* L.) and Hemp (*Cannabis sativa* L.) as fibre crops for phytoextraction of heavy metals: Biological, agro-technological and economical point of view. In *Plant-Based Remediation Processes*; Springer: Heidelberg/Berlin, Germany, 2013; pp. 199–237.
37. Mattina, M.I.; Lannucci-Berger, W.; Musante, C.; White, J.C. Concurrent plant uptake of heavy metals and persistent organic pollutants from soil. *Environ. Pollut.* **2003**, *124*, 375–378. [CrossRef]
38. Kos, B.; Leštan, D. Soil washing of Pb, Zn and Cd using biodegradable chelator and permeable barriers and induced phytoextraction by *Cannabis sativa*. *Plant Soil* **2004**, *263*, 43–51. [CrossRef]
39. Zehra, S.S.; Arshad, M.; Mahmood, T.; Waheed, A. Assessment of heavy metal accumulation and their translocation in plant species. *Afr. J. Biotechnol.* **2009**, *8*, 12.
40. Varun, M.; D'Souza, R.; Pratas, J.; Paul, M.S. Metal contamination of soils and plants associated with the glass industry in North Central India: Prospects of phytoremediation. *Environ. Sci. Pollut. Res.* **2012**, *19*, 269–281. [CrossRef] [PubMed]
41. Houben, D.; Evrard, L.; Sonnet, P. Mobility, bioavailability and pH-dependent leaching of cadmium, zinc and lead in a contaminated soil amended with biochar. *Chemosphere* **2013**, *92*, 1450–1457. [CrossRef] [PubMed]

© 2019 by the authors. Licensee MDPI, Basel, Switzerland. This article is an open access article distributed under the terms and conditions of the Creative Commons Attribution (CC BY) license (http://creativecommons.org/licenses/by/4.0/).

Article

Use of Bacteria and Synthetic Zeolites in Remediation of Soil and Water Polluted with Superhigh-Organic-Sulfur Raša Coal (Raša Bay, North Adriatic, Croatia)

Gordana Medunić [1,*], Prakash Kumar Singh [2], Asha Lata Singh [3], Ankita Rai [2], Shweta Rai [2], Manoj Kumar Jaiswal [4], Zoran Obrenović [5], Zoran Petković [6] and Magdalena Janeš [1]

1 Department of Geology, Faculty of Science, University of Zagreb, 10000 Zagreb, Croatia
2 Department of Geology, Institute of Science, Banaras Hindu University, Varanasi 221005, India
3 Department of Botany, Institute of Science, Banaras Hindu University, Varanasi 221005, India
4 Department of Earth Sciences, IISER, Kolkata 700073, India
5 Factory 'Alumina Ltd.', 71000 Zvornik, Bosnia and Herzegovina
6 Factory 'Zeochem Ltd.', 71000 Zvornik, Bosnia and Herzegovina
* Correspondence: gmedunic@geol.pmf.hr; Tel.: +385-1-460-5909

Received: 2 June 2019; Accepted: 8 July 2019; Published: 10 July 2019

Abstract: The Raša Bay (North Adriatic, Croatia) has been receiving various pollutants by inflowing streams laden with untreated municipal and coalmine effluents for decades. The locality was a regional center of coalmining (Raša coal), coal combustion, and metal processing industries for more than two centuries. As local soil and stream water were found to be contaminated with sulfur and potentially toxic trace elements (PTEs) as a consequence of weathering of Raša coal and its waste, some clean-up measures are highly required. Therefore, the aim of this study was to test the remediating potential of selected microorganisms and synthetic zeolites in the case of soil and coal-mine water, respectively, for the first time. By employing bacterial cultures of *Ralstonia* sp., we examined removal of sulfur and selected PTEs (As, Ba, Co, Cr, Cu, Ni, Pb, Rb, Se, Sr, U, V, and Zn) from soil. The removal of sulfur was up to 60%, arsenic up to 80%, while Se, Ba, and V up to 60%, and U up to 20%. By applying synthetic zeolites on water from the Raša coalmine and a local stream, the significant removal values were found for Sr (up to 99.9%) and Ba (up to 99.2%) only. Removal values were quite irregular (insignificant) in the cases of Fe, Ni, Zn, and Se, which were up to 80%, 50%, 30%, and 20%, respectively. Although promising, the results call for further research on this topic.

Keywords: coal; soil; water; bioremediation; *Ralstonia* sp.; sulfur; synthetic zeolite; removal

---

1. Introduction

Coal is a valuable resource in terms of cheap electricity production, and is also an economic source of strategically important elements (Ge, Ga, U, V, Se, rare earth elements, Y, Sc, Nb, Au, Ag, and Re) [1–3]. However, coal combustion emissions, enriched in S and potentially toxic trace elements (PTEs), such as Se, U, Cd, Hg, Pb, As, Ni, Cr, V, etc., could be hazardous for soil, water, air, and crop quality [4–7]. Coal combustion wastes (CCWs) are solid/mineral residues produced in huge quantities worldwide. If improperly disposed at unprotected landfills, their leachates, formed due to infiltrating rainwater, can contaminate local aquifers and terrestrial ecosystems with PTEs [8,9]. Soil and water degraded by the coal industry must be remediated and conserved using a strategy which needs to be achieved as cheaply and effectively as possible.

Various physical and chemical technologies of metal removal and/or stabilization from soils and sediments have been reported, each one with potential advantages and drawbacks. They include, for example, soil washing, thermal extraction, ion exchange, electrokinetic treatment, reverse osmosis, membrane technology, evaporation recovery, solidification, plasma vitrification, etc. [10]. They are mostly very expensive, and a viable alternative could be bioremediation, which is based on the microbially mediated transformation of the metal, and its subsequent biosorption and biomineralization [11]. It has been commonly used to remove organic matter and toxic chemicals from domestic and industrial waste by means of enzymatic attacks through the activities of living micro-organisms [12]. Environmental pollutants get biodegraded due to many processes, e.g., oxidation, mineralization, transformation to toxic or nontoxic compounds, and accumulation within an organism, or they get polymerized or otherwise bound to natural materials in soils, sediments, or waters [13]. For example, Park and colleagues [10] investigated the potential bioaugmentation in sediments to improve adsorption/biosorption of Cd and Zn. A batch experiment was performed in the lake sediments augmented with *Ralstonia* sp. HM-1. A reduction of metal concentrations in the liquid phase resulted due to adsorption in sediments, with 99.7% removal efficiency for both Cd and Zn in the aqueous solution after 35 days.

Regarding coal biodegradation, by employing biological processes, coal waste gets converted to a value-added liquid product in a process known as liquefaction [14]. Although a number of microorganisms are known to be actively involved in it, and for some, the extracellular enzymes involved are also known, the underlying mechanism(s) employed in the process of coal biodegradation are not yet fully established, partly due to the highly complex structure of coals [14]. It is a naturally complex process driven by an array of extracellular enzymes in the presence of various chelators and supporting enzymes released by different microorganisms that coinhabit the coal environment [15]. For example, bioremediation was successfully developed and applied on Indian and Indonesian coals using mixed bacterial consortium, which effectively removed more than 80% of Ni, Zn, Cd, Cu, and Cr, and 45% of Pb [16]. Also, the same research team has carried out desulfurization of coal using microorganisms *Ralstonia* sp. and *Pseudoxanthomonas* sp. [17], and found out the highest positive correlation between initial total S (2.2 Wt% dry ash free basis) and removal percentage (up to 45%) in the case of Indian Vastan lignite treated with *Ralstonia* sp. *Ralstonia* sp. are gram-negative, non-fermentative, rod-shaped bacteria ubiquitously present in soil and water [18]. They exhibit many advantages for the cleaning treatments due to their large range of environmental conditions, biodegradative abilities, and large metabolic diversity [19]. It was suggested [18] that *Ralstonia* sp. strain OR214 was tolerant to high concentrations of PTEs such as Cd, Co, Ni, and U.

Surface water pollution caused by coalmine discharges laden with toxic pollutants, especially PTEs, is a widespread problem and, occasionally, national standards for wastewater discharges from mines are violated [20]. Since PTEs are not biodegradable, they tend to accumulate in aquatic organisms, causing various diseases and disorders. Therefore, numerous treatment technologies have been developed to remove PTEs from wastewater, e.g., chemical precipitation, coagulation, ultra-filtration, biological systems, electrolytic processes, reverse osmosis, oxidation with ozone/hydrogen peroxide, membrane filtration, and ion exchange [21]. Due to their high cost and disposal problems, many of these conventional methods have not been widely applied on a large scale. Adsorption has been found to be superior compared to other techniques in terms of initial cost, flexibility and simplicity of design, ease of operation, and insensitivity to toxic pollutants [21]. Zeolites, natural as well as synthetic, represent low-cost materials with high adsorption selectivity for cationic contaminants [21–24]. Studies conducted on model solutions showed that synthetic zeolites, compared to natural ones, had higher cation-exchange capability, and consequently better PTE sorption performances [25].

The Raša Bay (North Adriatic, Croatia; Figure 1) has been receiving various pollutants by inflowing streams laden with untreated municipal and coalmine effluents for decades [20]. Locally mined superhigh-organic-sulfur (SHOS) Raša coal is a unique variety compared to other coal types worldwide based on its anomalously high levels of S, Se, and U [5]. The whole local area (about 600 km^2)

has experienced damaging environmental impacts of SHOS Raša coal due to mining, preparation, combustion, waste storage, and transport in the past [26]. Soil locations downwind from a local coal-fired power plant are severely polluted with S (up to 4.00%), Se (up to 6.80 mg/kg), Cd (up to 4.70 mg/kg), and organic contaminants, such as PAHs, up to 13,500 ng/g [4,27]. In particular, Selenium and selected PTEs were found to be increased in soil, sediment, surface water, locally grown vegetables, and local birds [5,28]. A local CCW, located in a Štrmac village (Figure 1, S4 and S8 locations), was found to have decreased Se values compared to relevant previous studies [29]. This indicates leaking problems, as the study area belongs to vulnerable karstic environment which promotes Se mobility due to high Eh/pH conditions [5].

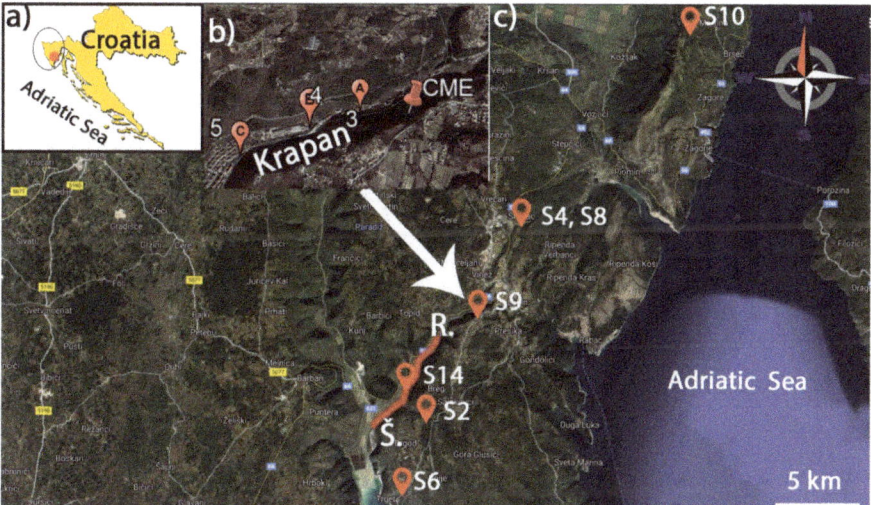

Figure 1. Map of the study area. (**a**) Red dot—study area (east side of the Istrian Peninsula). (**b**) Water sampling points: CME—coal-mine effluent (water samples no. 1 and 2, inside of a Raša coal-mine in Krapan village), water samples no. 3, 4, and 5 taken from the Krapan stream. (**c**) Soil sampling sites no. S2–S14. Š.—a former Štalije coal-separation unit, R.—the Raša town, S9—Krapan village.

The objectives of this study were as follows: (1) To apply bioremediation tests, by using *Ralstonia* sp. on SHOS Raša coal, as well as few selected soil samples polluted with SHOS coal and CCW, geochemically characterized in previous papers [4,5,20], and (2) to examine the ability of synthetic zeolites to remove PTEs from Raša coalmine water and stream water affected by coal leachates and nontreated municipal wastewater on laboratory scale.

2. Materials and Methods

2.1. Sampling and Sample Preparation

Soil and SHOS Raša coal sampling campaigns (Figure 1) were conducted on several occasions during the period 2013–2015 [4,20,28]. Soil samples were collected at: S2, a former small-scale coal-fired power plant Vlaška; S4, and S8, CCW locality in Štrmac village; S6, sampled close to a beach in the Trget village; S9, a former coal-mining town Krapan; S14, a former rail line (red-colored line in Figure 1), between the Raša town (R.) and a coal-separation unit Štalije (Š.); and S10, unpolluted soil [30] taken from a site 10–15 km NE from the study area. A summary of soil samples' (2 kg per sample) physicochemical properties (pH, LOI—loss on ignition, CEC—cation exchange capacity, and $CaCO_3$, acquired and described in previous papers [4,20,28]), along the levels of sulfur, PTEs in soil, and SHOS Raša coal, are shown in Table 1. Nonfiltered water samples (300 mL per sample) were collected from an old mining town Krapan and a city of Raša (Figure 1) in early May 2019. Raša coalmine water (n = 2) was sampled from the floor inside of an underground corridor in Krapan. Then, surface water (n = 2) was taken from a Krapan stream, which flows roughly parallel to the rail line (red line in Figure 1) and receives coa-mine water. The same stream was sampled (n = 1) downstream in the city of Raša, immediately after the point where municipal wastewater effluent (MWE) was being discharged into the stream. Samples were collected from the surface at a maximum depth of 10 cm in acid-cleansed plastic bottles. Zeolite treatments and PTEs measurements were carried out the following day.

Table 1. Basic physicochemical characteristics of soil ($CaCO_3$, and loss on ignition (LOI) in %; cation exchange capacity (CEC) in mEq/100), and values of S (%) and PTEs (mg/kg) in soil and Raša coal (na—not analyzed, DL—detection limit). Soil sample sites shown in Figure 1.

	Site	pH	$CaCO_3$	LOI	CEC	S	Se	V	U	Sr	Cr	Cu	Pb	Zn
S2	Vlaška	7.6	56	15	23	2.1	10	318	11	655	73	13	75	11
S4	CCW	na	na	na	na	0.7	3.3	118	5.7	285	89	1800	200	6580
S6	Trget	7.6	36	19	6	0.8	<DL	52	5.0	169	45	625	210	936
S8	CCW	7.2	39	25	22	2.5	1.8	72	1.8	277	52	111	69	334
S9	Krapan	7.1	12	19	12	0.9	<DL	560	0.2	74	1860	1847	72	953
S14	Rail	6.9	3	83	6	6.9	27	264	8.2	356	100	190	46	863
S10	Unpolluted	6.8	0	15	20	0.2	1.5	229	4.3	82	135	41	56	169
C	Coal	na	na	na	na	10.6	23	80	14	290	27	6	3	32

2.2. Bioremediation Tests

Soil and SHOS Raša coal samples were treated with a known bacterial species, *Ralstonia* sp., which was grown, separated, and mass cultivated under the optimum conditions, temperature of 35 °C, and growth pH of 7, in the Bioremediation lab, Department of Botany, Banaras Hindu University, and identified in the Institute of Microbial Technology (IMTECH), Chandigarh (India) on the basis of biochemical characteristics [16]. The exponential phase bacterial cells (1000 mg), representing the growth phase of bacteria where bacterial cells divide as fast as possible by intake of nutrients from GPY (i.e., glucose, peptone, and yeast extract) culture media, was obtained through the centrifugation at a rate of 10,000 rpm for 10 min, then washed two times with double-distilled water to remove any sort of contamination and extracted of growth media. The resultant bacterial biomass was mixed into 100 mL of 5% (w/v) solution of sodium alginate ($NaC_6H_7O_6$), prepared in the GPY medium. The mixture was pumped dropwise into the 0.2 M $CaCl_2$ solution, which served as a gelling agent to provide more stable beads (preferably in the laminar flow cabinet, to avoid contamination). The formed beads were harvested and resuspended in a 100 mL growth medium (GPY) in a 250 mL cotton plugged culture flask. Soil samples were initially sterilized under ultraviolet rays for up to 45 min in the laminar flow cabinet prior to the treatment. Then, 5.0 g of each sterilized sample was taken in a conical flask containing 75 mL of sterilized distilled water, and 100 beads (equivalents to 1000 mg dry weight of bacteria) were added in each flask. It was calculated that from 1 mL solution of biomass and sodium alginate, 25 beads were formed. For the creation of 100 beads, 4 mL of solution was required. Likewise, it was calculated that 1 mL of the solution of bacterial biomass in sodium alginate contained 250 mg of dry weight of bacteria. Therefore, 4 mL of the solution, which was required for 100 beads, had to contain 1000 mg of dry weight of bacteria. The treatment was carried out for 18 days, and analysis of sulfur content was performed after 6, 12, and 18 days, whereas analysis of PTEs was carried out on 12-day treatment samples. Bioremediation of SHOS Raša coal sample was also carried out for 7 days, and residual sulfur content was analyzed on 3-, 5-, and 7-day treatment samples.

2.3. Adsorption Experiment

Analysis and experiment were done on original, unfiltered samples, excluding sample no. 3, which was filtered due to high amount of suspended material. Water samples were prepared by 10-fold dilution, followed by the addition of 2% (v/v) HNO_3 s.p. and In (1 µg/L) as an internal standard.

Aliquots of each sample were used for the adsorption experiment. Two types of synthetic zeolites were used as follows: (1) (Z) Purmol 4ST, composed of zeolite (90%) and water (10%); white color, pH 9–11.5, and (2) (A) 3A-50% (activated zeolite), i.e., synthetic sodium potassium zeolite; pH 10.1–11.4. The adsorption experiment was carried out as follows: Aliquots (20 mL) of water samples were placed into plastic bottles containing accurately weighed amounts of the sorbents, i.e., (Z) Purmol 4ST (m = 1.0 g), and (A) 3A-50% (m = 1.0 g). The prepared suspensions were shaken using a mechanical shaker at 320 rpm. Following a contact time of 120 min, 5 mL of the suspension was taken by a syringe from each bottle and filtered through a 0.45-μm filter. Prior to analysis, the obtained solutions were diluted 10 times, acidified with 2% (v/v) HNO_3 (65%, suprapur, Fluka, Steinheim, Switzerland), and In (1 μg/L) was added as an internal standard.

2.4. Analysis of Sulfur and PTEs

Total sulfur contents (%) in original and bioremediated soil (fraction < 0.063 mm), and reference material ISE 979 (Rendzina soil from Wepal, the Netherlands) were determined using Eschka's mixture according to the standard test method [31]. Accuracy and precision were within 10% between analyzed and certified values.

The multi-element analysis of prepared nontreated soil samples was carried out with X-ray fluorescence (XRF). Prior to analysis, reference standard soils, MESS3 and NIST, were used for the data calibration [32]. All samples and standards were dried in an oven for two days at 60 °C, ground, and sieved through 200 ASTM sieve plate. The material (4 g) was mixed with 1 g of boric acid powder, and the mixture was homogenized. No trace of boric acid clump was left to avoid background scattering peaks. Samples were placed in press pellet using a KBr press pellet machine. Analysis was conducted with X-ray beam until the measurement was recorded by an automated detector inside the XRF (S8 TIGER, Bruker, Mannheim, Germany) analyzer.

Prior to high-resolution inductively coupled plasma mass spectrometry (HR-ICP-MS) analysis of bioremediated soil samples, subsamples (0.05 g) were subjected to total digestion in the microwave oven (Multiwave ECO, Anton Paar, Graz, Austria) in two-step procedure consisting of digestion with a mixture of 4 mL nitric acid (HNO_3)—1 mL hydrochloric acid (HCl)—1 mL hydrofluoric acid (HF), followed by addition of 6 mL of boric acid (H_3BO_3). Prior to analysis, sample digests were diluted 10-fold, acidified with 2% (v/v) HNO_3 (65%, supra pur, Fluka, Steinheim, Switzerland), and In was added (1 μg/L) as an internal standard. Details of the sample preparation protocols, description of instrument conditions, and measurement parameters are reported elsewhere [33]. Analysis was conducted with an Element 2 instrument (Thermo, Bremen, Germany). Standards for multi-element analysis were prepared by appropriate dilution of a multi-elemental reference standard (Analytika, Prague, Czech Republic) containing Al, As, Ba, Be, Bi, Cd, Co, Cr, Cs, Cu, Fe, Li, Mn, Ni, Pb, Se, Sr, Ti, Tl, V, and Zn, in which single-element standard solutions of Rb, Sb, Sn, and U (Analytika, Prague, Czech Republic) were added. All samples were analyzed for total concentration of following elements: As, Ba, Co, Cr, Cu, Ni, Pb, Rb, Se, Sr, U, V, and Zn. Quality control was performed by simultaneous analysis of the blank and the certified reference material for soil (NCS DC 73302, also known as GBW 07410, China National Analysis Center for Iron and Steel, Beijing, China). Good agreement (±10%) between analyzed and certified concentrations was obtained for all measured elements.

Multi-elemental analysis of prepared original and zeolite-treated solutions was performed by HR-ICP-MS using an Element 2 instrument (Thermo, Bremen, Germany). External calibration was used for the quantification. All samples were analyzed for total concentration of following elements: As, Ba, Cd, Cr, Cu, Fe, Mo, Ni, Pb, Se, Sr, U, V, and Zn. Typical instrument conditions and measurement parameters used throughout the work are reported elsewhere [33].

2.5. Data Analysis

Data analysis was conducted with free PAST software [34]. It included calculations of basic statistical parameters and Kendall's Tau correlation coefficients. Level of significance was 0.05.

3. Results and Discussion

3.1. Bioremediation of SHOS Raša Coal and Polluted Soil

3.1.1. Geochemical Characterization of Coal and Soil

Table 1 shows basic physicochemical characteristics of soil from the Raša Bay area, along a background (unpolluted) soil sample taken some 10–15 km away from the study area (Figure 1), and values of PTEs in soil and Raša coal.

Evidently, Raša soil is polluted with all analyzed PTEs and S compared to unpolluted soil collected far from the study area [4,5,20,28]. Briefly, S, Se, V, and U were elements indicative for Raša coal weathering, both from underground deposits and surface coal waste piles abandoned across the study area [1,5,26]. Strontium is typical for marine and karst environments, while Cr, Cu, Pb, and Zn commonly result from metal processing industries [20]. Samples S4 and S9 were collected from the Štrmac village, where various coalmining and foundry factories, which closed in the late 1950s, were previously active. Anomalously high levels of Cu, Pb, and Zn in the sample S4, collected above the CCW site (Figure 1), resulted from foundry waste, mixed with CCW. All the analyzed variables (Table 1) were used for the correlation analysis to see their mutual relations. Table 2 shows that S was positively correlated only with Se, but not significantly, while its relations with U and V were surprisingly very weak. Positive statistically significant relations were as follows: Se-U, Se-Sr, U-Sr, V-Cr, and Pb-Zn. The first three correlations indicate on weathering of Raša coal particles (Se and U), concomitant with weathering of karst bedrock (Sr), while V, Cr, Pb, and Zn are characteristic for metal processing activities [20]. Expectedly, LOI was positively correlated with S, i.e., coal particles, not significantly though. It is intertesting to note that two samples, one taken from a vicinity of the Trget beach, had lost all their Se (Table 1), presumably due to leaching processes [35].

Table 2. Kendall's Tau correlation coefficients (below the diagonal) of the variables shown in Table 1 (bold italic underlined significant at $p < 0.05$; p values displayed above the diagonal).

	pH	CaCO$_3$	LOI	CEC	S	Se	V	U	Sr	Cr	Cu	Pb	Zn
pH		0.07	0.10	0.27	0.19	0.58	0.43	0.43	0.79	0.19	0.43	0.07	0.43
CaCO$_3$	0.73		0.19	0.07	0.62	0.79	0.99	0.62	0.32	0.62	0.14	0.32	0.14
LOI	−0.66	−0.52		0.27	0.19	0.41	0.79	0.79	0.79	0.79	0.79	0.07	0.79
CEC	0.44	0.73	−0.44		0.79	0.58	0.43	0.43	0.79	0.19	0.79	0.19	
S	−0.52	−0.20	0.52	0.10		0.24	0.85	0.85	0.57	0.85	0.34	0.03	0.18
Se	−0.22	0.10	0.33	0.22	0.41		0.43	0.05	0.01	0.43	0.24	0.24	0.43
V	−0.31	0	−0.10	0.31	0.06	0.27		0.57	0.34	0.03	0.85	0.34	0.85
U	0.31	0.20	−0.10	0.10	0.06	**_0.69_**	0.20		0.01	0.85	0.18	0.85	0.34
Sr	0.10	0.40	0.10	0.31	0.20	**_0.82_**	0.33	**_0.86_**		0.85	0.09	0.57	0.18
Cr	−0.52	−0.20	0.10	0.10	0.06	0.27	**_0.73_**	−0.06	0.06		0.34	0.34	0.57
Cu	−0.31	−0.60	0.10	−0.52	−0.33	−0.41	0.06	−0.46	−0.60	0.33		0.85	0.01
Pb	0.73	0.40	−0.73	0.10	**_−0.73_**	−0.41	−0.33	−0.06	−0.20	−0.33	0.06		0.57
Zn	−0.31	−0.60	0.10	−0.52	−0.46	−0.27	−0.06	−0.33	−0.46	0.20	**_0.86_**	0.20	

3.1.2. Desulfurization of Coal and Soil

Figure 2 presents the results of desulfurization conducted on SHOS Raša coal. It shows that maximum removal of its sulfur, which was almost entirely in organic form [5,26], was up to 5% at best, i.e., not significant. However, it shows a general trend of decreased S concentrations in coal with time. Raša coal desulfurization was far less than S removal (9%) found in the case of Assam coal, which had 5–6% S, mainly in organic form, but also in pyritic form to some extent [36]. No dissolution of organic + sulfate S was found, as Thiobacillus ferrooxidans was not able to remove organically bound S [36]. Sulfur removal from Assam coal was only 9% compared to 91% from lignite, which had high content of pyritic S. It was also found that poor removal of S from Assam coal was due to extensive precipitation of jarosites, which was reflected in the maximum increase in the volatile matter in microbially treated Assam coal [36].

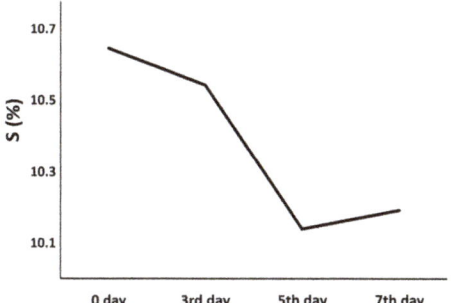

Figure 2. Sulfur concentrations (%) in original (0 day), and bioremediated (3-, 5-, and 7-day treatment periods) samples of SHOS Raša coal.

Microbial coal desulfurization, specifically the inorganic S, involves a complex combination of nonbiological and microbiologically catalyzed oxidations of sulfide minerals into sulfates, which are soluble in water. Herewith, desulfurized coal gets separated from liquid phase and washed with water [36]. The efficiency of bacterial bioremediation depends a lot on the type of sulfur species present in coal [37]. The sulfur-containing compounds in coal can be present as both aliphatic sulfur-containing chains (mercaptans, aliphatic sulfides, thiophenes) and heteroatoms in aromatic rings [38,39]. In high-sulfur coals, the relative proportion of aliphatic structures to total organic sulfur appears to be in the range of 30–50% [38]. Either in organic or pyritic form, sulfur amounts may depend on coal rank, where higher coal rank has higher amount of labile sulfur-containing compounds such as aliphatic thiols and sulfides [39,40]. High molecular weight compounds, such as aromatic ones, are usually more stable, and therefore heavier to break by bacteria [41,42]. Measurements made on Indonesian and Indian coals by means of *Ralstonia* sp. showed sulfur removal from 6% to 68% [17]. Also, investigations conducted on a same Indian coal sample aimed at comparison of desulfurization capacity of two bacteria, *Pseudoxanthomonas* sp. and *Ralstonia* sp., showed a better performance of the previous one [17,43]. Results from other investigations of coal remediation using other bacterial species showed sulfur removal values of 27%, 50%, and 31–51% [42,44,45].

The desulfurizing potential of *Ralstonia* sp. tested on soil polluted with SHOS Raša coal and CCW is presented in Figure 3. Values of removal of S (%) in samples S2, S4, S6, S8, S9, and S10 were up to 50, 30, 40, 60, 60, and 20, respectively. Similarly to desulfurization in SHOS Raša coal, desulfurization of soil increased with time. The highest removal was recorded in a sample collected from the CCW site (Štrmac locality), hosting a huge quantity of SHOS Raša coal combustion byproducts, as well as waste from a former foundry factory (closed in the late 1950s). The lowest removal was exhibited by the unpolluted soil sample collected away from the Raša locality (Figure 1). Hereby, the calculated Kendall's Tau correlation index between S removal values (%) and initial total S levels in soil (%) was 0.84. A similar finding was reported by [17], which determined a positive correlation between total S (2.2 Wt% dry ash free basis) and removal percentage levels (up to 45%).

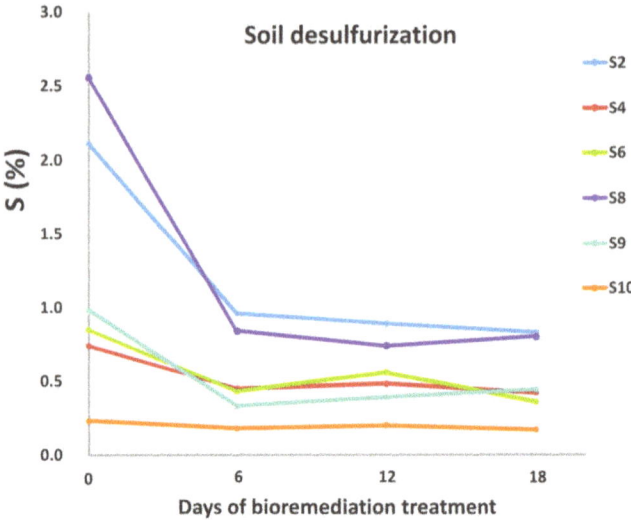

Figure 3. Soil total sulfur (%) in original (0 day), and bioremediated (6-, 12-, and 18-day treatment periods) samples. S10—unpolluted soil; S2, 6, and 9—soil dominantly polluted with SHOS Raša coal particles; S4, and 8—soil dominantly polluted with CCW particles.

3.1.3. Demineralization of Soil

Table 3 presents comparison of levels of PTEs in untreated and bioremediated soil samples with calculated removal values (%). Similar to soil desulfurization, lowest PTE removals were found for the unpolluted soil sample S10. Anomalously elevated Cu and Zn levels in the sample S4, both initial and bioremediated, are related to previously active practice of dumping foundry waste enriched in heavy metals [20]. Compared to initial Cu value, the bioremediated Cu was almost two-fold, the cause of which remains unclear, and it should be examined by future detailed microbiological studies. Almost all removal values in the case of two polluted soil samples (S4 and S14) were higher compared to the ones calculated for the unpolluted (S10) soil sample. The highest PTE removal values, except for Se and U, were observed for the sample S14. It was collected from a site of the former coal-train rail track Raša-Štalije (R.-Š., red-colored line in Figure 1). The Štalije location was a coal separation unit in the past. At the site S14, normally red-colored terra rossa soil is almost black due to dispersed coal particles fallen from coal-train coaches in the past [20]. The highest removal values in the case of the sample S10 are listed in descending order as follows: Rb > Ni = Zn > As = Ba = Cr = Cu = Sr = V > Pb > Co = U. The highest removal values in the case of the sample S4 were in following descending order: Ba = Se > V > Cr = Ni = Rb > Zn > As = Co = Pb > U. In the case of the sample S14, descending order of removal levels was following: As = > Ba = Co = Cu = Ni = Rb = Sr = V = Zn > Cr = Pb > U. Herewith, the lowest removal values were exhibited by U and Pb.

Similar to desulfurization, correlation indices between PTE removal levels (%) and initial total PTE levels (mg/kg) for soil samples S10, S4, and S14 were 0.60, 0.14, and 0.13, respectively. Herewith, soil desulfurization and demineralization trends had similar patterns, and they were also similar to previously found relevant relations for coal [16,17]. Evidently, bioremediation experiment carried out in this study was fairly successful. Moreover, its performance is comparable with coal PTE removals found by an earlier study [46], which were as follows: As (53%), Cu (39%), Co (42%), Cr (13%), Zn (31%), Ni (34%), and Pb (65%). The study [10] reported the potential of *Ralstonia* sp. HM-1, one of the bacteria resistant to potentially toxic metals, to improve adsorption onto lake sediments and biostabilization of Cd and Zn. Batch experiments were conducted using the spike of the synthetic Cd and Zn stock solution to investigate the effect of both the indigenous microorganism in sediment and the inoculation of *Ralstonia* sp. HM-1 in the bottle containing sediment and surface lake water. The reduction of the exchangeable fraction and the increase of bound organics and sulfide fractions were observed with the addition of *Ralstonia* sp. HM-1, showing its role in the prevention of metal elution from the sediment to water phase [10].

Table 3. Levels of PTEs (mg/kg) in nontreated (initial), and bioremediated (final, 12-day treatment) soil samples, with calculated removal values (na—not analyzed).

	S10			S4			S14		
	Initial	Final	Removal (%)	Initial	Final	Removal (%)	Initial	Final	Removal (%)
As	30.8	23.3	20	8.82	5.55	30	46.5	5.02	80
Ba	227	174	20	392	135	60	147	48.8	60
Co	19.3	18.2	0	9.05	6.22	30	9.30	3.05	60
Cr	135	106	20	89.7	47.7	40	100	42.6	50
Cu	41.4	31.3	20	1800	3000	-	190	71.7	60
Ni	123	80.6	30	52.3	26.5	40	60.3	19.5	60
Pb	56.4	48.9	10	200	132	30	46.3	20.3	50
Rb	122	67.7	40	30.4	18.0	40	23.7	9.40	60
Se	1.50	1.35	10	3.31	1.00	60	27.5	27.0	0
Sr	82.2	63.0	20	285	328	-	356	127	60
U	4.30	4.18	0	5.76	4.23	20	8.20	7.07	10
V	229	167	20	118	58.2	50	264	93.6	60
Zn	169	112	30	6580	3320	40	863	331	60

S10—unpolluted soil, S4—soil dominantly polluted with CCW ash, S14—soil dominantly polluted with Raša coal particles.

3.2. Zeolite Adsorptive Removal of PTEs from Coalmine and Stream Water Samples

Initial total levels of Se, U, V, and Mo in water samples, elements highly enriched in SHOS coals [5], are presented in Figure 4. Their close association, characteristic for SHOS Raša coal, was evidenced by statistically significant ($p < 0.05$), highly positive correlation coefficients (>0.99), reported earlier [5]. Selenium replaces S in organic complexes, and Se-bearing coals are exclusively high-S coals [47]. World stream water values [48] of Se, U, V, Mo, Cd, Ni, Sr, and Ba are as follows (µg/L): 0.2, 0.04, 0.9, 0.5, 0.02, 0.3, 70, and 20, respectively. Compared to them, Se was increased 35–45 times, U 45–55 times, V 3–5 times, Mo 25–35 times, Cd 6–9 times, Ni 3–6 times, Sr 10–13 times, and Ba 1.3–1.7 times. All the other elements were comparable to world stream water data [48]. By considering the downstream trend of sampling points, i.e., no. 1–5, only Pb, Fe, Cu, Zn, and As had increasing concentrations, while values of other PTEs were rather similar regardless of the sampling sites, shown in Figure 5 for Se, Mo, U, and V concentrations. Kendall's Tau correlation coefficients of the analyzed PTEs are shown in Table 4. They were predominantly positive, and several of them were significant. It should be noted that Se was not significantly correlated with either of analyzed PTEs, and it was even negatively correlated with few of them (Pb, Cu, Zn, and Ba). Selenium is known as an essential toxin due to its narrow range between dietary essentiality and toxicity for lifeforms. According to the study [35], waterborne Se levels of 2–5 µg/L pose concern to the aquatic life. Therefore, it is of interest to monitor Se levels in coalmine water and surface streams fed by it. As can be noticed from Figure 4, Se levels are comparable with previous studies [5,20,28], and are close to the Croatian regulatory limit value of 10 µg/L. In comparison to the world data, these values were much above the average total Se measured (µg/L) in wastewater from Spain (0.13), Belgium (0.35), Israel (0.44), Germany (0.12), Netherlands (0.12), and New York (<0.2) (references in [28]).

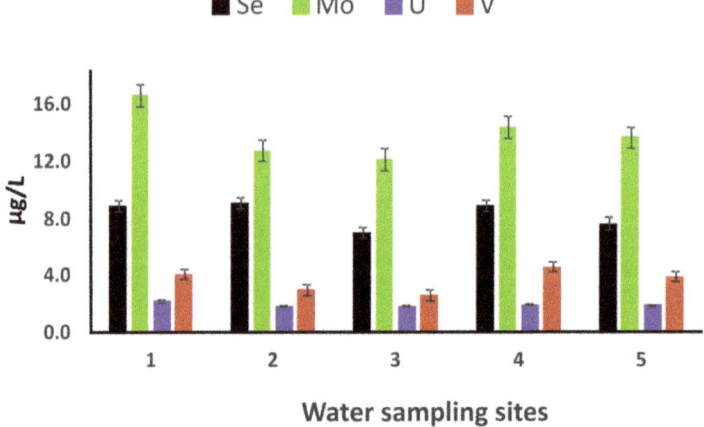

Figure 4. Initial Se, Mo, U, and V total concentrations (µg/L) in water samples collected as follows: 1, 2: coalmine water; 3, 4: Krapan stream, downstream of coalmine; and further downstream, sample no. 5: Krapan stream, downstream of municipal wastewater effluent.

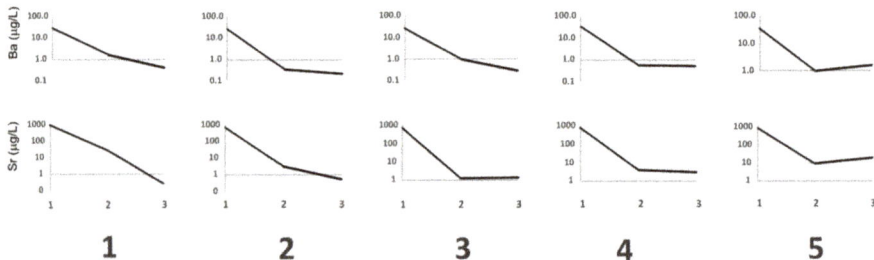

Figure 5. Water (1, 2: coalmine water; 3, 4: Krapan stream, downstream of coalmine; 5: Krapan stream, downstream of municipal wastewater effluent) Ba and Sr total values (vertical axis in log scale). 1—inital value, 2—value following the zeolite Purmol 4ST (Z) treatment, and 3—value following the activated zeolite 3A-50% (A) treatment.

Table 4. Kendall's Tau correlation coefficients (below the diagonal) of the variables measured in coalmine and surface water samples (bold italic underlined significant at $p < 0.05$; p values displayed above the diagonal).

	Cd	Pb	Cr	Fe	Ni	Cu	Zn	Sr	Ba	As	Se	Mo	U	V
Cd		0.79	0.07	0.79	0.19	0.79	0.43	0.07	0.43	0.79	0.27	0.02	0.07	0.07
Pb	−0.10		0.62	0.05	0.32	0.01	0.14	0.62	0.62	0.32	0.79	0.99	0.62	0.62
Cr	0.73	0.20		0.32	0.05	0.62	0.14	0.14	0.14	0.32	0.43	0.05	0.14	0.01
Fe	0.10	***0.80***	0.40		0.62	0.05	0.32	0.99	0.32	0.14	0.79	0.62	0.99	0.32
Ni	0.52	0.40	***0.80***	0.20		0.32	0.05	0.05	0.32	0.62	0.79	0.14	0.05	0.05
Cu	−0.10	***0.99***	0.20	***0.80***	0.40		0.14	0.62	0.62	0.32	0.79	0.99	0.62	0.62
Zn	0.31	0.60	0.60	0.40	***0.80***	0.60		0.14	0.14	0.32	0.79	0.32	0.14	0.14
Sr	0.73	0.20	0.60	0	***0.80***	0.20	0.60		0.62	0.99	0.79	0.05	0.01	0.14
Ba	0.31	0.20	0.60	0.40	0.40	0.20	0.60	0.20		0.05	0.79	0.32	0.62	0.14
As	0.10	0.40	0.40	0.60	0.20	0.40	0.40	0	***0.80***		0.79	0.62	0.99	0.32
Se	0.44	−0.10	0.31	0.10	0.10	−0.10	−0.10	0.10	−0.10	0.10		0.43	0.79	0.43
Mo	***0.94***	0	***0.80***	0.20	0.60	0	0.40	***0.80***	0.40	0.20	0.31		0.05	0.05
U	0.73	0.20	0.60	0	***0.80***	0.20	0.60	***0.99***	0.20	0	0.10	***0.80***		0.14
V	0.73	0.20	***0.99***	0.40	***0.80***	0.20	0.60	0.60	0.60	0.40	0.31	***0.80***	0.60	

The results of the zeolite adsorptive removal of PTEs from coalmine and surface water samples are presented in Table 5, and Figure 5.

Table 5. Water (Loc. (location)—1, 2: coalmine water; 3, 4: Krapan stream, downstream of coalmine; 5: Krapan stream, downstream of municipal wastewater effluent) PTE total values (µg/L). Tr. (treatment): 1—inital value, 2—zeolite Purmol 4ST (Z), and 3—activated zeolite 3A-50% (A).

Loc.	Tr.	Mo	Cd	Pb	U	V	Cr	Fe	Ni	Cu	Zn	As	Se	Ba	Sr
1	1	16.7	0.18	0.08	2.20	4.10	0.89	10.6	1.40	0.49	6.81	0.35	8.90	27.6	908
	2	18.4	0.16	0.07	2.31	5.30	10.1	15.9	1.43	2.26	4.64	0.84	11.4	1.57	27
	3	32.4	0.26	0.21	3.11	14.5	2.93	46.0	0.78	1.14	7.55	2.30	14.9	0.39	0.24
2	1	12.8	0.14	0.09	1.84	3.00	0.73	25.8	0.98	0.62	5.01	0.38	9.10	26.1	721
	2	14.0	0.12	0.06	1.91	4.20	10.1	19.6	0.89	0.73	3.46	0.99	8.90	0.33	3.00
	3	13.9	0.12	0.14	1.37	7.00	1.23	5.00	0.41	0.50	3.57	1.77	9.30	0.21	0.54
3	1	12.2	0.12	0.19	1.85	2.60	0.67	12.0	1.00	1.02	6.65	0.13	7.00	25.7	732
	2	13.3	0.13	0.24	2.12	8.50	9.50	168	1.24	4.60	11.3	1.21	9.00	0.95	1.21
	3	13.1	0.12	0.20	1.76	10.4	1.20	29.1	0.65	1.97	6.85	1.75	9.50	0.29	1.29
4	1	14.4	0.15	0.34	1.94	4.60	0.90	81.3	1.69	3.24	14.5	0.62	8.90	33.6	798
	2	15.1	0.14	0.36	1.95	5.20	9.92	33.6	2.63	3.23	13.8	1.36	9.10	0.55	3.84
	3	14.9	0.14	0.17	1.65	8.20	1.32	20.7	1.60	3.24	17.1	1.77	7.10	0.50	2.94
5	1	13.7	0.14	0.27	1.87	3.90	0.78	26.6	1.22	2.31	13.0	0.65	7.70	33.7	786
	2	15.0	0.14	0.20	1.96	4.90	9.80	34.0	2.63	2.97	10.5	1.09	8.20	1.02	9.34
	3	15.0	0.14	0.89	1.72	8.10	1.39	32.2	1.29	2.50	10.9	1.82	7.30	1.65	18.6

The best removal efficiencies were found for Ba and Sr. The results presented in Figure 5 demonstrate the decrease of Ba and Sr for the both zeolite treatments in all water samples except of the sample no. 5. In the case of Ba, removal values (%) of the zeolite Purmol 4ST (Z), and the activated zeolite 3A-50% (A) for the samples 1–5 were as follows, respectively: 94.3 and 98.6; 98.7 and 99.2; 96.3 and 98.9; 98.4 and 98.5; and 97.0 and 95.1. Excluding sample no. 5, the A zeolites exhibited better performances compared to the Z ones. In the case of Sr, removal values (%) of the Z, and A for the samples 1–5 were as follows, respectively: 97.0 and 99.9; 99.6 and 99.9; 99.8 and 99.8; 99.5 and 99.6; and 98.8 and 97.6. Again, excluding sample no. 5, the A zeolites exhibited better performances compared to the Z ones. It seems that the sample no. 5, impacted by both coalmine and MWE effluents, was more challenging in terms of purification with the applied zeolites compared to the other four water samples. Possibly, it could be ascribed to increased levels of Pb, Fe, Cu, Zn, and As in the sample no. 5, as lower removal efficiencies have been referred [25] in conditions of increased PTE concentrations. However, at very low PTE concentrations, the effects of zeolites likely become negligible, as was probably the case with Pb levels in this study. By converting 2.5 mmol Pb/L, reported in [25], it is about 518 mg/L, which is much higher than Pb data in this study (Table 5). Commonly, tests of water treatment with zeolites operate with PTE concentrations in the range 100x µg/L to 100x mg/L. It was not possible to compare the results of this study (Table 5) with literature values. To the best of our knowledge, there are no published similar studies focused on karst coalmine effluents treated with zeolites. Herewith, the removal effects of the applied zeolites were observed in the case of Sr and Ba only, which can be ascribed to their highest concentrations in the initial samples compared to the rest of analyzed PTEs. Also, there is no single process capable of adequate water treatment, mainly due to the complex nature of the effluents [21]. A combination of different processes is commonly necessary to achieve the desired water quality in the most economical way. The paper [21] reviewed adsorption capacities of various industrial wastes in terms of heavy metal removal. It demonstrated that adsorption capacities of the adsorbents varied depending on the characteristics of the adsorbents, the extent of chemical modification, and the concentration of adsorbates. Generally, percent adsorption increased with increased adsorbent dose, contact time, and agitation speed. It should be noted that for every investigated element, there was a favorable pH range in which maximum adsorption occurred [21]. Differently to Ba and Sr, the results are specific to few samples, as in several situations, post-treatment concentrations were comparable, or even higher, than the initial ones. In conclusion, by considering a +/−10% variation of the values in Table 5, typical for many analytical methods, the effect of zeolites was significant for Ba and Sr only.

4. Conclusions

The present study applied bacterial cultures of *Ralstonia* sp. on SHOS Raša coal and soil polluted with coal and coal-combustion waste. The removal of organically bound sulfur from coal was negligible (up to 5%), while soil desulfurization was up to 60%. Values of removal of PTEs from soil were as follows: Up to 80% for As, and Mo, up to 60% for Se, Ba, and V, and up to 20% for U. By applying synthetic zeolites on SHOS Raša coalmine water and municipal wastewater, the significant removal was found for Sr (99.9%) and Ba (99.2%) only. The activated zeolites (3A-50%) were slightly more efficient sorbents for PTEs compared to the Purmol 4ST zeolites. The overall conclusion is as follows: The examined microorganisms could be used for soil cleanup in terms of sulfur and the abovementioned PTEs, which is a cost-effective alternative compared to various chemical and physical methods. Regarding PTE removal from SHOS Raša coalmine water, synthetic zeolites proved excellent sorbents for Ba and Sr, while Se, which is most important among PTEs at the study locality, should be examined by future multidisciplinary research.

Author Contributions: Conceptualization—G.M., P.K.S., A.L.S.; Methodology—G.M., P.K.S., A.L.S.; Validation—G.M., P.K.S., A.L.S., Z.O., Z.P.; Formal analysis—A.R., S.R., M.J.; Resources—P.K.S., A.L.S., M.K.J., Z.O., Z.P.; Data curation—G.M., M.J.; Visualization—G.M.; Writing (original draft preparation)—G.M.; Writing (review & editing)—G.M.

Funding: This research received no external funding.

Acknowledgments: The first author (G.M.) is acknowledging that this collaborative study, specifically the bioremediation experiment, entitled 'Decontamination of SHOS Raša coal and soil polluted by Raša coal using bacterial biomass: a case study of coal and soil from the Labin city area (North Adriatic, Croatia)', was carried out at BHU (Varanasi, India) thanks to the TECO grant ("EU-India project TECO ICI+/2014/342-817-Technological eco-innovation for the quality control and the decontamination of polluted waters and soils"). All co-authors thank Mladen Bajramović (IUR) and Glorija Paliska Bolterstein (the Raša city mayor) for the help with field-work. Following persons are greatly acknowledged for measurements, help with chemical equipment, laboratory analyses, and assistance: Željka Fiket (IRB, Zagreb, Croatia), Ankica Rađenović (Faculty of Metallurgy, Sisak, Croatia), Štefica Kampić (PMF, Zagreb, Croatia), Vladimir Damjanović and Željko Ostojić (Alumina Ltd, Zvornik, BiH). Nikola Medunić created Figure 1. Students Vanja Geng and Laura Kozjak are acknowledged for assistance in the field and lab. The critical and constructive feedback made by three anonymous reviewers immensely improved the clarity and quality of this paper, which is highly appreciated by all co-authors.

Conflicts of Interest: The authors declare no conflict of interest.

References

1. Dai, S.; Yan, X.; Ward, C.R.; Hower, J.C.; Zhao, L.; Wang, X.; Zhao, L.; Ren, D.; Finkelman, R.B. Valuable elements in Chinese coals. *Int. Geol. Rev.* **2016**. [CrossRef]
2. Rađenović, A. Inorganic constituents in coal. *Kem. Ind.* **2006**, *55*, 65–71.
3. Anggara, F.; Amijaya, D.H.; Harijoko, A.; Tambaria, T.N.; Sahri, A.A.; Asa, Z.A.N. Rare earth element and yttrium content of coal in the Banko coalfield, South Sumatra Basin, Indonesia: Contributions from tonstein layers. *Int. J. Coal Geol.* **2018**, *196*, 159–172. [CrossRef]
4. Medunić, G.; Ahel, M.; Božičević Mihalić, I.; Gaurina Srček, V.; Kopjar, N.; Fiket, Ž.; Bituh, T.; Mikac, I. Toxic airborne S, PAH, and trace element legacy of the superhigh-organic-sulphur Raša coal combustion: Cytotoxicity and genotoxicity assessment of soil and ash. *Sci. Total Environ.* **2016**, *566–567*, 306–319. [CrossRef] [PubMed]
5. Medunić, G.; Kuharić, Ž.; Krivohlavek, A.; Fiket, Ž.; Rađenović, A.; Gödel, K.; Kampić, Š.; Kniewald, G. Geochemistry of Croatian superhigh-organic-sulphur Raša coal, imported low-S coal and bottom ash: Their Se and trace metal fingerprints in seawater, clover, foliage and mushroom specimens. *Int. J. Oil Gas Coal Technol.* **2018**, *18*. [CrossRef]

6. Saikia, J.; Saikia, P.; Boruah, R.; Saikia, B.K. Ambient air quality and emission characteristics in and around a non-recovery type coke oven using high sulphur coal. *Sci. Total Environ.* **2015**, *530–531*, 304–313. [CrossRef]
7. Zacchini, M.; Pietrini, F.; Mugnozza, G.S.; Iori, V.; Pietrosanti, L.; Massacci, A. Metal tolerance, accumulation and translocation in poplar and willow clones treated with cadmium in hydroponics. *Water Air Soil Pollut.* **2009**, *197*, 23–34. [CrossRef]
8. Pandey, V.C.; Singh, J.S.; Singh, R.P.; Singh, N.; Yunus, M. Arsenic hazards in coal fly ash and its fate in Indian scenario. *Res. Conserv. Recycl.* **2011**, *55*, 819–835. [CrossRef]
9. Voltaggio, M.; Spadoni, M.; Sacchi, E.; Sanam, R.; Pujari, P.R.; Labhasetwar, P.K. Assessment of groundwater pollution from ash ponds using stable and unstable isotopes around the Koradi and Khaperkheda thermal power plants (Maharashtra, India). *Sci. Total Environ.* **2015**, *518–519*, 616–625. [CrossRef]
10. Park, Y.-J.; Ko, J.-J.; Yun, S.-L.; Lee, E.Y.; Kim, S.-J.; Kang, S.-W.; Lee, B.-C.; Kim, S.-K. Enhancement of bioremediation by Ralstonia sp. HM-1 in sediment polluted by Cd and Zn. *Bioresour. Technol.* **2008**, *99*, 7458–7463. [CrossRef]
11. Ojuederie, O.B.; Babalola, O.O. Microbial and Plant-Assisted Bioremediation of Heavy Metal Polluted Environments: A Review. *Int. J. Environ. Res. Public Health* **2017**, *14*, 1504. [CrossRef] [PubMed]
12. Vimal, S.R.; Singh, J.S.; Arora, N.K.; Singh, S. Soil-Plant-Microbe Interactions in Stressed Agriculture Management: A Review. *Pedosphere* **2017**, *27*, 177–192. [CrossRef]
13. Crawford, R.L.; Crawford, D.L. *Bioremediation: Principles and Applications*; Cambridge University Press: New York, NY, USA, 2005; 406p.
14. Ralph, J.P.; Catcheside, D.E.A. Transformations of low rank coal by Phanerochaete chrysosporium and other wood-rot fungi. *Fuel Process. Technol.* **1997**, *52*, 79–93. [CrossRef]
15. Sekhohola, L.M.; Igbinigie, E.E.; Cowan, A.K. Biological degradation and solubilisation of coal. *Biodegradation* **2013**, *24*, 305–318. [CrossRef] [PubMed]
16. Singh, P.K.; Singh, A.L.; Kumar, A.; Singh, M.P. Mixed bacterial consortium as an emerging tool to remove hazardous tracemetals from coal. *Fuel* **2012**, *102*, 227–230. [CrossRef]
17. Singh, A.L.; Singh, P.K.; Kumar, A.; Singh, M.P. Desulfurization of selected hard and brown coal samples from India and Indonesia with *Ralstonia* sp. and *Pseudoxanthomonas* sp. *Energy Explor. Exploit.* **2012**, *30*, 985–998. [CrossRef]
18. Utturkar, S.M.; Bollmann, A.; Brzoska, R.M.; Klingeman, D.M.; Epstein, S.E.; Palumbo, A.V.; Brown, S.D. Draft Genome Sequence for Ralstonia sp. Strain OR214, a Bacterium with Potential for Bioremediation. *Microbiol. Soc. Announc.* **2013**. [CrossRef]
19. Ryan, M.P.; Pembroke, J.T.; Adley, C.C. Ralstonia pickettii in environmental biotechnology: Potential and applications. *J. Appl. Microbiol.* **2007**. [CrossRef]
20. Medunić, G.; Kuharić, Ž.; Krivohlavek, A.; Đuroković, M.; Dropučić, K.; Rađenović, A.; Oberiter, B.L.; Krizmanić, A.; Bajramović, M. Selenium, sulphur, trace metal, and BTEX levels in soil, water, and lettuce from the Croatian Raša Bay contaminated by superhigh organic sulphur coal. *Geosciences* **2018**, *8*, 408. [CrossRef]
21. Ahmaruzzaman, M. Industrial wastes as low-cost potential adsorbents for the treatment of wastewater laden with heavy metals. *Adv. Colloid Interface Sci.* **2011**, *166*, 36–59. [CrossRef]
22. Fiket, Ž.; Galović, A.; Medunić, G.; Furdek Turk, M.; Ivanić, M.; Dolenec, M.; Biljan, I.; Šoster, A.; Kniewald, G. Adsorption of rare earth elements from aqueous solutions using geopolymers. *Proceedings* **2018**, *2*, 567. [CrossRef]
23. Jha, V.K.; Matsuda, M.; Miyake, M. Sorption properties of the activated carbon-zeolite composite prepared from coal fly ash for Ni^{2+}, Cu^{2+}, Cd^{2+} and Pb^{2+}. *J. Hazard. Mater.* **2008**, *160*, 148–153. [CrossRef] [PubMed]
24. Munthali, M.W.; Johan, E.; Aono, H.; Matsue, N. Cs+ and Sr2+ adsorption selectivity of zeolites in relation to radioactive decontamination. *J. Asian Ceram. Soc.* **2015**, *3*, 245–250. [CrossRef]
25. Ćurković, L.; Cerjan-Stefanović, Š.; Filipan, T. Metal ion exchange by natural and modified zeolites. *Water Res.* **1997**, *31*, 1379–1382. [CrossRef]
26. Medunić, G.; Rađenović, A.; Bajramović, M.; Švec, M.; Tomac, M. Once grand, now forgotten: What do we know about the superhigh-organic-sulphur Raša coal? *Min. Geol. Pet. Eng. Bull.* **2016**, *34*, 27–45. [CrossRef]
27. Dvoršćak, M.; Stipičević, S.; Mendaš, G.; Drevenkar, V.; Medunić, G.; Stančić, Z.; Vujević, D. Soil burden by persistent organochlorine compounds in the vicinity of a coal-fired power plant in Croatia: A comparison study with urban-industrialized area. *Environ. Sci. Pollut. Res.* **2019**. [CrossRef] [PubMed]

28. Medunić, G.; Kuharić, Ž.; Fiket, Ž.; Bajramović, M.; Singh, A.L.; Krivovlahek, A.; Kniewald, G.; Dujmović, L. Selemium dan other potentially toxic elements in vegetables and tissues of three non-migratory birds exposed to soil, water and aquatic sediment contaminated with seleniferous Raša coal. *Min. Geol. Pet. Eng. Bull.* **2018**, *33*, 53–62. [CrossRef]
29. Valković, V.; Makjanić, J.; Jakšić, M.; Popović, S.; Bos, A.J.J.; Vis, R.D.; Wiederspahn, K.; Verheul, H. Analysis of fly ash by X-ray emission spectroscopy and proton microbeam analysis. *Fuel* **1984**, *63*, 1357–1362. [CrossRef]
30. Miko, S.; Durn, G.; Adamcová, R.; Čović, M.; Dubíková, M.; Skalský, R.; Kapelj, S.; Ottner, F. Heavy metal distribution in karst soils from Croatia and Slovakia. *Environ. Geol.* **2003**, *45*, 262–272. [CrossRef]
31. ASTM International. Standard Test Methods for Total Sulfur in the Analysis Sample of Refuse Derived Fuel. Available online: https://www.astm.org/Standards/E775.htm (accessed on 7 September 2018).
32. Kodom, K.; Preko, K.; Boamah, D. X-ray fluorescence (XRF) analysis of soil heavy metal pollution from an industrial area in Kumasi, Ghana. *Soil Sediment Contam.* **2012**, *21*, 1006–1021. [CrossRef]
33. Fiket, Ž.; Mikac, N.; Kniewald, G. Mass Fractions of Forty-Six Major and Trace Elements, Including Rare Earth Elements, in Sediment and Soil Reference Materials Used in Environmental Studies. *Geostand. Geoanal. Res.* **2017**, *41*, 123–135. [CrossRef]
34. Hammer, Ø.; Harper, D.A.T.; Ryan, P.D. PAST: Paleontological statistics software package for education and data analysis. *Palaeontol. Electron.* **2001**, *4*, 1–9. Available online: http://palaeo-electronica.org/2001_1/past/issue1_01.htm (accessed on 30 May 2019).
35. Lemley, A.D. Guidelines for evaluating selenium data from aquatic monitoring and assessment studies. *Environ. Monit. Assess.* **1993**, *28*, 83–100. [CrossRef] [PubMed]
36. Acharya, C.; Kar, R.N.; Sukla, L.B. Bacterial removal of sulfur from three different coals. *Fuel* **2001**, *80*, 2207–2216. [CrossRef]
37. Berthelin, J. Microbial Weathering Processes in Natural Environments. In *Physical and Chemical Weathering in Geochemical Cycles*; Lerman, A., Meybeck, M., Eds.; Springer: Dordrecht, The Netherlands, 1988; Volume 251.
38. Calkins, W.H. The chemical forms of sulfur in coal: A review. *Fuel* **1994**, *73*, 475–484. [CrossRef]
39. Rađenović, A. Sulfur in coal. *Kem. Ind.* **2004**, *53*, 557–565.
40. Boudou, J.; Boulègue, J.; Maléchaux, L. Identification of some sulphur species in a high organic sulphur coal. *Fuel* **1987**, *66*, 1558–1569. [CrossRef]
41. Wrenn, B.A.; Venosa, A.D. Selective enumeration of aromatic and aliphatic hydrocarbon degrading bacteria by a most-probable-number procedure. *Can. J. Microbiol.* **1995**, *42*, 252–258. [CrossRef]
42. Tripathy, S.S.; Kar, R.N.; Kumar Mishra, S.; Twardowska, I.; Sukla, L.B. Effect of chemical pretreatment on bacterial desulphurisation of Assam coal. *Fuel* **1998**, *77*, 859–864. [CrossRef]
43. Singh, P.K.; Singh, A.L.; Kumar, A.; Singh, M.P. Control of different pyrite forms on desulfurization of coal with bacteria. *Fuel* **2013**, *106*, 876–879. [CrossRef]
44. Aller, A.; Martinez, O.; de Linaje, J.A.; Moran, A. Biodesulphurisation of coal by microorganisms isolated from the coal itself. *Fuel Process. Technol.* **2001**, *69*, 45–57. [CrossRef]
45. Cardona, I.C.; Márquez, M.A. Biodesulfurization of two Colombian coal with native microorganisms. *Fuel Process. Technol.* **2009**, *90*, 1099–1106. [CrossRef]
46. Singh, A.L.; Singh, P.K.; Kumar, A.; Singh, M.P. Demineralization of Rajmahal Gondwana coals by bacteria: Revelations from X-ray diffraction (XRD) and Fourier Transform Infra Red (FTIR) studies. *Energy Explor. Exploit.* **2015**, *33*, 755–767. [CrossRef]
47. Yudovich, Y.E.; Ketris, M.P. Selenium in coal: A review. *Int. J. Coal Geol.* **2006**, *67*, 112–126. [CrossRef]
48. Reimann, C.; de Caritat, P. *Chemical Elements in the Environment. Factsheets for the Geochemist and Environmental Scientist*; Springer-Verlag: Berlin/Heidelberg, Germany, 1998; 398p.

© 2019 by the authors. Licensee MDPI, Basel, Switzerland. This article is an open access article distributed under the terms and conditions of the Creative Commons Attribution (CC BY) license (http://creativecommons.org/licenses/by/4.0/).

Article

Removal and Ecotoxicity of 2,4-D and MCPA in Microbial Cultures Enriched with Structurally-Similar Plant Secondary Metabolites

Elżbieta Mierzejewska [1], Agnieszka Baran [2], Maciej Tankiewicz [3] and Magdalena Urbaniak [1,4,5,*]

1. Department of Applied Ecology, Faculty of Biology and Environmental Protection, University of Lodz, Banacha 12/16, 90-237 Lodz, Poland
2. Department of Agricultural and Environmental Chemistry, Faculty of Agriculture and Economics, University of Agriculture in Krakow, Mickiewicza 21, 31-120 Cracow, Poland
3. Department of Environmental Toxicology, Faculty of Health Sciences, Medical University of Gdansk, Dębowa 23 A, 80-204 Gdansk, Poland
4. European Regional Centre for Ecohydrology of the Polish Academy of Sciences, Tylna 3, 90-364 Lodz, Poland
5. Department of Biochemistry and Microbiology, Faculty of Food and Biochemical Technology, University of Chemistry and Technology in Prague, Technicka 3, 166 28 Prague, Czech Republic
* Correspondence: magdalena.urbaniak@vscht.cz

Received: 20 June 2019; Accepted: 8 July 2019; Published: 13 July 2019

Abstract: The removal of contaminants from the environment can be enhanced by interactions between structurally-related plant secondary metabolites (PSMs), selected xenobiotics and microorganisms. The aim of this study was to investigate the effect of selected PSMs (ferulic acid—FA; syringic acid—SA) on the removal of structurally-similar phenoxy herbicides (PHs): 2,4-dichlorophenoxyacetic acid (2,4-D) and 2-methyl-4-chlorophenoxyacetic acid (MCPA). The study also examines the biodegradation potential of soil bacteria, based on the occurrence of functional *tdfA*-like genes, and the ecotoxicity of the samples against two test species: *Sinapis alba* L. and *Lepidium sativum* L. The microbial cultures spiked with the PSMs demonstrated higher phenoxy acid removal: 97–100% in the case of 2,4-D and 99%–100% for MCPA. These values ranged from 5% to 100% for control samples not amended with FA or SA. The higher herbicide removal associated with PSM spiking can be attributed to acceleration of the microbial degradation processes. Our findings showed that the addition of SA particularly stimulated the occurrence of the total number of *tfdA* genes, with this presence being higher than that observed in the unamended samples. PSM spiking was also found to have a beneficial effect on ecotoxicity mitigation, reflected in high (102%) stimulation of root growth by the test species.

Keywords: 2,4-D; MCPA; plant secondary metabolites; ferulic acid; syringic acid; biodegradation; ecotoxicity

1. Introduction

Two of the most commonly-used herbicides in agriculture and the home/garden market sector are phenoxy herbicides (PHs), 2,4-D (2,4-dichlorophenoxyacetic acid) and MCPA (2-methyl-4-chlorophenoxyacetic acid) [1]. These compounds are able to selectively control the growth of dicotyledonous weeds [2]. All PHs are constructed from ring-like structures, with at least one chlorine atom attached to the ring at different positions [3]. Their mode of action is similar to that of phytohormones (auxins), in that they disturb the physiological processes of plants and their growth regulation [4].

Around 6.5 million kg of the active ingredients of PHs (i.e., 2,4-D and MCPA) were sold in the EU in 2016, with around two million kg being sold in Poland [5]. However, such profligate usage runs the risk of misuse: incorrect storage and application practices may widen the dispersal of the compound and its metabolites throughout the environment, especially in soil and water ecosystems, thus disturbing the ecological sustainability of habitats. Environmental studies have shown the concentrations of 2,4-D and MCPA to fluctuate seasonally, e.g., from 0 to 150 µg/L in the Narew River (Poland) [6], and can reach a level of 329.42 µg/L in water coming from rice fields [7]. This is of particular concern, as EU Directive E98/83/EC specifies the maximum permissible concentration of pesticide residues in drinking water to be 0.50 µg/L. Both 2,4-D and MCPA exhibit high toxicity to soil and water organisms, causing severe malformations and cell death [8,9]. The high toxicity of the PHs, together with their increasing persistence in acidic soil, creates a need to identify nature-based solutions that can enhance their degradation by autochthonous microorganisms, particularly bacteria.

One significant way of removing PHs from soil, and mitigating their toxic effects, is by the use of indigenous soil microbiota harboring desirable catabolic genes, typically those from the *tfdA* cluster. The first step in the bacterial phenoxy herbicide degradation pathway is initiated by α-ketoglutarate-dependent dioxygenase, which is encoded by *tfdA* or *tfdA-like* genes [10,11] such as *tfdAα* (detected in α-Proteobacteria) and *tfdA* Class I, II and III (identified in γ- and β-Proteobacteria) [12]. The *tfdA* genes encode aromatic ring hydroxylation dioxygenases (RHDO), which are widely distributed among a number of microorganisms. *TfdA-like* genes can also be transferred through horizontal gene transfer [12].

However, this degradation activity by indigenous soil microbiota in the environment can be restricted by unfavorable conditions, one of which being the limited availability of carbon. Hence, the biodegradative potential of soil microbiota can be increased by the addition of plant secondary metabolites (PSMs) to the soil, these being natural organic compounds frequently resembling the chemical structures of xenobiotics.

The presence of PSMs such as flavonoids, coumarins, terpenes and phenolic compounds in the soil plays an important role in the ecological relationships between plants and microorganisms. PSMs can influence the chemical and physical properties of soil, protect plants against pathogens, serve as substrates or enhance the catabolic pathway of soil microorganisms in the presence of xenobiotics [13]. They can enhance the activity of degradation processes in three key ways: by serving as primary substrates in cometabolism and providing energy for microorganism growth, by acting as inducers of degradative enzymes due to their structural similarities to xenobiotics, and by enhancing the degree of contamination removal by increasing the bioavailability of pollutants in soil [13]. The structural similarity between certain xenobiotics and PSMs may have a profound impact on the biodegradation of a given structurally-related compound (Table 1) [14,15], insofar that PH removal can be enhanced by the application of PSMs with similar chemical structures [16,17].

Table 1. Examples of PSMs and Their Effect on the Biodegradation of Selected Contaminants.

Contaminant.	Studied Plant	Plant Secondary Metabolite	Observed Effect	Literature
Polychlorinated biphenyls (PCBs)	n.a.	flavonoid, naringin	enhancement of PCBs reduction	[15]
Trichloroethene (TCE)	n.a.	cumen	enhancement of the biodegradation of TCE by *R. gordoniae* up to 75% within 24 h	[18]
Dichlorodiphenyldichloroethylene (DDE)	*Cucurbita pepo*	low molecular weight acids (e.g., citric acid)	enhancement of p,p'-DDE bioavailability in soil	[19]
Polycyclic aromatic hydrocarbons (PAH)	*Apium graveolens*	linoleic acid	enhancement of benzo[α]pyrene removal	[20]
2,4-dichlorophenol (2,4-DCP)	soil samples taken from sites located under *Pinus sylvestris*, *Quercus robur*	limonene and α-pinene	induction of 2,4-DCP degradation by indigenous soil microbiota	[21]

n.a.—not analyzed.

The present microcosm study examines the effect of the application of two PHs, *viz.* 2,4-D and MCPA, on the degradation of two structurally-similar PSMs, *viz.* FA and SA, to confirm whether such similarity influences the rate of removal [16]. The study examines the influence of the PSMs addition on

(1) changes in PH concentration over time, (2) the presence of PH-degrading genes (3) and changes in ecotoxicity occurring throughout the experiment, the latter being measured using two dicotyledonous plant species: *Lepidium sativum* and *Sinapis alba*.

2. Materials and Methods

2.1. Microcosm Setup

The study was conducted in bacterial cultures containing liquid Mineral Salt Medium (MSM) (1 g/L of KNO_3, 0.5 g/L of K_2HPO_4 and $MgSO_4$ 7 H_2O, 0.05 g/L of NaCl and $CaCl_2$, and 0.01 g/L of $FeCl_3$). MSM was filtered through a microbiological Corning™ (New York, NY, USA) Disposable Vacuum Filter (0.22 μL) and enriched with soil microorganisms derived from the agricultural soil extract (SE) 50%:50%, v/v (see Supplementary Materials: Texts S1 and S2; Table S1).

Prepared MSM or MSM+SE samples were amended with 2,4-D (≥95.0% technical purity, molecular weight 220.04 g/mol, 677 ppm water solubility at 25 °C, pKa value 2.73, Sigma Aldrich® (St. Louis, MS, U.S.) or MCPA (≥95.0% technical purity, molecular weight 200.61 g/mol, 0.825 g/L water solubility at 20 °C, pKa value 3.07, Sigma Aldrich® (St. Louis, MS, U.S.), at doses of 0.1 and 0.5 mM.

FA (≥95.0% purity, molecular weight 194.186 g/mol, 780 mg/mL water solubility at 25° C, pKa value 4.58, Sigma Aldrich® (St. Louis, MS, U.S.), or SA (≥95.0% purity, molecular weight 198.17 g/mol, 5.78 mg/mL water solubility at 25 °C, pKa value 4.34, Sigma Aldrich® (St. Louis, MS, U.S.) were applied at a dose of 0.25 mM [22].

Samples containing only sterile MSM, and sterile MSM + sterile soil extract (SSE) were used as controls to assess the degree of physicochemical degradation. The samples were incubated in darkness at 25 °C for 24 days.

2.2. 2,4-D and MCPA Concentration Measurement and Their Ppercentage Removal

Subsamples were collected at the beginning of the experiment and after the 24-day incubation period, and these were used for monitoring changes in 2,4-D and MCPA concentration.

The concentration of 2,4-D was assessed using an enzyme immunoassay (ELISA) RaPID Assay® (A00082, Guildford, UK) 2,4-D Test Kit according to the manufacturer's instructions [23]. Each analytical batch contained a sample blank, a control sample of known concentration, four calibration standards (0, 1, 10 and 50 μg/L) and the test samples. As the initial concentrations of 2,4-D (0.1 mM and 0.5 mM) were outside the range of the standard curve, the samples were appropriately diluted. The precision was verified by duplicate analyses, and the test reproducibility was measured using coefficients of variation (CVs); these CVs should be lower than 10% for the calibration standards, and lower than 15% for the samples. If the CVs exceeded the above values, the whole procedure was repeated to achieve good quality of the obtained results. The minimum detection level of the kit was 0.7 μg/L. Samples showing a concentration lower than the minimum detection level were considered to be negative.

The concentration of MCPA was determined using a GC-MS TQ 8040 (Shimadzu Corp., Kyoto, Japan) gas chromatograph equipped with split/splitless injector operating in a splitless mode at 250 °C, and a triple quadrupole mass spectrometer (MS) connected to "LabSolutions" software (version 4.45, Shimadzu Corp., Kyoto, Japan) extended with Pesticide Smart Database (version 1.03, Shimadzu Corp., Kyoto, Japan) (see Supplementary Materials: Text S3).

The working standard solutions for the calibration study were prepared by spiking the tested samples with the standard solution in the concentration range of 0.01–100 μg/mL. The linear range for MCPA was studied by replicate analysis of the standard stock solutions. The linear regression value was calculated with the mean peak areas of five replicate injections. The linear regression was in the range of 0.19–100 μg/mL with a coefficient of determination of 0.9996. The coefficient of variability, i.e., the percentage of relative standard deviation (CV%), calculated as the mean value of the concentrations across the linear range, was found to indicate good precision (1.1%). The sensitivity of the method was 62.5 ng/mL, considered in terms of the limit of detection (LOD), calculated from calibration functions. The limit of quantization (LOQ), defined as three times the LOD value, was 0.19 µg/mL. After the basic validation parameters had been determined, the environmental samples were analyzed in nine replications. For each sample, the mean value was calculated together with the relative standard deviation.

The percentage removal of 2,4-D and MCPA was calculated in reference to the initial concentration of herbicide used (0.1/0.5 mM). The percentage removal of studied phenoxy herbicides was calculated according to Equation (1):

$$PR = \frac{C_i - C_f}{C_i} \times 100\% \qquad (1)$$

Equation (1) PR—percentage removal of phenoxy herbicides after 24 days of incubation, C_i—initial concentration of phenoxy herbicide used in the microcosm preparation; C_f—final concentration of phenoxy herbicide measured after 24 days of incubation.

2.3. Molecular Analysis

Subsamples for molecular analysis were collected every six days, i.e., four times during the incubation period. The collected subsamples were examined for the occurrence of bacterial *16S rRNA* gene fragments and five functional phenoxy acid degradative genes (*tfdA*, *tfdA*α and *tfdA* Class I, *tfdA* Class II and *tfdA* Class III). DNA was extracted from the soil bacteria using a *GeneMATRIX Soil DNA Purification Kit EurX®* (Gdansk, Poland), according to the manufacturer's instructions. Polymerase Chain Reaction (PCR) conditions were applied according to literature (see Supplementary Materials: Table S2) with minor modifications of annealing temperatures. The 20 µL reaction mixture contained: sterile H_2O, 1xbuffer, 3.5 mM $MgCl_2$, 0.2 mM dNTP, 0.5 µM primers, 0.1 mg/mL BSA, 1–2.5 U/µL Taq Polymerase and 15–30 ng of template DNA. Five sets of *tfdA* gene primers (see Supplementary Materials: Table S2) were used for the amplification of given bacterial DNA fragments. The conditions for the PCR were optimized for each studied gene. PCR products were checked using 1.5% agarose gel electrophoresis and stained with ethidium bromide.

The *16S rRNA* gene fragments (1300–1400 bp) obtained from the variants amended with MCPA and SA were additionally amplified by PCR using thermostable Pfu DNA polymerase (ThermoScientific® (Waltham, MA, U.S.) and bacterium specific forward primer (27F5′-AGAGTTTGATCCTGGCTCAG-3′) and a universal reverse primer (1492R5-GGTTACCTTGTTACGACTT-3′) [24]. The amplified *16S rRNA* gene fragments were purified using a QIAGEX® II Gel Extraction Kit (Qiagen) (Hilden, Germany) and subjected to sequencing. Homology searches were performed using the National Center for Biotechnology Information microbial and nucleotide BLAST network service (http://blast.ncbi.nlm.nih.gov/Blast.cgi) and Vector NTI AdvanceTM 9 software (Invitrogen) (Carlsbad, CA, USA).

2.4. Ecotoxicity Assessment

Ecotoxicity measurements were performed twice during the experiment: at the beginning and after a 24-day incubation period, using *Phytotoxkit* Test (Microbiotest Inc., Nazareth, Belgium) [25], a commercial toxicity bioassay. The test compares the degree of inhibition of root length of certain test species after three days of exposure to a control sample with that of an uncontaminated control containing only distilled water. For the purpose of this experiment, the dicotyledons *Lepidium sativum* (L.) and *Sinapis alba* (L.) were used as test plants. The samples were classified as non-toxic when the percent effect of root growth inhibition (PE%) was ≥20%, slightly toxic for 20% ≤ PE < 50%, toxic for 50% ≤ PE < 100%, and highly toxic for PE = 100% [26].

The percentage effect on root growth inhibition of the studied soil was calculated according to Equation (2):

$$PE = \frac{A - B}{A} \times 100\% \qquad (2)$$

Equation (2) PE—percentage effect of root growth inhibition, A—plant root length in control soil; B—plant root length in studied soil.

3. Results

3.1. Changes in 2,4-D and MCPA Concentration

Table 2 indicates the results of 2,4-D and MCPA degradation during 24 days of incubation. Regarding the samples amended with 0.1 mM 2,4-D, between 83% and 93% degradation was observed for sterile untreated controls; however, this increased to 98% in samples with soil microorganisms enriched with SE, and to 100% for samples with MSM + SE + FA. For sterile samples amended with 0.5 mM 2,4-D, the samples treated with MSM + SE, MSM and MSM + FA demonstrated a removal rate of 100%, while those treated with SSE alone demonstrated 99% removal. Those treated with FA displayed slightly lower removal (97%).

Table 2. Percentage Removal (PR) of PH (2,4-D and MCPA) in Studied Samples.

PH	Concentration (mM)	Sample *	PR
2,4-D	0.1	MSM	86
		MSM + FA	93
		MSM + FA + SSE	83
		MSM + SE	98
		MSM + SE + FA	100
	0.5	MSM	100
		MSM + FA	100
		MSM + FA + SSE	99
		MSM + SE	100
		MSM + SE + FA	97
MCPA	0.1	MSM	40
		MSM + SA	11
		MSM + SA + SSE	19
		MSM + SE	53
		MSM + SE + SA	99
	0.5	MSM	27
		MSM + SA	5
		MSM + SA + SSE	12
		MSM + SE	99
		MSM + SE + SA	100

* MSM—mineral salt medium; SSE—sterile soil extract; SE—soil extract; FA—ferulic acid; SA—syringic acid.

In contrast, lower removal rates were observed for the samples amended with 0.1 mM MCPA: 11–40% removal was observed for the sterile, untreated samples, 53% for those treated with SE and 99% for those treated with MSM + SE + SA. Similar removal rates were observed for samples amended with 0.5 mM MCPA: MCPA removal ranged from 5% to 27% for sterile untreated samples, which increased to 99% for the SE samples and 100% for the SE + SA.

3.2. Molecular Analysis

Table 3 and Figure 1 show the results of the molecular analysis for studied samples. The bacterial *16S rRNA* gene fragment was detected in all studied samples containing SE. In the MSM + SE variant, the *tfdAα* gene fragment was detected after 6, 12 and 18 days of incubation and the *tfdA* gene fragment after 18 and 24 days. In the samples amended with 0.1 mM 2,4-D, the *tfdA* Class III gene was present during the whole experiment; however, the gene detection pattern changed after the addition of FA: the *tfdA* Class I gene was detected after 6, 12, 18 days and the *tfdAα* gene was detected after 18 and 24 days. The use of 0.5 mM 2,4-D, i.e., the higher concentration, shifted the frequency and timing of gene fragment detection: in samples with SE, *tfdAα* was detected only once after 24 days and *tfdA* Class III was observed after 12 and 18 days of incubation. The genes were found to be less apparent in the MSM + SE + FA variant (Table 4): only *tfdA* Class I was observed, on one occasion, after 6 days, *tfdAα* was observed after 6, 18 and 24 days.

Table 3. PCR Results for Target Genes; "+"Presence of PCR Product on 1.5% Agar Gel Electrophoresis; CI—Class I, CII—Class II, CIII—Class III.

		Target Gene	16S rRNA	tfdA alfa	tfdA CI	tfdA CII	tfdA CIII	tfdA	16S rRNA	tfdA alfa	tfdA CI	tfdA CII	tfdA CIII	tfdA
PH	Days of Incubation	2,4-D/MCPA Concentration (mM)			MSM + SE						MSM + SE + FA			
	6		+	+					+	+	+			
	12	0	+	+					+					
	18		+	+				+	+	+				
	24		+					+	+	+				
2,4-D	6		+			+			+	+				
	12	0.1	+			+			+	+				
	18		+			+			+	+	+			
	24		+			+			+	+				
	6		+						+					
	12	0.5	+			+			+		+			
	18		+			+			+					
	24		+	+					+					
					MSM + SE						MSM + SE + SA			
	6		+	+					+				+	
	12	0	+	+					+				+	
	18		+	+				+	+	+			+	
	24		+					+	+				+	
MCPA	6		+						+	+		+	+	
	12	0.1	+						+				+	
	18		+						+				+	
	24		+						+			+		
	6		+	+	+	+			+					
	12	0.5	+			+			+	+	+	+		
	18		+			+			+	+	+	+	+	
	24		+	+	+	+			+	+		+	+	

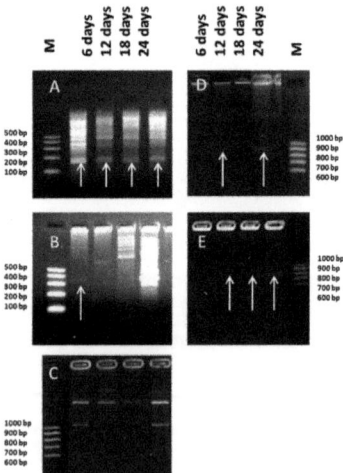

Figure 1. PCR Results Visualized on Gel Electrophoresis for Samples with MCPA (0.1 mM), SE and SA after 6, 12, 18, 24 days of incubation; A—*16S rRNA*; B—*tfdAα*; C—*tfdA* Class I; D—*tfdA* Class II; E—*tfdA* Class III M—marker, bp—base pairs.

Table 4. Number of Different Functional Genes Present in Studied Samples After 6, 12, 18, 24 Days of Incubation and Total Number of Detected Functional Genes.

PH	Concentration	Sample Variant *	Nr of Detected Functional Genes				
			Days of Incubation				Total
			6	12	18	24	
-	0	MSM + SE	1	1	2	1	5
		MSM + SE + FA	2	0	1	1	3
		MSM + SE + SA	0	1	2	1	4
2,4-D	0.1	MSM + SE	1	1	1	1	4
		MSM + SE + FA	1	1	2	1	5
	0.5	MSM + SE	0	1	1	1	3
		MSM + SE + FA	0	1	0	0	1
MCPA	0.1	MSM + SE	0	0	0	0	0
		MSM + SE + SA	3	1	1	1	6
	0.5	MSM + SE	3	1	1	3	8
		MSM + SE + SA	0	3	4	3	10

* MSM—mineral salt medium; SSE—sterile soil extract; SE—soil extract; FA—ferulic acid; SA—syringic acid.

Among the variants amended with 0.1 mM MCPA, no functional genes were detected in the SE samples. However, the SA samples demonstrated *tfdAα* on day 6 of incubation, *tfdA* Class II on day 6 and day 24, and *tfdA* Class III on days 6, 12 and 18. In the samples with 0.5 mM MCPA, all tested gene fragments were observed (*tfdAα* after 6 and 24 days, *tfdA* Class I after 24 days, Class II after 6 and 24 days and Class III after 6, 12, 18 days). In the MSM + SE + SA variant, *tfdAα* was present after 12, 18, 24 days, *tfdA* Class I after 12 and 18 days, *tfdA* Class II after 12, 18 and 24 days, and *tfdA* Class III after 18 and 24 days. In comparison to the sample amended with SE, the addition of SA was associated with a twofold increase in gene detection (Table 4). After 12 days, *tfdA* Class III was detected throughout the whole incubation period in samples amended with MSM + SE + SA. When SA was applied, the detection of the studied genes was found to double for both the *tfdAα* and the *tfdA* Class I genes.

In addition, most of the samples treated with 0.1 mM MCPA did not demonstrate the presence of any *tfdA*-like genes; however, 30% of the 0.1 mM MCPA samples treated with SA were found to contain three of the five tested genes: *tfdAα*, *tfdA* Class II and *tfdA* Class III (Table 4). Greater detection rates were observed in the case of samples treated with 0.5 mM MCPA: the genes were detected in 45% of all samples, i.e., 40% when PSM was not added and 50% when it was (Table 3).

Our analysis confirmed that SA spiking increased the number of detected functional genes and MCPA removal efficiency. Hence, the sample containing MCPA, SE and SA was selected for further analysis of the biodiversity of the microbial populations which were present in samples amended with herbicide and selected PSMs. Analyses of the *16S rRNA* gene sequences within the samples revealed widespread enrichment of several sequence variants associated with known microbial degraders of many classes of environmental organic pollutants (Table 5). These degraders include members of the genera *Rhodoferax*, *Achromobacter*, *Burkolderia* and *Cupriavidus* (see Supplementary Materials: Figures S1 and S2).

Table 5. The Results of Sequence Alignments for Samples with MCPA, SE and SA.

Class of Microorganisms	Identified Strain	Homology	E-value
β-proteobacteria	*Rhodoferax saidenbachensis* strain OX0321	100%	0
	Achromobacter dolens strain BFHC1 5	99%	1.0×10^{-44}
	Burkholderia sp. strain A5	99%	1.0×10^{-44}
	Cupriavidus sp. strain CI099	99%	4.0×10^{-44}

3.3. Changes in Ecotoxicity

Table 6 shows the results of the ecotoxicity assay based on two dicotyledonous species: *L. sativum* and *S. alba*; measurements were taken at the beginning of the experiment and after 24 days of incubation. At the beginning of the experiment, the MSM + SE sample demonstrated root growth inhibition of 51% (toxic) for *L. sativum* and 57% (toxic) for *S. alba*. All samples treated with 0.1 mM or 0.5 mM 2,4-D displayed 100% inhibition (highly toxic) for *L. sativum*; however, the 0.1 mM 2,4-D samples displayed 98% inhibition (toxic) for *S. alba* and the 0.5 mM 2,4-D samples 94% inhibition (toxic). All samples amended with MCPA (0.1 and 0.5 mM) demonstrated 100% (highly toxic) root growth inhibition in both studied plant species.

Table 6. Root Growth Inhibition (PE) in Samples Before and After 24 Days of Incubation.

PH	2,4-D/MCPA Concentration (mM)	*L. sativum*					*S. alba*				
		MSM	MSM + PSM *	MSM + SSE	MSM + SE	MSM + SE+ PSM *	MSM	MSM + PSM *	MSM + SSE	MSM + SE	MSM + SE + PSM *
					PE at the Beginning of the Experiment						
-	-	n.a.	n.a.	n.a.	51	n.a.	n.a.	n.a.	n.a.	57	n.a.
2,4-D	0.1	100	n.a.	n.a.	n.a.	n.a.	98	n.a.	n.a.	n.a.	n.a.
	0.5	100	n.a.	n.a.	n.a.	n.a.	94	n.a.	n.a.	n.a.	n.a.
MCPA	0.1	100	n.a.	n.a.	n.a.	n.a.	100	n.a.	n.a.	n.a.	n.a.
	0.5	100	n.a.	n.a.	n.a.	n.a.	100	n.a.	n.a.	n.a.	n.a.
					PE After 24 Days of Incubation						
2,4-D	0.1	100	98	100	100	100	96	95	97	94	98
	0.5	10	100	100	100	100	96	96	96	97	97
MCPA	0.1	100	100	100	−102 *	−69 *	100	100	100	−34 *	−21 *
	0.5	100	100	100	−47 *	−55 *	100	100	100	−38 *	−35 *

n.a.—not analyzed; * PSM—plant secondary metabolite (FA for 2,4-D samples and SA for MCPA-amended samples); Negative values indicate stimulation of root growth by the test plants with regard to the control sample.

After 24 days, in sterile samples amended with 0.1 mM 2,4-D, the ecotoxicity still remained very high. Samples treated with MSM showed 100% growth inhibition (highly toxic) for *L. sativum* and 96% (toxic) for *S. alba*. Likewise, MSM+FA demonstrated 98% inhibition (toxic) for *L. sativum* and 95% (toxic) for *S. alba*. The MSM + SSE variant demonstrated 100% (highly toxic) inhibition for *L. sativum* and 97% (toxic) inhibition for *S. alba*.

At the higher initial concentration of 2,4-D (0.5 mM), sterile samples demonstrated 100% (highly-toxic) inhibition for *L. sativum* and 96% (toxic) for *S. alba*. Both samples with SE and those with MSM + SE + FA displayed 100% (highly-toxic) inhibition for *L. sativum* and 97% (toxic) for *S. alba*.

The samples treated with MCPA also demonstrated high ecotoxicity, with the sterile samples being 100% (highly toxic) for both plants at both PH concentrations. Nevertheless, the application of soil extract ameliorated this highly toxic effect. Among the samples amended with 0.1 mM MCPA, the SE samples showed a stimulation effect of −102% (non-toxic) for *L. sativum* and −34% (non-toxic) for *S. alba*, while the SE + SA samples demonstrated −69% (non-toxic) for *L. sativum* and −21% (non-toxic) for *S. alba*. For the samples dosed with 0.5 mM MCPA, the PE% values for the MSM + SE samples were −47% (non-toxic) for *L. sativum* and −38% (non-toxic) for *S. alba*, while those for the MSM + SE + SA samples were −55% (non-toxic) for *L. sativum* and −35% (non-toxic) and *S. alba*.

4. Discussion

To improve crop yields and mitigate the risk of economic loss caused by weeds, it is often necessary to apply herbicides, e.g., PHs, to soil; however, inappropriate storage practices and usage can result in PHs being dispersed throughout the environment. Following dispersal, their residues can accumulate in indigenous plant and aquatic organisms, where they can adversely affect their metabolism. Although they are initially released into the environment in the form of commercial products containing phenoxy acids salts or esters, these products immediately hydrolyze to their corresponding anionic or neutral forms upon exposure to environmental conditions [27]. Residual forms of active ingredients (2,4-D,

MCPA) can be adsorbed by soil particles and transported to terrestrial and water ecosystems (surface- and groundwater) with runoff and in the soil profile [28]. Hence, there is a need to identify methods that enable effective elimination of these pollutants from the environment; of these, nature-based approaches offer the greatest potential, as these are cheaper and less likely to be harmful to the environment. One such approach is based on the stimulation of biodegradation: a promising, environmentally-friendly and cost-effective method of enhancing naturally-occurring processes of contaminant removal. The present study compares the potential of selected PSMs as stimulators of biodegradation of certain PHs which form the basis of widely-used herbicides: FA against 2,4-D, and SA against MCPA. The method employed a tripartite approach: phenoxy acid removal efficiency was assayed, the bacterial communities involved in the biodegradation processes were identified by bacterial gene analysis, and changes in sample ecotoxicity were determined.

4.1. The Influence of Selected PSMs on 2,4-D/MCPA Removal

Various PSMs appear to stimulate the biodegradation of organic pollutants in soil (Table 1). However, as the sterile samples and those amended with SE demonstrated similar efficiency in reducing 2,4-D concentration, it is possible that such degradation is driven mainly by physicochemical processes. Recent research by McMartin et al. [29] found the half-life values of 2,4-D in water samples to range between 30 and 40 hours, which is consistent with our present results obtained in sterile samples (Table 2). McMartin et al. also reported that the microorganisms required an acclimation period of 18 days before commencing the biodegradation of 2,4-D.

In contrast to our results, it has previously been suggested that samples amended with *Pseudomonas cepacia* populations displayed significant reductions of 2,4-D in comparison to controls, and that biodegradation of 2,4-D is determined by biotic factors [30]. McLoughlin et al. [21] also note that PSMs such as α-pinen and limonene play a significant role in enhancing the removal of 2,4-DCP (2,4-D metabolite: 2,4-dichlorophenol) from soil matrices. Our present findings indicated that both 2,4-D and FA, and their metabolites, exerted a toxic effect on the bacterial consortia in the amended samples; hence, it is more likely that the removal of 2,4-D was mainly driven by physicochemical processes.

After 24 days of incubation, MCPA removal in the sterile variants ranged from 19% to 40% when applied at 0.1 mM, and from 5% to 27% when applied at 0.5 mM; the addition of soil microbiota enhanced the removal rate to 53% (0.1 mM) and 99% (0.5 mM). The application of both SE and SA boosted the process of MCPA removal to 99% for 0.1 mM MCPA and to 100% for 0.5 mM MCPA. However, regarding the soil amended with 0.5 mM 2,4-D, no significant difference was observed between the sterile samples and those amended with soil extract.

These results indicated that two related compounds, in this case, 2,4-D and MCPA, can behave very differently in the presence of structurally-related PSMs: although the biodegradation of MCPA was enhanced by SA, no such effect was observed on 2,4-D by FA. Two critical factors determining the biostimulation effect of PSM appear to be the selection of structurally-related PHs and PSMs, and the initial concentration of the contaminant.

4.2. The Influence of Selected PSMs on the Degradation Potential of SE Bacteria toward 2,4-D/MCPA

Previous studies indicated that the biodegradation processes taking place in the rhizosphere are stimulated solely by PSMs. They can serve as cometabolites and provide the energy needed by microorganisms to carry out biodegradation processes. It is worth noting that PSMs can stimulate the expression of desirable genes present in environmental matrices if they bear a similar structural similarity to certain xenobiotics. Additionally, PSMs can be used as a primary source of carbon for bacterial communities to support their growth and increase their tolerance to higher concentrations of toxic compounds [21,30]. The presence of functional genes indicates that the microorganisms present in environment have the potential to degrade xenobiotics such as PHs.

PH degradative *tfdA*-like genes belong to the catabolic gene cluster *tfdABCDFE*, which is thought to be widely distributed in the environment among *Proteobacteria*. These genes have been extensively studied since the early use of 2,4-D-based herbicides in the 1950s. These *tfdA*-like genes have been used as markers for the presence of 2,4-D catabolism for two key reasons: they encode the dioxygenase enzyme, which plays a critical role in the initiation of PH biodegradation, and they are characterized by a unique nucleotide sequence and function [12]. Among the 2,4-D degraders, the bacteria harboring *tfdA*-like genes can be divided into four distinct groups, according to their base nucleotide sequence. The first group, the α-*Proteobacteria*, harbor the *tfdA*α gene; this group comprises bacteria closely related to *Bradyrhizobium* spp. [31]. The remaining three groups, harboring *tfdA* Class I, II and III, include various bacterial strains belonging to γ- and β-*Proteobacteria* [11,32]. Class I was isolated from contaminated sites and is known to be present in β- and γ- *Proteobacteria* such as *Cupriavidus pinatubonensis JMP134*; in contrast, *tfdA* Class II is less widely distributed, being found in *Burkholderia* spp., while *tfdA* Class III was found in *Comamonas acidovorans* [33]. The location of the *tfdA* gene also differs between classes: *tfdA* Class I and III *tfdA* are located on self-transmissible plasmids [34], while *tfdA* Class II genes are located on bacterial chromosomes [12].

Therefore, the presence and type of *tfdA*-like genes can be used as indicators to identify the biodegradation potential of bacteria present in studied soil. Our findings confirm the presence of bacterial DNA, including the conservative *16S rRNA* gene fragment, in all the samples amended with SE. PCR amplification revealed almost all selected functional gene fragments to be abundant in the studied samples; however, significant differences in gene profile were observed between samples amended with 2,4-D and MCPA at the two initial concentrations.

These functional genes were present in the sample spiked with SE (Table 3), which indicates that PH-degrading bacteria were already present in the studied soil. These results are consistent with those of our previous studies, where *tfdA* Class III and *tfdA*α genes were detected in control soil [35]; however, although the soil extract was prepared from the same soil material as in the previous studies, only the *tfdA* and *tfdA*α genes were detected [35]. It is important to underline that these two experiments were conducted in different media: soil and liquid MSM. The presence of functional genes suggests that the soil used for soil extract preparation might have been exposed to PHs for a longer time prior to collection. This observation is consistent with ecotoxicity results, where samples with the soil extract exhibited ecotoxicity levels as high as 57%.

In samples amended with 2,4-D, the initial concentration had a significant effect on the presence of the studied genes, with *tfdA* Class III genes being detected later at the higher dose of 2,4-D (0.5 mM), which may be due to the cytotoxic action of 2,4-D. In addition, higher herbicide concentrations were associated with lower numbers of the present functional genes. Following supplementation with a selected PSM (FA), the pattern of genes in the samples changed to *tfdA* Class I and *tfdA*α, indicating that FA influenced the structure of the microbial communities. The same amplicons were detected in the sample treated with SE + FA. FA has a strong effect on the microorganisms in the local environment and their degradative activities; however, its presence does not appear to lower the ecotoxicity effect of 2,4-D or its removal. Additionally, the enrichment of the sample with 0.5 mM 2,4-D and FA suppressed detection of the gene, suggesting that the two substances may have a synergistic negative effect on ecotoxicity.

Different results were obtained for samples amended with MCPA than those treated with 2,4-D. No PCR products were observed in samples treated with 0.1 mM of MCPA. This might be due to the presence of bacteria which possess other genes responsible for PH degradation. Previous studies have identified the presence of other RHDO genes in *Bradyrhizobium* sp., these being the *cadA* family [31,33]. These amplicons have also been detected in the environment (e.g., activated sludge) [12]. It is important to note that various fungi have also been found to metabolize PHs and to play the sole role in their biodegradation e.g., *Streptomyces* sp. (isolated from forest soil in Vietnam), *Serratia marcescens* and *Penicillium* sp. (isolated from contaminated soil in Brazil) [36]. Therefore, it is possible that the observed absence of bacterial functional genes might be associated with the action of fungi, indeed, the extract used for the experiment would have contained both bacteria and fungi; however, this was not investigated in the present study. Nevertheless, fungal strains are known to produce several non-specific enzymes which take part in the degradation of phenolic compounds [37].

Following the addition of SA, increased numbers of bacterial MCPA-degradative genes were detected. Samples spiked with both MCPA and MSM + SE + SA displayed twice the numbers of detected genes than those amended only with MCPA. Such an increase in *tfdA*-like gene numbers suggest that SA can enhance the biodegradation of MCPA, a substance with close structural similarity. As noted above, the samples amended with a higher dose of MCPA demonstrated higher numbers of degradative *tdfA*-like genes; this is also consistent with the results observed for the degradation and mitigation of ecotoxicity of 2,4-D (Table 6). The presence of the *tdfA*-like genes in the soil indicates that of microorganisms capable of degrading MCPA.

These findings confirm those of Bælum et al. (2006) [38], who showed that bacteria harboring *tfdA* Class III genes were active during the degradation of MCPA; interestingly, they also recorded *tfdA* Class III and *tfdA*α in control soil, but not in the MCPA spiked soil samples [38], which is consistent with our previous findings [35]. Similar findings were observed by Poll et al. 2010 [32], who compared the abundance of *tfdA* and *tfdA*α in control and MCPA-amended samples. This indicates that the soil was probably enriched in a natural substrate enabling the growth of bacteria harboring the *tfdA*α gene, not necessarily phenoxy acids [31]. The presence of *tfdA* Class III amplicons in the SA-amended sample indicates that the addition of PSM influences the metabolic properties of the present microbial communities.

By identifying the dominant microorganisms in a sample, it is possible to recognize which bacteria are capable of PH biodegradation. In order to identify the dominant bacterial strains in samples amended with MCPA and SA, the *16S rRNA* sequences were analyzed. The greatest homology (100%) was achieved with the *16S rRNA* gene of *Rhodoferax saidenbachensis* strain OX0321 (accession number MG576020.1) (Table 5); in addition, the *16S rRNA* genes of *Achromobacter dolens* strain BFHC1 5 (accession number MG897148.1), *Burkholderia* sp. strain A5 (accession number KY623377.1) and *Cupriavidus* sp. strain CI099 (accession number MG798754.1) were found to display 99% homology. These results are consistent with other studies which have revealed the presence of MCPA-degrading strains belonging to the β-proteobacteria: *Rhodoferax* sp. [39], *Rhodoferax fermentans* TFD23, AF049536 [40], *Cupriavidus necator* JMP134 [41], *Cupriavidus* sp. [42], *Burkholderia* sp. [43,44] and *Achromobacter* sp. [45]. It is also worth noting that microbial communities isolated from rhizosphere soil contaminated with mecocrop (MCPP) differ substantially from those isolated from bulk soil, comprising mainly bacteria belonging to the γ-Proteobacteria (*Pseudomonas* spp. and *Acinetobacter calcoaceticus*) in the presence of MCPP [46].

4.3. The Influence of Selected PSMs on the Ecotoxicity of 2,4-D/MCPA-Enriched Samples

Previous studies have examined the ecotoxicity of 2,4-D and MCPA against organisms from various trophic levels. MCPA application was found to increase of soil ecotoxicity towards buckwheat (*Fagospyrum esculentum* var. Kora), causing stem deformation and discoloration of leaves [47]. Polit et al. [48] demonstrated that MCPA has a toxic effect on seed germination and seedling development of winter oilseed rape (*Brassica napus*). The application of 1 μM and 23 mM 2,4-D inhibited root/hypocotyl elongation and disturbed mesophyll cell structure in *Sinapsis arvensis* (wild mustard) and *Pisum Sativum* (pea), respectively [42,49]. Both, MCPA and 2,4-D, exhibit high toxicity against water organisms, e.g., *Microcystis aeruginosa* (toxigenic cyanobacteria); *Danio rerio* (zebrafish); *Daphnia magna*, *Thamnocephalus platyurus*, *Artemia franciscana* (planktonic crustaceans) and *Selenastrum capricornutum* (green algae), causing severe malformations and cell death [8,9]. Studies by Sarikaya and Yilmaz [50] showed that 2,4-D (66 mg/L) cause internal hemorrhage and behavioral changes in *C. carpio*. Higher concentrations of 2,4-D caused lysis of human erythrocytes under laboratory conditions [51]. The spread of 2,4-D in the environment can also result in the contamination of water ecosystems, leading to cellular deformation of green algae [52], abnormal cellular proliferation in amphibians (*Rhinella arenarum*) [53] and the development of non-viable embryos in invertebrates (*Biomphalaria glabrata*) [54].

The present study used an assay based on two dicotyledonous plants (L. sativum and S. alba) to confirm the ecotoxic effect of studied PHs and their mitigation as a result of ongoing physicochemical and biological processes. The ecotoxicity of pure SE at the start of the experiment exceeded the limit of 20% for both L. sativum (51%) and S. alba (57%): as the soil used for the SE preparations was collected from an agricultural field, it is possible that it had been exposed to contaminants prior to collection. After incubation, it was found that although the % removal of 2,4-D in the studied cases ranged between 83–100%, the ecotoxic effect did not change over time, remaining above the toxicity limit of 20%. Of the two plants used in the assay, L. sativum was more sensitive to the treatment, displaying 100% root growth inhibition in all samples with 2,4-D, which is consistent with our previous study [35]. This ecotoxicity might have been caused by the formation of 2,4-D metabolites, as the catabolism of 2,4-D leads to the subsequent formation of 2,4-DCP and 3,5-DCC. These intermediates, formed during the degradation processes, have sometimes been found to be more toxic than the original compound [55]. 2,4-DCP has previously been observed to exhibit a strong cytotoxic effect and exert high affinity to the structures of plan cells [56].

The ecotoxicity of 2,4-D was not mitigated by addition of 0.1 or 0.5 mM FA. FA can be excreted to the root rhizosphere in various amounts under stress conditions e.g., when plants are exposed to contaminants that disturb their metabolism [57]. FA also has strong antimicrobial properties [58] and can inhibit seed germination, root and shoot growth and cell division, and exert a negative influence on the physiological parameters of the plants [59]. To conclude, FA was not found to exert a positive effect on ecotoxicity mitigation by increasing the rate of 2,4-D biodegradation, probably due to the formation of metabolites of 2,4-D exhibiting higher toxicity than the original compound. This suggests that both 2,4-D and FA had a toxic effect on the proliferation and catabolic activity of bacterial consortia in the amended samples.

The opposite situation was observed for MCPA: the physicochemical and microbial degradation processes not only reduced the ecotoxicity of the samples but even stimulated the root growth of the test plants. This finding suggests that soil microorganisms introduced to the samples play a major role in the process of herbicide detoxification. Furthermore, although both test plants responded similarly to the initial concentration of PH, higher root growth stimulation was observed for L. sativum after incubation. This is consistent with other studies, where S. alba has been found to be more sensitive to MCPA [35].

The above findings indicate that PSM application may have a positive influence on mitigating the ecotoxicity of the samples; however, this influence depends on the selected compounds and concentration of contaminants. In this case, at the higher initial concentration of MCPA used in the study (0.5 mM), higher stimulation of root growth was observed in samples treated with the PSM, i.e., SA (−55%) than those which were not (−47%).

5. Conclusions

The study presents an interdisciplinary approach to the problem of PH contamination in soil. It focuses not only on the removal rate of the herbicide, but also examines its influence on soil ecotoxicity and the potential for PH degradation using indigenous soil bacteria. The present findings reveal that measurements of removal percentage can be misleading when estimating ecotoxicity and biodegradation potential. In this case, selected PSMs had different effects on 2,4-D and MCPA biodegradation. Two critical factors were identified for the removal of contaminants and mitigation of the ecotoxicity: the choice of structurally-related phenoxy acid and PSM, and the initial concentration of phenoxy acid. Although the application of the chosen PSM (FA), contributed to the depletion of 2,4-D, a high ecotoxic effect was still observed at the end of incubation. In contrast, SA treatment enhanced the biodegradation of MCPA and encouraged the development of beneficial bacteria harboring a wide array of *tfdA*- like genes.

This study not only identifies changes in the presence of functional, bacterial degradative genes following PSM application, but also indicates that the two chosen xenobiotics, 2,4-D and MCPA, exert different effects on the abundance of degradative potential and ecotoxicity mitigation. The obtained molecular, instrumental and ecotoxicity assessment results demonstrate that the application of PSM can positively influence the removal of structurally-related herbicides; however, the final effect is highly selective and needs further, more elaborate investigation of the molecular mechanisms behind biostimulation processes.

Supplementary Materials: The following are available online at http://www.mdpi.com/2073-4441/11/7/1451/s1, Table S1: Mean and Standard Deviation for Physical and Chemical Properties of Soil; * Significant Differences at α ≤ 0.05 According to the Mann-Whitney U-Test, Table S2: Target Genes, PCR Rimers and Their Optimal Annealing Temperature, Figure S1: The Alignment Analysis of *16S rRNA* Gene Fragment (1300–1400 bp) Amplified in Samples Enriched with MCPA and Siringic Acid (Query Sequence) and Nucleotide Sequence of *Rhodoferax Saidenbachensis* OX00321, Figure S2: The Alignment Analysis of *16s rRNA* Gene Fragment (1300–1400 bp) Amplified in Samples Enriched with Mcpa and Syringic Acid (Query Sequence) and Nucleotide Sequence of *Achromobacter Dolens, Burkholderia* sp. Strain a5 and *Cupriavidus* sp. Strain ci099.

Author Contributions: Conceptualization, M.U., E.M.; methodology, M.U., E.M., M.T., A.B.; validation, M.U., E.M., M.T., A.B..; investigation M.U., E.M.; resources, M.U., E.M.; data curation, M.U., E.M.; writing—original draft preparation, M.U., E.M.; writing—review and editing, M.U., E.M.; visualization, M.U., E.M.; supervision, M.U.; project administration, M.U., E.M.; funding acquisition, M.U., E.M.

Funding: This work was supported by the University of Lodz Student Research Grant "Plant Secondary Metabolites as stimulators of bacterial degradation of 2,4-D and MCPA" and the European Structural and Investment Funds, OP RDE-funded project 'CHEMFELLS4UCTP' (No. CZ.02.2.69/0.0/0.0/17_050/0008485).

Conflicts of Interest: The authors declare no conflict of interest. The funders had no role in the design of the study; in the collection, analyses, or interpretation of data; in the writing of the manuscript, or in the decision to publish the results.

Abbreviations

The abbreviations with proper definitions used in the study:

PSM	plant secondary metabolite
FA	ferulic acid
SA	syringic acid
PH	phenoxy herbicides
2,4-D	2,4-dichlorophenoxy acid
MCPA	2-methyl-4-chlorophenoxyacetic acid
SE	soil extract
SSE	sterile soil extract
PR	percentage removal
PE	percentage effect

References

1. 2008–2012 Market Estimates. *Pesticides Industry Sales and Usage*; U.S. Environmental Protection Agency: Washington, DC, USA, 2017.
2. Smith, A.E.; Mortensen, K.; Aubin, A.J.; Molloy, M.M. Degradation of MCPA, 2,4-D, and Other Phenoxyalkanoic Acid Herbicides Using an Isolated Soil Bacterium. *J. Agric. Food Chem.* **1994**, *42*, 401–405. [CrossRef]
3. Michael, A. Kamrin Phenoxy and Benzoic Acid Herbicides. In *Pesticide Profiles*, 1st ed.; CRC Press: Boca Raton, FL, USA, 1997; ISBN 978-1-56670-190-7.
4. Skiba, E.; Wolf, W.M. Commercial Phenoxyacetic Herbicides Control Heavy Metal Uptake by wheat in a divergent way than pure active substances alone. *Environ. Sci. Eur.* **2017**, *29*, 26. [CrossRef] [PubMed]
5. Eurostat Sales of pesticides by type of pesticide. Available online: https://ec.europa.eu/eurostat/web/products-datasets/product?code=tai02 (accessed on 11 March 2019).
6. Ignatowicz, K.; Struk-Sokołowska, J. Sezonowe wahania zanieczyszczeń agrotechnicznych w rzece Narwi ze szczególnym uwzględnieniem herbicydów fenoksyoctowych. *Środkowo-Pomorskie Tow. Nauk. Ochr. Środowiska.* **2004**, *4*, 189–205.

7. Ismail, B.S.; Prayitno, S.; Tayeb, M.A. Contamination of rice field water with sulfonylurea and phenoxy herbicides in the Muda Irrigation Scheme, Kedah, Malaysia. *Environ. Monit. Assess.* **2015**, *187*, 406. [CrossRef] [PubMed]
8. Li, K.; Wu, J.Q.; Jiang, L.L.; Shen, L.Z.; Li, J.Y.; He, Z.H.; Wei, P.; Lv, Z.; He, M.F. Developmental toxicity of 2,4-dichlorophenoxyacetic acid in zebrafish embryos. *Chemosphere.* **2017**, *171*, 40–48. [CrossRef] [PubMed]
9. Caux, P.-Y.; Ménard, L.; Kent, R.A. Comparative study of the effects of MCPA, butylate, atrazine, and cyanazine on Selenastrum capricornutum. *Environ. Pollut.* **1996**, *92*, 219–225. [CrossRef]
10. Batogliu-Pazarbasi, M.; Milosevic, N.; Malaguerra, F.; Binning, P.J.; Albrechtsen, H.J.; Bjerg, P.L.; Aamand, J. Discharge of landfill leachate to streambed sediments impacts the mineralization potential of phenoxy acid herbicides depending on the initial abundance of *tfd*A gene classes. *Environ. Pollut.* **2013**, *176*, 275–283. [CrossRef] [PubMed]
11. Mcgowan, C.; Fulthorpe, R.; Wright, A.; Tiedje, J.M.; Gowan, C.M.C. Evidence for Interspecies Gene Transfer in the Evolution of Evidence for Interspecies Gene Transfer in the Evolution of 2,4-Dichlorophenoxyacetic Acid Degraders. *Appl. Environ. Microbiol.* **1998**, *64*, 4089–4092.
12. Kitagawa, W.; Kamagata, Y. Diversity of 2,4-Dichlorophenoxyacetic Acid (2,4-D)-Degradative Genes and Degrading Bacteria. In *Biodegradative Bacteria: How Bacteria Degrade, Survive, Adapt, and Evolve*; Nojiri, H., Fukuda, M., Tsuda, M., Kamagata, Y., Eds.; Springer: Japan, 2014; pp. 43–57. ISBN 978-4-431-54519-4.
13. Musilova, L.; Ridl, J.; Polivkova, M.; Macek, T.; Uhlik, O. Effects of Secondary Plant Metabolites on Microbial Populations: Changes in Community Structure and Metabolic Activity in Contaminated Environments. *Int. J. Mol. Sci.* **2016**, *17*, 1205. [CrossRef]
14. Hu, C.; Zhang, Y.; Tang, X.; Luo, W. PCB Biodegradation and bphA1 Gene Expression Induced by Salicylic Acid and Biphenyl with Pseudomonas fluorescence P2W and Ralstonia eutropha H850. *Pol. J. Environ. Stud.* **2014**, *23*, 1591–1598.
15. Uhlik, O.; Musilova, L.; Ridl, J.; Hroudova, M.; Vlcek, C.; Koubek, J.; Holeckova, M.; Mackova, M.; Macek, T. Plant secondary metabolite-induced shifts in bacterial community structure and degradative ability in contaminated soil. *Appl. Microbiol. Biotechnol.* **2013**, *97*, 9245–9256. [CrossRef] [PubMed]
16. Kruczek, M. Pumpkin (Cucurbita sp.) as a source of health-beneficial compounds with antioxidant properties Dynia (Cucurbita sp.) jako źródło prozdrowotnych związków o charakterze antyoksydacyjnym. *Przem. Chem.* **2015**, *1*, 86–90. [CrossRef]
17. Yoon, J.-Y.; Chung, I.-M.; Thiruvengadam, M. Evaluation of phenolic compounds, antioxidant and antimicrobial activities from transgenic hairy root cultures of gherkin (Cucumis anguria L.). *South. African. J. Bot.* **2015**, *100*, 80–86. [CrossRef]
18. Suttinun, O.; Lederman, P.B.; Luepromchai, E. Application of terpene-induced cell for enhancing biodegradation of TCE contaminated soil. *Songklanakarin, J. Sci. Technol.* **2004**, *26*, 131–142.
19. White, J.C.; Mattina, M.I.; Lee, W.Y.; Eitzer, B.D.; Iannucci-Berger, W. Role of organic acids in enhancing the desorption and uptake of weathered p,p-DDE by Cucurbita pepo. *Environ. Pollut.* **2003**, *124*, 71–80. [CrossRef]
20. Yi, H.; Crowley, D.E. Biostimulation of PAH degradation with plants containing high concentrations of linoleic acid. *Environ. Sci. Technol.* **2007**, *41*, 4382–4388. [CrossRef] [PubMed]
21. McLoughlin, E.; Rhodes, A.H.; Owen, S.M.; Semple, K.T. Biogenic volatile organic compounds as a potential stimulator for organic contaminant degradation by soil microorganisms. *Environ. Pollut.* **2009**, *157*, 86–94. [CrossRef] [PubMed]
22. Zhou, X.; Wu, F. Effects of amendments of ferulic acid on soil microbial communities in the rhizosphere of cucumber (Cucumis sativus L.). *Eur. J. Soil Biol.* **2012**, *50*, 191–197. [CrossRef]
23. Modern Water RaPID Assay ® 2,4-D. Available online: https://www.modernwater.com/pdf/MW_Factsheet_Rapid-Assay_2-4-D.pdf (accessed on 11 September 2017).
24. Orphan, V.J.; Sylva, S.P.; Hayes, J.M.; Delong, E.F. Comparative Analysis of Methane-Oxidizing Archaea and Sulfate-Reducing Bacteria in Anoxic Marine Sediments Comparative Analysis of Methane-Oxidizing Archaea and Sulfate-Reducing Bacteria in Anoxic Marine Sediments. *Appl. Environ. Microbiol.* **2001**, *67*, 1922–1934. [CrossRef] [PubMed]
25. MicroBioTests Inc. Standard Operational Procedure, Phytotoxkit. Seed Germination and Early Growth Microbiotest with Higher Plants. Available online: https://www.microbiotests.com/SOPs/Phytotestkit%20(complete%20test)%20SOP%20-%20A5.pdf (accessed on 11 September 2017).

26. Persoone, G.; Marsalek, B.; Blinova, I.; Törökne, A.; Zarina, D.; Manusadzianas, L.; Nalecz-Jawecki, G.; Tofan, L.; Stepanova, N.; Tothova, L.; et al. A practical and user-friendly toxicity classification system with microbiotests for natural waters and wastewaters. *Environ. Toxicol.* **2003**, *18*, 395–402. [CrossRef]
27. Paszko, T.; Muszyński, P.; Materska, M.; Bojanowska, M.; Kostecka, M.; Jackowska, I. Adsorption and degradation of phenoxyalkanoic acid herbicides in soils: A review. *Environ. Toxicol. Chem.* **2016**, *35*, 271–286. [CrossRef] [PubMed]
28. Gavrilescu, M. Fate of Pesticides in the Environment and its Bioremediation. *Eng. Life Sci.* **2005**, *5*, 497–526. [CrossRef]
29. McMartin, D.W.; Gillies, J.A.; Headley, J.V.; Peterson, H.G. Biodegradation Kinetics of 2,4-Dichlorophenoxyacetic Acid (2,4-D) in South Saskatchewan River Water. *Can. Water Resour. J.* **2000**, *25*, 81–92. [CrossRef]
30. Urbaniak, M.; Mierzejewska, E.; Tankiewicz, M. The stimulating role of syringic acid, a plant secondary metabolite, in the microbial degradation of structurally-related herbicide, MCPA. *PeerJ* **2019**, *7*, e6745. [CrossRef] [PubMed]
31. Itoh, K.; Tashiro, Y.; Uobe, K.; Kamagata, Y.; Suyama, K.; Yamamoto, H. Root Nodule Bradyrhizobium spp. Harbor tfdAα and cadA, Homologous with Genes Encoding 2,4-Dichlorophenoxyacetic Acid-Degrading Proteins. *Appl. Environ. Microbiol.* **2004**, *70*, 2110–2118. [CrossRef] [PubMed]
32. Poll, C.; Pagel, H.; Devers-Lamrani, M.; Martin-Laurent, F.; Ingwersen, J.; Streck, T.; Kandeler, E. Regulation of bacterial and fungal MCPA degradation at the soil-litter interface. *Soil Biol. Biochem.* **2010**, *42*, 1879–1887. [CrossRef]
33. Kitagawa, W.; Takami, S.; Miyauchi, K.; Masai, E.; Kamagata, Y.; Tiedje, J.M.; Fukuda, M. Novel 2,4-Dichlorophenoxyacetic Acid Degradation Genes from Oligotrophic. *Society* **2002**, *184*, 509–518.
34. Bælum, J.; Jacobsen, C.S.; Holben, W.E. Comparison of 16S rRNA gene phylogeny and functional tfdA gene distribution in thirty-one different 2,4-dichlorophenoxyacetic acid and 4-chloro-2-methylphenoxyacetic acid degraders. *Syst. Appl. Microbiol.* **2010**, *33*, 67–70. [CrossRef]
35. Mierzejewska, E.; Baran, A.; Urbaniak, M. The influence of MCPA on soil phytotoxicity and the presence of genes involved in its biodegradation. *Arch. Environ. Prot.* **2017**, *44*, 58–64.
36. Silva, T.M.; Stets, M.I.; Mazzetto, A.M.; Andrade, F.D.; Pileggi, S.A.V.; Fávero, P.R.; Cantú, M.D.; Carrilho, E.; Carneiro, P.I.B.; Pileggi, M. Degradation of 2,4-D Herbicide by Microorganisms Isolated from Brazilian Contaminated Soil. *Braz. J. Microbiol.* **2007**, *38*, 522–525. [CrossRef]
37. Del Pilar Castillo, M.; Andersson, A.; Ander, P.; Stenström, J.; Torstensson, L. Establishment of the white rot fungus Phanerochaete chrysosporium on unsterile straw in solid substrate fermentation systems intended for degradation of pesticides. *World J. Microbiol. Biotechnol.* **2001**, *17*, 627–633. [CrossRef]
38. Bælum, J.; Henriksen, T.; Christian, H.; Hansen, B.; Jacobsen, C.S. Degradation of 4-Chloro-2-Methylphenoxyacetic Acid in Top- and Subsoil Is Quantitatively Linked to the Class III tfdA Gene. *Appl. Environ. Microbiol.* **2006**, *72*, 1476–1486. [CrossRef] [PubMed]
39. Ehrig, A.; Müller, R.H.; Babel, W. Isolation of phenoxy herbicide-degrading Rhodoferax species from contaminated building material. *Acta Biotechnol.* **1997**, *17*, 351–356. [CrossRef]
40. Lee, T.H.; Kurata, S.; Nakatsu, C.H.; Kamagata, Y. Molecular analysis of bacterial community based on 16S rDNA and functional genes in activated sludge enriched with 2,4-dichlorophenoxyacetic acid (2,4-D) under different cultural conditions. *Microb. Ecol.* **2005**, *49*, 151–162. [CrossRef] [PubMed]
41. Streber, W.R.; Timmis, K.N.; Zenk, M.H. Analysis, cloning, and high-level expression of 2,4-dichlorophenoxyacetate monooxygenase gene tfdA of Alcaligenes eutrophus JMP134. *J. Bacteriol.* **1987**, *169*, 2950–2955. [CrossRef] [PubMed]
42. Pazmiño, D.M.; Rodríguez-Serrano, M.; Romero-Puertas, M.C.; Archilla-Ruiz, A.; del Río, L.A.; Sandalio, L.M. Differential response of young and adult leaves to herbicide 2,4-dichlorophenoxyacetic acid in pea plants: Role of reactive oxygen species. *Plant. Cell Environ.* **2011**, *34*, 1874–1889. [CrossRef] [PubMed]
43. Suwa, Y.; Wright, A.D.; Fukumori, F.; Nummy, K.A.; Hausinger, R.P.; Holben, W.E.; Forney, L.J. Characterization of a chromosomally encoded 2,4-dichlorophenoxyacetic acid (2,4-D)/alpha-ketoglutarate dioxygenase from Burkholderia sp. RASC. *Appl. Environ. Microbiol.* **1996**, *62*, 2464–2469. [PubMed]
44. Fulthorpe, R.R.; McGowan, C.; Maltseva, O.V.; Holben, W.E.; Tiedje, J.M. 2,4-Dichlorophenoxyacetic acid-degrading bacteria contain mosaics of catabolic genes. *Appl. Environ. Microbiol.* **1995**, *61*, 3274–3281.

45. Xia, Z.Y.; Zhang, L.; Zhao, Y.; Yan, X.; Li, S.P.; Gu, T.; Jiang, J.D. Biodegradation of the Herbicide 2,4-Dichlorophenoxyacetic Acid by a New Isolated Strain of Achromobacter sp. LZ35. *Curr. Microbiol.* **2017**, *74*, 193–202. [CrossRef] [PubMed]
46. Lappin, H.M.; Greaves, M.P.; Slatert, J.H. Degradation of the Herbicide Mecoprop [2-(2-Methyl-4-Chlorophenoxy) Propionic Acid] by a Synergistic Microbial Community. *Appl. Environ. Microbiol.* **1985**, *49*, 429–433. [PubMed]
47. Podolska, G. The effectiveness and phytotoxicity of herbicide in buckwheat cv. Kora. *Polish, J. Agron.* **2014**, *19*, 17–24.
48. Polit, J.T.; Praczyk, T.; Pernak, J.; Sobiech, Ł.; Jakubiak, E.; Skrzypczak, G. Inhibition of germination and early growth of rape seed (Brassica napus L.) by MCPA in anionic and ester form. *Acta Physiol. Plant.* **2014**, *36*, 699–711. [CrossRef]
49. Wei, Y.D.; Zheng, H.G.; Hall, J.C. Role of auxinic herbicide-induced ethylene on hypocotyl elongation and root/hypocotyl radial expansion. *Pest. Manag. Sci.* **2000**, *56*, 377–387. [CrossRef]
50. Sarikaya, R.; Yilmaz, M. Investigation of acute toxicity and the effect of 2,4-D (2,4-dichlorophenoxyacetic dichlorophenoxyacetic acid) herbicide on the behavior of the common carp (Cyprinus carpio L., 1758; Pisces, Cyprinidae). *Chemosphere* **2003**, *52*, 195–201. [PubMed]
51. Bukowska, B. Effects of 2,4-D and its metabolite 2,4-dichlorophenol on antioxidant enzymes and level of glutathione in human erythrocytes. *Comp. Biochem. Physiol. C Toxicol. Pharmacol.* **2003**, *135*, 435–441. [CrossRef]
52. Martínez-Ruiz, E.B.; Martínez-Jerónimo, F. Exposure to the herbicide 2,4-D produces different toxic effects in two different phytoplankters: A green microalga (Ankistrodesmus falcatus) and a toxigenic cyanobacterium (Microcystis aeruginosa). *Sci. Total Environ.* **2018**, *619–620*, 1566–1578. [CrossRef] [PubMed]
53. Aronzon, C.M.; Sandoval, M.T.; Herkovits, J.; Pérez-Coll, C.S. Stage-dependent toxicity of 2,4-dichlorophenoxyacetic on the embryonic development of a South American toad, Rhinella arenarum. *Environ. Toxicol.* **2011**, *26*, 373–381. [CrossRef] [PubMed]
54. Estevam, E.C.; Nakano, E.; Kawano, T.; de Bragança Pereira, C.A.; Amancio, F.F.; de Albuquerque Melo, A.M.M. Dominant lethal effects of 2,4-D in Biomphalaria glabrata. *Mutat. Res. Genet. Toxicol. Environ. Mutagen.* **2006**, *611*, 83–88. [CrossRef] [PubMed]
55. Schweigert, N.; Hunziker, R.; Escher, B.; Eggen, R. Acute toxicity of (chloro-)catechol-copper combinations in Escherichia coil corresponds to their membrane toxicity in vitro. *Environ. Toxicol. Chem.* **2001**, *20*, 239–247. [PubMed]
56. Lurquin, P.F. Production of a toxic metabolite in 2,4-D-resistant GM crop plants. *3 Biotech.* **2016**, *6*, 4–7. [CrossRef] [PubMed]
57. Piaia, B.; Alves, C.; Gularte, O.; Teixeira, D.; Cristofari, M.; Ricardo, M.; Carriço, S.; Chimelo, M.; Luiz, R.; Luis, E.; et al. Chemosphere The phytoremediation potential of Plectranthus neochilus on 2,4-dichlorophenoxyacetic acid and the role of antioxidant capacity in herbicide tolerance. *Chemosphere* **2017**, *188*, 231–240.
58. Shi, C.; Sun, Y.; Zheng, Z.; Zhang, X.; Song, K.; Jia, Z.; Chen, Y.; Yang, M.; Liu, X.; Dong, R.; et al. Antimicrobial activity of syringic acid against Cronobacter sakazakii and its effect on cell membrane. *Food Chem.* **2016**, *197*, 100–106. [CrossRef] [PubMed]
59. Singh, H.P.; Kaur, S.; Batish, D.R.; Kohli, R.K. Ferulic acid impairs rhizogenesis and root growth, and alters associated biochemical changes in mung bean (Vigna radiata) hypocotyls. *J. Plant. Interact.* **2014**, *9*, 267–274. [CrossRef]

© 2019 by the authors. Licensee MDPI, Basel, Switzerland. This article is an open access article distributed under the terms and conditions of the Creative Commons Attribution (CC BY) license (http://creativecommons.org/licenses/by/4.0/).

Article

Removal of Diesel Oil in Soil Microcosms and Implication for Geophysical Monitoring

Francesca Bosco [1], Annalisa Casale [1], Fulvia Chiampo [1,*] and Alberto Godio [2]

[1] Department of Applied Science and Technology, Politecnico di Torino, Corso Duca degli Abruzzi 24, 10129 Torino, Italy
[2] Department of Environment, Land and Infrastructure Engineering, Politecnico di Torino, Corso Duca degli Abruzzi 24, 10129 Torino, Italy
* Correspondence: fulvia.chiampo@polito.it; Tel.: +39-011-090-4685

Received: 30 May 2019; Accepted: 8 August 2019; Published: 11 August 2019

Abstract: Bioremediation of soils polluted with diesel oil is one of the methods already applied on a large scale. However, several questions remain open surrounding the operative conditions and biological strategies to be adopted to optimize the removal efficiency. This study aimed to investigate the environmental factors that influence geophysical properties in soil polluted with diesel oils, in particular, during the biodegradation of this contaminant by an indigenous microbial population. With this aim, aerobic degradation was performed in soil column microcosms with a high concentration of diesel oil (75 g kg^{-1} of soil); the dielectric permittivity and electrical conductivity were measured. In one of the microcosms, the addition of glucose was also tested. Biostimulation was performed with a Mineral Salt Medium for Bacteria. The sensitivity of the dielectric permittivity versus temperature was analyzed. A theoretical approach was adopted to estimate the changes in the bulk dielectric permittivity of a mixture of sandy soil-water-oil-gas, according to the variations in the oil content. The sensitivity of the dielectric permittivity to the temperature effects was analyzed. The results show that (1) biostimulation can give good removal efficiency; (2) the addition of glucose as a primary carbon source does not improve the diesel oil removal; (3) a limited amount of diesel oil was removed by adsorption and volatilization effects; and (4) the diesel oil efficiency removal was in the order of 70% after 200 days, with different removal percentages for oil components; the best results were obtained for molecules with a low retention time. This study is preparatory to the adoption of geophysical methods to monitor the biological process on a larger scale. Altogether, these results will be useful to apply the process on a larger scale, where geophysical methods will be adopted for monitoring.

Keywords: bioremediation; biostimulation; diesel oil; indigenous microorganisms; kinetics; dielectric permittivity

1. Introduction

Soil pollution has an anthropogenic origin, often due to accidental or improper industrial spills, stockpile leakages, improper waste disposal, mining, and military activities, to name just a few examples. Looking at the contaminant classes, most of them are mineral oils, heavy metals, metalloids or organic compounds, due to hydrocarbon leakages in the subsoil and groundwater.

Diesel oil is widely used in many industrial sectors, mainly for transport and energy plants, and therefore it is one of the common soil pollutants, with huge impacts on human health, the environment, and economy. For its removal, in situ bioremediation is considered an environmentally friendly and cost-effective solution, due to the metabolic capability of microorganisms to degrade the pollutants. So far, bioremediation is performed by means of

- biostimulation: The addition of macro- and/or micronutrients to enhance indigenous biomass growth and pollutant degradation [1,2];
- bioaugmentation: The addition of enriched microbial cultures (autochthonous or allochthonous degraders) to the soil [3];
- combined biostimulation and bioaugmentation [4,5].

Several studies were carried out in different operative conditions, adopting one of the aforesaid strategies, and aimed to optimize hydrocarbon removal [3,6–9]. It is widely known that biostimulation is an easier process to carry out [2,5,10,11]. However, when biostimulation does not give satisfactory results, bioaugmentation is adopted, or coupled methods are applied.

In a previous study carried out with the same soil contaminated with a commercial diesel oil (7.5% w/w soil) [12], the effect of biostimulation was investigated. The results clearly showed that the medium favoring bacterial growth was more effective in promoting indigenous microbial activity (in terms of CO_2 production and biomass dry weight). In a subsequent work [13], the biodegradation efficiencies of biodegradation with and without the addition of a primary carbon source (e.g., glucose) were compared. The results in the biostimulated and bioaugmented microcosms were very similar. It was confirmed that biostimulation can be adopted as a strategy to remove aerobically high concentrations of diesel oil.

In the present study, which is part of a project aiming to apply geophysical methods to monitor the biodegradation of diesel oil in soils, biostimulation was used to remove a high concentration of pollutant (75 g kg^{-1} of soil). The process was performed by aerobic indigenous microorganisms with and without the addition of a primary carbon source (glucose). The study aimed to better understand if diesel oil removal could be improved by biostimulation of indigenous bacteria; moreover, the contribution of the soil itself to the overall removal was tested.

Kinetic modelling was also performed in order to predict the process performance when transferring the system to a field scale, where geophysical monitoring could be adopted as a fast tool to monitor the behavior of biostimulation.

The challenging issue was to characterize and monitor changes in hydrological and biogeochemical properties/processes, using geophysical measurements at both the laboratory column and field scales. This issue impacts the design of remediation schemes, and it is relevant to check the remediation performances at many contaminated sites. Particularly, at the field scale, the most commonly adopted methods are electrical resistivity, induced polarization, and georadar [14,15]. However, research in bio-geophysics focusses on how microbial growth and biofilm formation could have a potential direct impact on changes in geophysical properties, such as in frequency dependent dielectric permittivity [16]. Moreover, rock texture, surface area, porosity, pore size and shape geometry, tortuosity, formation factor, cementation, and mechanical properties could affect the electromagnetic properties of the soil.

In designing laboratory experiments and geophysical monitoring, a crucial issue is to evaluate how environmental factors could affect geophysical properties; particularly, the dielectric permittivity and electrical conductivity are sensitive to temperature effects. Many authors discussed the effect of temperature both on the dielectrical permittivity and the electrical conductivity of soil; Or and Wraith [17] discussed the role of temperature in complex dielectric permittivity, observed by using time domain reflectometry. The temperature coefficients for both dielectric permittivity and conductivity depend on the mixture composition, the frequency, and the temperature range, and they are useful to compensate for the effect of temperature change during measurements.

At a given frequency and for small temperature changes, the dielectric properties of a mixture vary according to linear temperature coefficients, defined as the percent change in either permittivity or conductivity per Celsius degree. The linear temperature coefficients are limited to a number of specific discrete frequencies and temperatures; outside of these ranges, the temperature impact on the dielectric properties may no longer be linear [18].

Generally, the relative dielectric permittivity and electrical conductivity trends with temperature differ in terms of frequency: In the microwave frequency range, the change in relative permittivity is 2% per Celsius degree and the change in conductivity is between 1 and 2% per Celsius degree, depending on the mixture and on the frequency and temperature range considered.

In such a context, one of the aims is to study/define the sensitivity of dielectric permittivity to the biodegradation of diesel oil in a laboratory column. Starting from the results of microcosm experiments, the analysis of the expected response of the dielectric permittivity during the diesel oil removal, over time, can be carried out.

Dielectric permittivity is a basic electromagnetic property, controlling the radio-wave propagation into the soil, and it is the parameter involved in georadar (GPR) investigation or in Time Domain Reflectometry (TDR) monitoring. Laboratory measurements on sandy soil saturated in water or in diesel oil are found in the literature (e.g., [19]), while time-lapse monitoring of the dielectric permittivity changes caused by oil degradation are less common.

Theoretical models were proposed to describe the contaminant fluid behavior and its effects on dielectric permittivity; for instance, Endres and Redman [20] developed a pore-scale fluid model for clay-free granular soils. In order to predict the characteristics of the hydrocarbon spill, Carcione and Seriani [21] proposed a model for the complex permittivity of a soil composed of sandy grains, clay and silt, partially saturated with gas (air), water, and hydrocarbon. The theory is based on the self-similar model [22,23]. At radar frequencies of 10 MHz–2 GHz, the interfacial and electrochemical mechanisms, such as surface effects, associated with the soil/water interface can be neglected [24]. More recently, multiphase models were implemented for estimating the dielectric permittivity of a mixture of soil-water-gas and oil; the models were validated by using an experimental set-up based on Time Domain Reflectometry (TDR) [25], which is very similar to the approach adopted in our study.

In this study, we focus on the dielectric permittivity of a soil-water-diesel, oil-gas system by evaluating the sensitivity of state of-art geophysical tools to monitor the degradation effect over time.

2. Materials and Methods

2.1. Soil Properties

The soil was the same as that used for a previous study [13] and was collected in Trecate (Northern Italy), near a site polluted with a crude oil spill. Previous analyses established that the presence of crude oil was not detectable [14].

The soil was sieved and material within the range 0.2–2 mm was used for the study, after oven drying at 70 °C. It should be noted that in its original condition, this soil has a very low water content, which is negligible.

The chemical and physical soil properties are shown in Table 1.

Table 1. Chemical and geophysical soil properties.

Chemical Parameter	Value
pH	7.32 ± 0.04
Soluble bicarbonate (mg kg^{-1} of dry soil)	66.9 ± 10.8
Soluble chlorides (mg kg^{-1} of dry soil)	26.2 ± 0.3
Soluble sulphates (mg kg^{-1} of dry soil)	211 ± 3
Ammonia (mg kg^{-1} of dry soil)	2.18 ± 0.11
Nitrates (mg kg^{-1} of dry soil)	68.0 ± 0.4
Geophysical Parameter	**Value**
Porosity (% volume)	40–42
Density (kg m^{-3})	2700
Dielectric permittivity for dry soil	2.5–3
Dielectric permittivity for saturated soil	25–30

The analysis of the Total Organic Carbon of the soil evidenced a negligible content (<0.1% by weight). The external carbon source was due to diesel oil or to diesel oil and glucose.

2.2. Soil Microcosms

Microcosms were set up in four closed Plexiglas columns (diameter = 0.05 m; height = 0.4 m), each filled with 200 g of soil (layer height = 0.07 m) and operated with different conditions, namely:

1. Control abiotic microcosm (A): Sterilized soil, spiked with 15 g of commercial diesel oil and hydrated with 37 mL of sterilized water to obtain a soil moisture level of about 15 % by weight; the sterilization was carried out in an autoclave, at 120 °C for 2 h;
2. control biotic microcosm (C): Soil hydrated with 37 mL of Mineral Salt Medium for Bacteria (MSMB);
3. biostimulated microcosm (BIOS): Soil treated with 15 g of diesel oil and hydrated with 37 mL of MSMB;
4. biostimulated microcosm with added glucose (BIOS-G): The addition of 0.74 g of glucose to 200 g of soil treated with 15 g of diesel oil and hydrated with 37 mL of MSMB; after 143 days, the same amount of glucose was added again.

Soil water content was monitored at time t = 0, 15, 68, 110, 135, 143, and 175 days. After 143 days, 37 mL of MSMB was added to microcosms C, BIOS, and BIOS-G, and the same amount of sterilized water was added to microcosm A.

Each column was opened every 3–4 days to aerate the microcosm. For microcosm A, all the operations were carried out under biological hood.

2.3. Respirometric Measurements

The respirometric activity of the soil microbial population, as CO_2 evolution, was measured by CO_2 absorption in NaOH solution, using the method described by Bosco et al. [13].

In each microcosm, the measurement was performed every 3–4 days until 218 days.

2.4. Microbial Counts

Soil samples of each microcosm were taken at t = 34, 68, 110, and 175 days to perform the viable microbial count on Malt Extract Agar (MEA) plates, in line with the method used by Bosco et al. [13]. The colony counting was performed after 3 days, and the results were expressed as the number of Colony Forming Units (CFU) per gram of soil.

2.5. Residual Diesel Oil Concentration

In each microcosm, the residual diesel oil concentration was measured by gas-chromatographic analysis of the extract, according to the EPA method 8015. The gas chromatograph (GF) was equipped with a flame ionization detector (FID) and a DB-5 fused silica capillary column, operated with helium gas as the carrier. For the oven, the following temperature program was adopted: Maintaining at 50 °C for 1 min, heating by 8 °C min^{-1} up to 320 °C and maintaining at 320 °C for 10 min, with a total retention time of 45 min. The injector and detector were maintained at 220 and 280 °C, respectively. The injected extract volume was 1 µL in splitless mode. The residual diesel oil concentration was calculated using a calibration curve obtained with the same commercial diesel oil.

The analyses were performed at different times (0, 15, 68, 110, 135, 175, and 203 days). For each sampling, one sample was taken, and two extracts were prepared and analyzed in triplicate with the aforesaid GC-FID procedure. The gas chromatograms were analyzed as follows:

- The chromatogram from 0 to 8 min was not considered due to the presence of solvent peaks.
- The final part, from 35 to 45 min, was also excluded due to the presence of very small and not well-defined peaks.
- The residual retention time, from 8 to 35 min, was divided into three equal periods, and the corresponding species were classified as:

- Low retention (LR species), in the period 8–17 min.
- Medium retention (MR species), in the period 17–26 min.
- High retention (HR species), in the period 26–35 min.

In each period, the cumulative concentration was calculated as for the total residual one.

2.6. Kinetic Modeling

As for common chemical reactions, the kinetics of hydrocarbon biodegradation can be described with the general model of differential rate law:

$$r = dC/dt = -kC^n \tag{1}$$

where r is the reaction rate, C is the residual reagent concentration, t is time, k is the reaction rate constant, and n is the reaction order.

Notwithstanding the broad spectrum of possibilities for the reaction order, for bioremediation the most commonly used models are the first-order (n = 1) and the second-order (n = 2), since they give satisfactory results, and their use is rather simple. Therefore, these two models were tested with the data.

2.6.1. First-Order Reaction Rate

In the bioremediation of hydrocarbon-polluted soil, several authors showed the reliability of the first-order reaction rate model [8,26–28]. For this model, Equation (1) becomes

$$r = dC/dt = -kC \tag{2}$$

Adapting this model to diesel oil biodegradation, it can be assumed that the diesel oil is the key reagent. By integration, the following expression is achieved:

$$C(t) = C_0 \exp(-kt) \tag{3}$$

where $C(t)$ is the residual diesel oil concentration at time t (mg kg^{-1} of soil), C_0 is the initial diesel oil concentration (mg kg^{-1} of soil), k is the reaction rate constant (day^{-1}), and t is time (day).

The kinetic rates are often expressed in terms of half-life time $t_{1/2}$, meaning the time by which the starting concentration is halved: At $t = t_{1/2}$, $C = C_0/2$. Therefore, rearranging Equation (3), it is possible to write and define the half-life time as

$$t_{1/2} = \ln 2/k = 0.693/k \tag{4}$$

2.6.2. Second-Order Reaction Rate

Some authors modeled their experimental data for hydrocarbon bioremediation with the second-order reaction rate [29,30].

In this case, with n = 2, Equation (1) becomes

$$r = dC/dt = -kC^2 \tag{5}$$

The solution is

$$1/C(t) = 1/C_0 + kt \tag{6}$$

with the half-life time, $t_{1/2}$:

$$t_{1/2} = 1/(kC_0). \tag{7}$$

2.7. Geophysical Features for Future Monitoring and Preliminary Measurements

The sensitivity of the bulk dielectric permittivity to changes in the mass of each single element of the materials in the column is checked with the Complex Refractive Index Model (CRIM) (e.g., [25]). This model allows us to predict the bulk dielectric permittivity by accounting for the contribution of each fraction; it was widely adopted to predict the dielectric permittivity of multiphase systems (e.g., Reference [19]). The general formulation of the CRIM model, modified after Knigths and Endres [24], is given by the following formula:

$$\epsilon_{bulk}^{\alpha} = (1-\phi)\,\epsilon_{grain}^{\alpha} + (\phi S_w)\,\epsilon_{water}^{\alpha} + (\phi S_o)\,\epsilon_{oil}^{\alpha} + \phi\,(1-S_o-S_w)\,\epsilon_{water}^{\alpha} \qquad (8)$$

where ϕ is the soil porosity, and S_w and S_o are the water and oil saturation, respectively, while ϵ defines the dielectric permittivity of the different materials. The α-exponent accounts for non-linear effects of the interaction between different phases of the mixture. The typical range of the α-exponent is 0.25–0.6.

The simulation was carried out on the same sandy soil adopted for the experiment, with the following assumptions:

- Average porosity: 0.38–0.4;
- soil density: 2700 kg m^{-3};
- diesel oil density: 800 kg m^{-3};
- water saturation: 0.4;
- diesel oil concentration equal to 76,000 mg kg^{-1} of soil, equivalent to diesel oil volume of 0.25 m^3 m^{-3} of the total volume (the mixture oil-water-soil-air);
- water relative electrical permittivity equal to 78 (the water salinity was not considered).

The values and the adopted methodology to estimate the dielectric permittivity of each single component of the mixture are given in Table 2.

Table 2. Reference values of relative dielectric permittivity for the materials adopted in the sensitivity analysis.

Phase	Relative Dielectric Permittivity	Method
Solid (grains)	3.2 +/− 0.2	Estimated according to Complex Refractive Index Model (CRIM) formula on saturated specimen (Time Domain Reflectometry (TDR) probe) [24]
Water	78.5 +/− 0.2 (at 25 °C)	Measured with TDR probe [25]
Diesel oil	2.2 +/− 0.1	Measured with Open-Ended-Coaxial cable [19]

3. Results

3.1. Respirometric Measurements

The respirometric activity, expressed as the cumulative amount of CO_2 produced in all the tested microcosms, is reported in Figure 1. The main results are the following:

1. The control abiotic microcosm (A) had no respirometric activity, as expected.
2. The control biotic microcosm (C) had the highest respirometric activity in the first 35–40 days, with an exponential trend, then, the daily CO_2 production continued but with a reduced amount (Figure 2). After 40 days, the cumulative CO_2 quantity grew with an almost linear tendency, demonstrating that the daily production was almost constant and in the order of 0.014 g CO_2 kg^{-1} of soil day^{-1}. In both cases, the regression coefficient R^2 had values very close to 1.
3. For the BIOS-G microcosm, a slight deviation from linearity occurred at the beginning of the run, when the microbial activity was probably enhanced by the glucose presence, due to the quick use of this primary carbon source with respect to the use of more complex molecules.

4. Both biostimulated microcosms, namely without (BIOS) and with glucose (BIOS-G), had cumulative production of CO_2 with linear growth, around 0.058 g CO_2 kg^{-1} of soil day^{-1} and 0.080 g CO_2 kg^{-1} of soil day^{-1}, respectively. These amounts contained the biotic quantity measured in microcosm C, to say the amount naturally produced by the microbial activity in the absence of diesel oil and glucose.
5. Comparing the CO_2 production of the BIOS and BIOS-G microcosms, the system with added glucose (BIOS-G) produced a higher amount, especially at the beginning of the test, and the difference increased until about 110 days, when the value was in the order of 2.4–2.5 g CO_2 kg^{-1} of soil (Figure 3). This trend was also found after the second addition of glucose to BIOS-G at t = 143 days, when the BIOS-G microcosm started to produce CO_2 more quickly than the BIOS microcosm (this feature ended at 163–165 days).
6. Without any external addition of carbon sources (microcosm C), the soil had relevant microbial activity: Compared to the cumulative amount achieved in the others, its weight was around 25% of the amount produced when just diesel oil was present (BIOS) and around 17% of the cumulative quantity produced in BIOS-G.

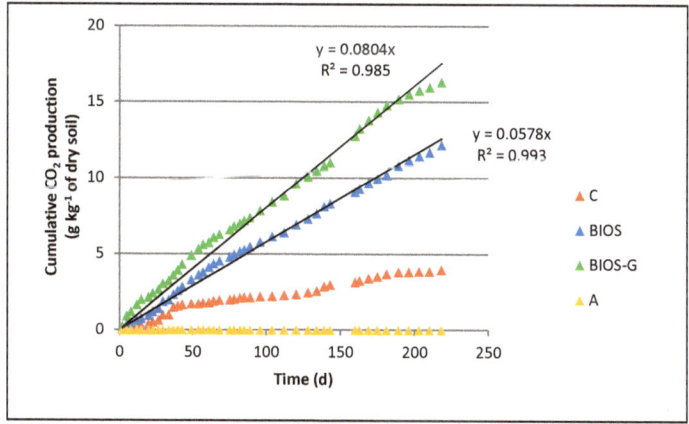

Figure 1. Cumulative amount of CO_2 production due to microbial respiration (A = abiotic microcosm; BIOS = biostimulated microcosm; BIOS-G = biostimulated microcosm and glucose addition; C = control microcosm).

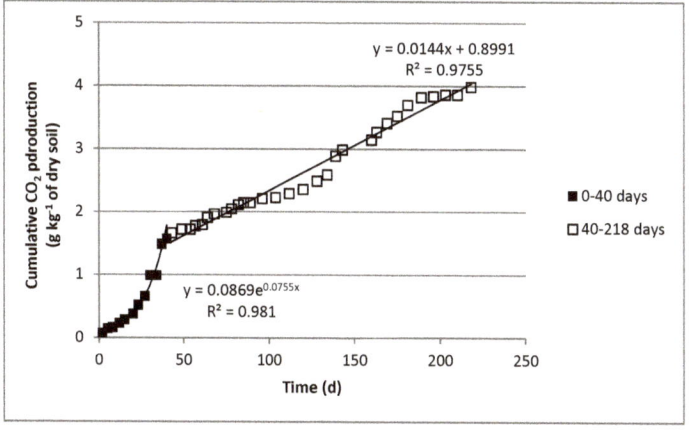

Figure 2. Cumulative amount of CO_2 production of a biotic microcosm (natural soil).

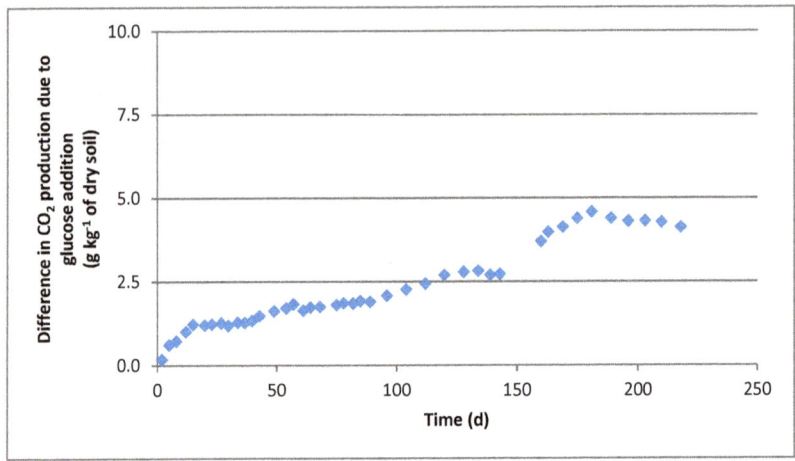

Figure 3. Difference in CO_2 production due to glucose addition.

3.2. Microbial Counts

Figure 4 shows the microbial counts obtained on MEA, at different times, in the C, BIOS, and BIOS-G microcosms. We observed that the BIOS-G microcosm had the highest number of colonies during the first two months. The population growth was enhanced by glucose, which as the primary carbon source was metabolized rapidly. For the BIOS-G microcosm, the addition of glucose at t = 143 days apparently influenced the microbial population, which at t = 175 days was more than twice that found at t = 110 days. However, growth also occurred for the C and BIOS microcosms, therefore the addition of MSMB seemed to enhance the growth more than glucose itself.

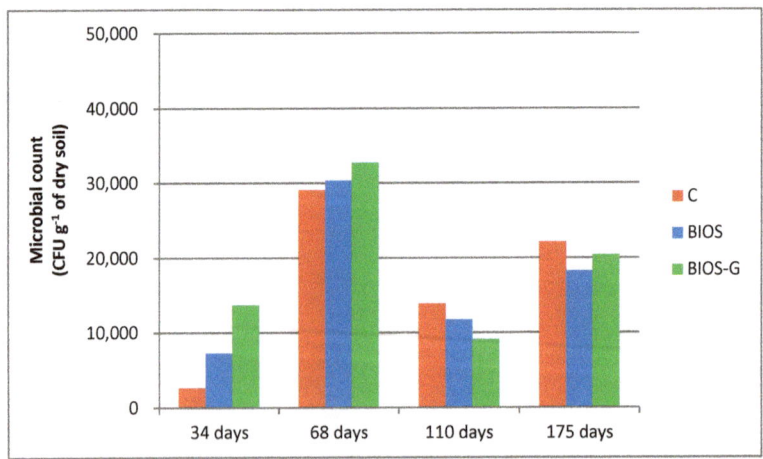

Figure 4. Microbial count at different times in the experimental tests' respiration (BIOS = biostimulated microcosm; BIOS-G = biostimulated microcosm and glucose addition; C = control microcosm).

Among the microcosms, the control one (C) showed the highest value both at 110 days and at 175 days (at 175 days, the population was 60% higher than that found at 110 days). A very similar population growth (around 60%) was achieved in the BIOS microcosm from 110 to 175 days.

3.3. Diesel Oil Removal

The monitoring of diesel oil content is presented in Figure 5 for the microcosms treated with diesel oil at an initial concentration equal to 75,926 mg diesel oil kg^{-1} of soil.

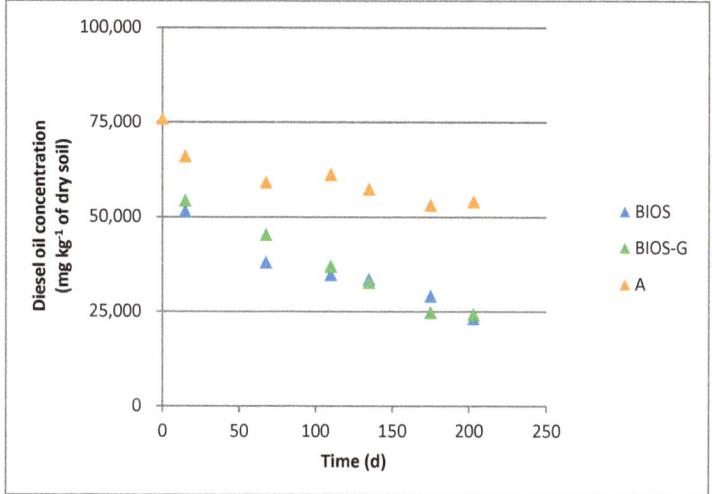

Figure 5. Residual diesel oil concentration in microcosms' respiration (A = abiotic microcosm; BIOS = biostimulated microcosm; BIOS-G = biostimulated microcosm and glucose addition).

During the biodegradation tests, the residual contaminant concentration was measured after 15, 68, 110, 135, 175, and 203 days.

As far as the residual diesel oil concentration in soil is concerned, the following considerations can be pointed out:

- At the end of the test (t = 203 days), in the abiotic microcosm the residual concentration was around 50,000 mg kg^{-1} of soil, corresponding to a removal efficiency in the order of 30%.
- After 200 days, for the BIOS and BIOS-G microcosms the removal efficiency was very similar and almost 70%, suggesting that the presence of a primary carbon source had no effect on the diesel oil biodegradation. This confirmed the results obtained in a previous work [13] carried out in the same conditions (i.e., biostimulation) but with a different microcosm volume.
- The data shown for the BIOS and BIOS-G microcosms included the abiotic contribution (value obtained in microcosm A); therefore, the net diesel oil removal due to biodegradation was the difference between the overall and abiotic values.
- For the BIOS and BIOS-G microcosms, the trend of residual diesel oil concentration was still decreasing after 200 days; however, this became almost negligible and not substantial in real-case applications.

To better understand the biodegradation and removal process as a whole, in each microcosm, the composition of the residual diesel oil was analyzed and the results compared to the initial one. This information is useful to roughly estimate the removal capability of the process carried out in similar operative and soil conditions.

The results are reported in Figures 6–8, for the abiotic, BIOS, and BIOS-G microcosms, respectively.

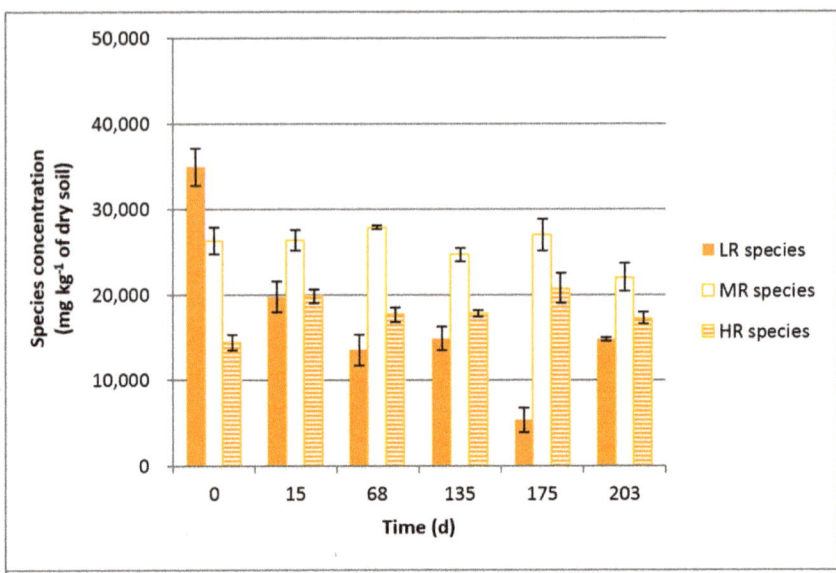

Figure 6. Change in residual diesel oil composition during the process for the abiotic microcosm.

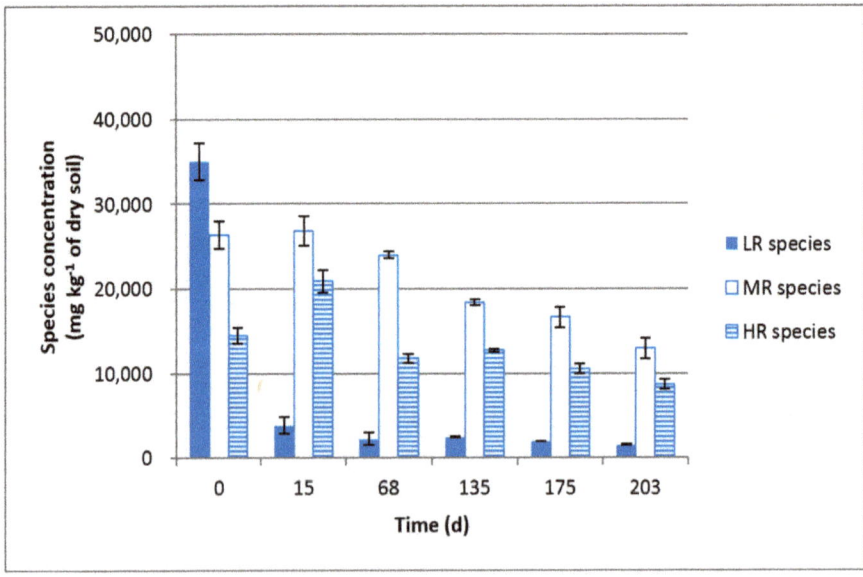

Figure 7. Change in residual diesel oil composition during the process for the BIOS microcosm.

The removal of LR species was around 57% of the initial content (Figure 6). This occurred rather rapidly, and after about two months the LR species concentration remained at around 15,000 mg kg^{-1} of soil.

For MR and HR species, very limited removal was achieved. This means that the removal in the abiotic microcosm was mainly for LR species, representative of simple molecules, like aliphatic ones.

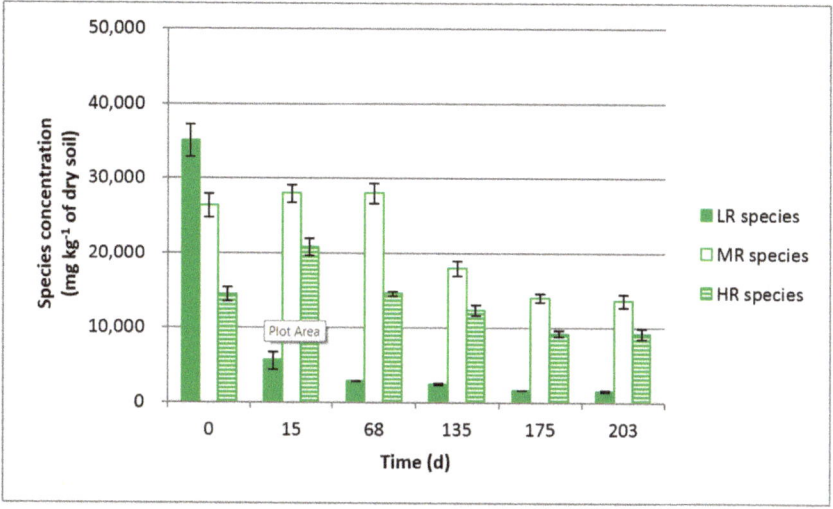

Figure 8. Change in residual diesel oil composition during the process for the BIOS-G microcosm.

For the BIOS microcosm, the results are shown in Figure 7. In this case, it is possible to observe a different behavior:

- For LR species, after 15 days their concentration was greatly reduced, and after 68 days, the value remained about constant, at around 2000 mg kg^{-1} of soil. The trend was similar to that shown by the abiotic system, but the constant value was much lower, evidencing that part of these compounds were metabolized by the diesel oil-degrading microorganisms (the LR species contained more quickly biodegradable molecules).
- The removal of MR species was slower and with a lag phase, which could be estimated at two months, since after only 68 days the MR species concentration started to slowly decrease with a trend still continuing at t = 203 days.
- Regarding HR species, the removal was not consistent and seemed to occur slowly with long time durations.

The monitoring of the BIOS-G microcosm (Figure 8) shows results very similar to those obtained with the BIOS microcosm, coherent with the overall removal (Figure 5).

The starting diesel oil composition can be described as:

- LR species: About 46% by weight.
- MR species: About 35% by weight.
- HR species: About 19% by weight.

The results achieved in both the biostimulated microcosms show that diesel oil removal gave good overall results with the adopted operative conditions. However, the content decrease was mainly due to LR species removal, thanks to the simultaneous abiotic and biotic contributions, and this should be taken into account for subsequent remediation applications.

3.4. Kinetic Modeling

3.4.1. First-Order Reaction Rate

The data of residual diesel oil concentration were used to model the process kinetics. As discussed in paragraph 2.6.1, the most widely adopted model is the first-order one.

Looking at the data presented in Figure 5 and as discussed in paragraph 3.3, the diesel oil removal process was not influenced by the presence of glucose; therefore, the results obtained for BIOS and BIOS-G were modeled together. Figure 9 shows these data and the fitting line describing the first-order kinetics, stressed to have C_0 = 75926 mg kg^{-1} of soil. The fitting was performed with the least-square method with uncertain data.

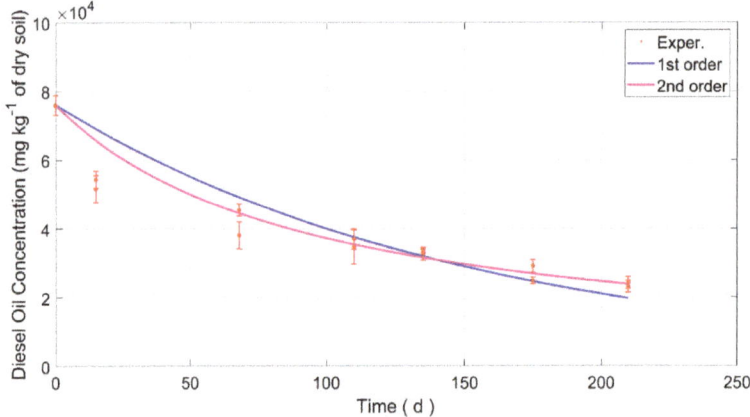

Figure 9. Kinetic modelling for the BIOS and BIOS-G microcosms.

This line has an equation equal to:

$$C(t) = 75926 \exp(-0.0064\, t), \text{ with } R^2 = 0.79.$$

The reaction rate constant was equal to 0.0064 day^{-1}. Using Equation (4), the half-life time $t_{1/2}$ was calculated: $t_{1/2}$ = 108 days.

As is evident from Figure 9, the model did not properly fit the experimental data in the first phase of biodegradation (t = 15 and 68 days); for later times, the fitting became adequate. Longer runs could support this reliability, even if it must be noted that for real-scale biodegradation, the processing duration influences the treatment costs, and shorter times are preferable.

3.4.2. Second-Order Reaction Rate

The line modeling the second-order reaction rate is also shown in Figure 9. As for the first-order model, the data fitting was stressed to obtain concentration C = 75926 mg kg^{-1} of soil at t = 0; as in the previous case, the fitting was performed by the least-square method with uncertain data.

The interpolating equation is $1/C(t) = 1.316 \times 10^{-5} + 1.373 \times 10^{-7} \times t$, and the value of the correlation coefficient is R^2 = 0.93. The coefficient was higher than the first-order one, showing that this model seems more suitable to fit the experimental data. Assuming the fitting parameters, the calculation of the half-life time with equation (7) gave $t_{1/2}$ = 96 days.

3.5. Geophysical Features for Future Monitoring and Preliminary Measurements

A preliminary test to check the behavior of the relative dielectric permittivity and the temperature is shown in Figure 10. The measurements were collected in a sample of the tested system, i.e., sandy soil, partially saturated with water, diesel oil, and gas. A Water Content Reflectometer (WCR) probe was adopted to measure the dielectric permittivity and the sample temperature. We observed a slight increase in the dielectric permittivity when the temperature increased and vice versa.

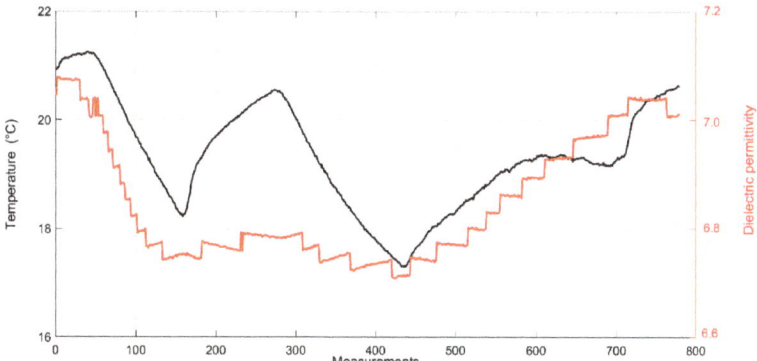

Figure 10. Relative dielectric permittivity (red) and temperature (black) measured in a mixture of sandy grain, diesel oil, water and gas (measurements performed with a Water Content Probe). The measures refer to a time-window of 24 h.

We applied formula (8) to model the behavior of dielectric permittivity of the same mixture of sandy soil, partially saturated with water, oil and gas. The results of the modeling are depicted in Figure 11. The simulation considered a decrease in oil saturation from 0.4 to 0.2, due to the combined effect of evaporation, adsorption and degradation. We assumed that the diesel oil was displaced by gas during the degradation. We plotted the trend of the α-exponent for values of α-exponent in the range 0.2–0.6.

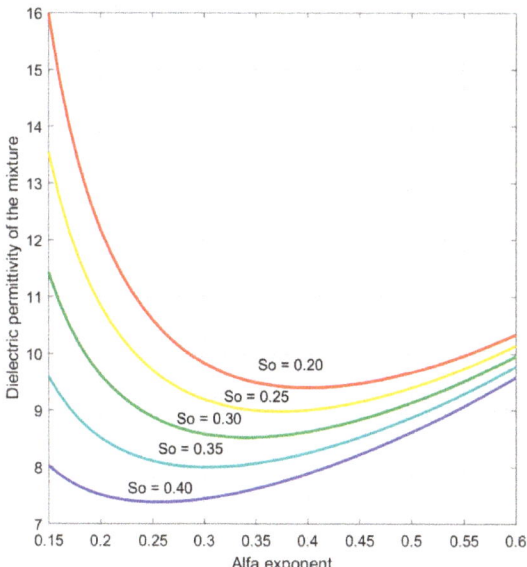

Figure 11. Effect of α-exponent of the CRIM formula on relative dielectric permittivity at different diesel oil contents (S_o), considering a relative water content of 0.4.

The preliminary measurements showed that the best α-coefficient to match the experimental result is close to being 0.4, considering the uncertainty because of the probe and all the measurements' inaccuracies.

4. Discussion

Here, we discuss the main findings in terms of biological, kinetic, and geophysical issues. As a biological strategy, biostimulation was adopted, exploiting indigenous microorganisms. The results showed that the overall removal efficiency could be high in a time compatible with common bioremediation treatments, that is to say around 50% in about four months. As a whole, the contribution was the synergic result due to biological and physical removal. Particularly, the abiotic system showed an evident removal of LR species, around 57% of the initial content. Probably, some part of these compounds evaporated and/or was adsorbed; this effect occurred very rapidly. A similar result was obtained by Chen et al. [31]. They studied the bioremediation effect on Total Petroleum Hydrocarbons (TPHs) with biostimulation tests and documented a reduction of TPH mass in the order of 80%, basically due to the removal of aliphatic compounds. A different kind of soil (clay loam soil) was used, and this could be a relevant issue if comparing with our results in sandy soil. Ani and Ochin [32] studied the use of wastes, such as goat manure and palm oil mill effluent, as additional nutrients for the bioremediation of soils polluted with TPHs. Their findings showed the positive effect of these additions when the soil is poor. Therefore, future studies are required to evaluate the effect of soil composition and possible nutrients on the reliability of biostimulation and soil degradation.

The biological contribution can be useful to predict the performance of similar systems, exploiting the indigenous microorganism and enhancing the growth of bacteria populations. The overall effects of the environment can be evaluated in terms of evaporation and soil adsorption.

The sum of these two terms, the biological and the environmental, can estimate the amount of removable pollutant.

The results of the biostimulated microcosms showed that glucose addition did not improve the removal efficiency, and this is a positive issue in terms of real-scale application.

The analysis of the residual diesel oil species showed that, as expected, the simple molecules were removed at a great extent in a short time (15 days), with the synergic contribution of soil and biological process. For the other species, slow removal was shown, completely due to the microbial activity.

In terms of kinetics, the first-order reaction rate was applied with good success to predict the overall amount of removed pollution. This was previously ascertained by other researchers, as shown in paragraph 2.6. Recently, Ortega et al. [33] showed the opportunity to simulate and predict the diesel oil removal from polluted soils by a first-order kinetic modeling able to consider the addition of soil amendments.

In our study, the second order of the reaction rate was also adopted, and it provided a better data fitting with a correlation coefficient equal to 0.93.

The biodegradation results were used to study their impact on an optimal design of future geophysical monitoring. We analyzed the impact on the design of geophysical experiments, both at the laboratory scale and at the field scale. Particularly, if a second order of kinetic is suitable for estimating the behavior of degradation versus time, the time-lapse geophysical experiments must be designed in order to sample with a higher repetition rate the first stage of the degradation, while a first order of kinetic asks for a more regular sampling of the time-lapse experiments.

We focused on the main environmental factors that affect the reliability of laboratory and in-field geophysical monitoring; particularly, we pointed out how the dielectric permittivity depends on the temperature. The monitoring of the temperature is therefore necessary in order to compensate the data for the temperature drift. Particularly, we observed a gradient of $dT/d\varepsilon = 0.1$ °C F^{-1} m, which agrees with similar trends observed in previous works [34]. We did not discuss the temperature effect on the electrical conductivity; this was well documented by several authors [17,34].

We simulated the expected behavior of the dielectric permittivity according to the change in diesel oil saturation, by performing a sensitive analysis (Figure 11). The model reliability had to be checked for every experiment; especially, the optimum value of the α-exponent had to be calibrated. We pointed out that the most favorable condition was yielded when the α-exponent was lower than 0.3, because a higher sensitivity of the dielectric permittivity to the variations of the diesel content was observed. For instance, when α is equal to 0.25, an increase in the dielectric permittivity from 7.5 up to 10.5 is predicted; this is equivalent to a variation of more than 40%. Otherwise, if the α-exponent is equal to 0.5, the simulation predicts an increase in the dielectric permittivity of only about 20%.

According to our sensitivity analysis, the model seemed to be able to predict the degradation rate starting from the geophysical measurements, even if with some uncertainty due to the accuracy of geophysical parameter measurements and reliability of the model. The removal of contaminant inside soil pores produced an increment of soil dielectric permittivity. Therefore, it was directly related to diesel oil reduction, since the contaminant was replaced by the displacing fluid. As also pointed out by Comegna et al. [25], the amount of contaminant in soil can be inferred if the total volume of pore fluid is known in advance.

At this stage of the research, the physical properties of the system are mainly related to the physical and electro-magnetic behaviors of solid media and the chemical and electrical characteristics of fluids. We are not yet considering the interactions between the media, the fluids, and microbes and their metabolic processes. New tests and more detailed analysis are requested in order to understand the geophysical sensitivity of the electromagnetic devices to microbial effects. Particularly, we need to explore how microbial processes could affect the propagation and attenuation terms of the dielectrical permittivity: This could occur at the scale of the cell surface, as suggested by Atenkawa and Slater [35].

Electromagnetic devices, such as WCR and TDR, combine measures of the real part of dielectrical permittivity, which is related to the propagation of the electromagnetic signal, and conductivity, which refers to the attenuation: This integration offers a great opportunity to not only monitor the degradation effects, but also to analyze the microbial activity at the laboratory scale. A good correlation between the presence of hydrocarbon contamination and a substantial increase in electrical conductivity, caused by microbiological activities, was reported in previous studies [36]. Moreover, the long-term degradation of hydrocarbons (in the field) determines a marked attenuation of the electromagnetic signal [37]: In such a context, the monitoring of electrical conductivity, combined with the propagation term (the real part of the dielectrical permittivity), takes on an important role in assessing microbial activity.

This study revealed some findings that are useful for the future application of bioremediation to remove diesel oil from polluted soil, and for planning a strategy for monitoring degradation effects by using a geophysical approach.

Author Contributions: Conceptualization, F.B. and A.G.; methodology, F.B.; software, F.C. and A.G.; validation, F.C.; formal analysis, A.G.; investigation, F.B.; resources, F.B.; data curation, A.C.; writing—original draft preparation, F.C. and A.G.; writing—review and editing, F.C. and A.G.; visualization, F.B.; supervision, F.C.; project administration, F.C.; funding acquisition, F.C.

Funding: This research was funded by the project "GEOPHYSICAL METHODS TO MONITOR SOIL BIOREMEDIATION", funded by the Italian Ministry of Foreign Affairs and International Cooperation in the frame of the Executive Programme of Scientific and Technological Cooperation between the Republic of India and the Italian Republic for the years 2017–2019—SIGNIFICANT RESEARCH.

Conflicts of Interest: The authors declare no conflict of interest.

References

1. Kalantari, R.R.; Mohseni-Bandpi, A.; Esrafili, A.; Nasseri, S.; Ashmagh, F.R.; Jorfi, S.; Ja'fari, M. Effectiveness of biostimulation through nutrient content on the bioremediation of phenanthrene contaminated soil. *J. Environ. Health Sci. Eng.* **2014**, *12*, 143. [CrossRef]
2. Simpanen, S.; Dahli, M.; Gerlach, M.; Mikkonen, A.; Malk, V.; Mikola, J.; Romantschuk, M. Biostimulation proved to be the most efficient method in the comparison of in situ soil remediation treatments after a simulated oil spill accident. *Environ. Sci. Pollut. Res.* **2016**, *23*, 25024–25038. [CrossRef] [PubMed]
3. Poi, G.; Aburto-Medina, A.; Mok, P.C.; Ball, A.S.; Shahsavari, E. Large scale bioaugmentation of soil contaminated with petroleum hydrocarbons using a mixed microbial consortium. *Ecol. Eng.* **2017**, *102*, 64–71. [CrossRef]
4. Fan, M.Y.; Xie, R.-J.; Qin, G. Bioremediation of petroleum-contaminated soil by a combined system of biostimulation–bioaugmentation with yeast. *Environ. Technol.* **2014**, *35*, 391–399. [CrossRef] [PubMed]

5. Wu, M.; Dick, W.A.; Li, W.; Wang, X.; Yang, Q.; Wang, T.; Xu, L.; Zhang, M.; Chen, L. Bioaugmentation and biostimulation of hydrocarbon degradation and the microbial community in a petroleum-contaminated soil. *Int. Biodeterior. Biodegrad.* **2016**, *107*, 158–164. [CrossRef]
6. Guarino, C.; Spada, V.; Sciarrillo, R. Assessment of three approaches of bioremediation (Natural attenuation, Landfarming and Bioagumentation-Assistited Landfarming) for a petroleum hydrocarbons contaminated soil. *Chemosphere* **2017**, *170*, 10–16. [CrossRef] [PubMed]
7. Liu, X.; Selonen, V.; Steffen, K.; Surakka, M.; Rantalainen, A.-L.; Romantschuk, M.; Sinkkonen, A. Meat and bone meal as a novel biostimulation agent in hydrocarbon contaminated soils. *Chemosphere* **2019**, *225*, 574–578. [CrossRef]
8. Safdari, M.-S.; Kariminia, H.-R.; Rahmati, M.; Fazlollahi, F.; Polasko, A.; Mahendra, S.; Wilding, W.V.; Fletcher, T.H. Development of bioreactors for comparative study of natural attenuation, biostimulation, and bioaugmentation of petroleum-hydrocarbon contaminated soil. *J. Hazard. Mater.* **2018**, *342*, 270–278. [CrossRef]
9. Wu, M.; Li, W.; Dick, W.A.; Ye, X.; Chen, K.; Kost, D.; Chen, L. Bioremediation of hydrocarbon degradation in a petroleum-contaminated soil and microbial population and activity determination. *Chemosphere* **2017**, *169*, 124–130. [CrossRef]
10. Lopes Julio, A.D.; Rocha Fernandes, R.d.C.; Dutra Costa, M.; Lima Neves, J.C.; Montes Rodrigues, E.; Rogerio Totola, M. A new biostimulation approach based on the concept of remaining P for soil bioremediation. *J. Environ. Manag.* **2018**, *207*, 417–422. [CrossRef]
11. Polyak, Y.M.; Bakina, L.G.; Chugunova, M.V.; Mayachkina, N.; Gerasimov, A.; Bure, V.M. Effect of remediation strategies on biological activity of oil-contaminated soil—A field study. *Int. Biodeterior. Biodegrad.* **2018**, *126*, 57–68. [CrossRef]
12. Casale, A.; Bosco, F.; Chiampo, F.; Franco, D.; Ruffino, B.; Godio, A. Soil microcosm set up for a bioremediation study. *Int. J. Appl. Sci. Environ. Eng.* **2018**, *1*, 277–280.
13. Bosco, F.; Casale, A.; Mazzarino, I.; Godio, A.; Ruffino, B.; Mollea, C.; Chiampo, F. Microcosm evaluation of bioaugmentation and biostimulation efficacy on diesel-contaminated soil. *J. Chem. Technol. Biotechnol.* **2019**. [CrossRef]
14. Arato, A.; Weher, M.; Bíró, B.; Godio, A. Integration of geophysical, geochemical and microbiological data for a comprehensive small-scale characterization of an aged LNAPL-contaminated site. *Environ. Sci. Pollut. Res.* **2014**, *1*, 8948–8963. [CrossRef]
15. Koroma, S.; Arato, A.; Godio, A. Analyzing geophysical signature of a hydrocarbon-contaminated soil using geoelectrical surveys. *Environ. Earth Sci.* **2015**, *74*, 2937–2948. [CrossRef]
16. Atekwana, E.A.; Werkema, D.D.; Atekwana, E.A. Biogeophysics: The effects of microbial processes on geophysical properties of the shallow subsurface. *Appl. Hydrogeophys.* **2006**, *71*, 161–193. [CrossRef]
17. Or, D.; Wraith, J.M. Temperature effects on soil bulk dielectric permittivity measured by time domain reflectometry: A physical model. *Water Resour. Res.* **1999**, *35*, 371–383. [CrossRef]
18. Lazebnik, M.; Converse, M.; Booske, J.H.; Hagness, S.C. Ultra-wideband Temperature-Dependent Dielectric Properties of Animal Liver Tissue in the Microwave Frequency Range. *Phys. Med. Biol.* **2006**, *51*, 1941–1955. [CrossRef]
19. Godio, A. Open ended-coaxial Cable Measurements of Saturated Sandy Soils. *Am. J. Environ. Sci.* **2007**, *3*, 175–182. [CrossRef]
20. Endres, A.L.; Redman, D.J. Modeling the electrical properties of porous rocks and soils containing immiscible contaminants. *J. Environ. Eng. Geophys.* **2009**, 105–112. [CrossRef]
21. Carcione, J.M.; Seriani, G. An electromagnetic modelling tool for the detection of hydrocarbons in the subsoil. *Geophys. Prospect.* **2000**, *48*, 231–256. [CrossRef]
22. Sen, P.N.; Scala, C.; Cohen, M.H. A self-similar model for sedimentary rocks with applications to the dielectric constant of fused glass beads. *Geophysics* **1981**, *46*, 781–795. [CrossRef]
23. Feng, S.; Sen, P.N. Geometrical model of conductive and dielectric properties of partially saturated rocks. *J. Appl. Phys.* **1985**, *58*, 3236–3243. [CrossRef]
24. Knight, R.; Endres, A. A new concept in modeling the dielectric response of sandstones: Defining a wetted rock and bulk water system. *Geophysics* **1990**, *55*, 586–594. [CrossRef]
25. Comegna, A.; Coppola, A.; Dragonetti, G.; Sommella, A. Dielectric response of a variable saturated soil contaminated by Non-Aqueous Phase Liquids (NAPLs). *Procedia Environ. Sci.* **2013**, *19*, 701–710. [CrossRef]

26. Adesodun, J.K.; Mbagwu, J.S.C. Biodegradation of waste-lubricating petroleum oil in a tropical alfisol as mediated by animal droppings. *Bioresour. Technol.* **2008**, *99*, 5659–5665. [CrossRef]
27. Agarry, S.E.; Aremu, M.O.; Aworanti, A. Kinetic Modelling and Half-Life Study on Bioremediation of Soil Co-Contaminated with Lubricating Motor Oil and Lead Using Different Bioremediation Strategies. *Soil Sediment Contam. Int. J.* **2013**, *22*, 800–816. [CrossRef]
28. Komilis, D.P.; Vrohidou, A.E.K.; Voudrias, E.A. Kinetics of aerobic bioremediation of a diesel-contaminated sandy soil: Effect of nitrogen addition. *Water Air Soil Pollut.* **2010**, *208*, 193–208. [CrossRef]
29. Nwankwegu, A.S.; Onwosi, C.O. Bioremediation of gasoline contaminated agricultural soil by bioaugmentation. *Environ. Technol. Innov.* **2017**, *7*, 1–11. [CrossRef]
30. Sarkar, D.; Ferguson, M.; Datta, R.; Birnbaum, S. Bioremediation of petroleum hydrocarbons in contaminated soils: Comparison of biosolids addition, carbon supplementation, and monitored natural attenuation. *Environ. Pollut.* **2005**, *136*, 187–195. [CrossRef]
31. Chen, F.; Li, X.; Zhu, Q.; Ma, J.; Hou, H.; Zhang, S. Bioremediation of petroleum-contaminated soil enhanced by aged refuse. *Chemosphere* **2019**, *222*, 98–105. [CrossRef]
32. Ani, K.A.; Ochin, E. Response surface optimization and effects of agricultural wastes on total petroleum hydrocarbon degradation. *Beni-Suef Univ. J. Basic Appl. Sci.* **2018**, *7*, 564–574. [CrossRef]
33. Ortega, M.F.; García-Martínez, M.-J.; Bolonio, D.; Canoira, L.; Llamas, J.F. Weighted linear models for simulation and prediction of biodegradation in diesel polluted soils. *Sci. Total Environ.* **2019**, *686*, 580–589. [CrossRef]
34. Seyfried, S.S.; Grant, L.E. Temperature Effects on Soil Dielectric Properties Measured at 50 MHz. *Vadose Zone J.* **2007**, *6*, 759–765. [CrossRef]
35. Atekwana, E.A.; Slater, L. Biogeophysics: A new frontier in Earth science research. *Rev. Geophys.* **2009**, *47*, RG4004. [CrossRef]
36. Cassiani, G.; Binley, A.; Kemna, A.; Wehrer, M.; Flores Orozco, A.; Deiana, R.; Boaga, J.; Rossi, M.; Dietrich, P.; Werban, U.; et al. Noninvasive characterization of the Trecate (Italy) crude-oil contaminated site: Links between contamination and geophysical signals. *Environ. Sci. Pollut. Res.* **2014**, *21*, 8914–8931. [CrossRef]
37. Godio, A.; Arato, A.; Stocco, S. Geophysical characterization of a non aqueous-phase liquid-contaminated site. *Environ. Geosci.* **2010**, *17*, 141–162. [CrossRef]

© 2019 by the authors. Licensee MDPI, Basel, Switzerland. This article is an open access article distributed under the terms and conditions of the Creative Commons Attribution (CC BY) license (http://creativecommons.org/licenses/by/4.0/).

Article

Gradual Exposure to Salinity Improves Tolerance to Salt Stress in Rapeseed (*Brassica napus* L.)

Michael Santangeli [1,2], Concetta Capo [1], Simone Beninati [1], Fabrizio Pietrini [3] and Cinzia Forni [1,*]

1. Dipartimento di Biologia, Università di Roma Tor Vergata, Via della Ricerca Scientifica, 00133 Rome, Italy
2. Department of Forest and Soil Sciences, University of Natural Resources and Life Sciences, Institute of Soil Research (IBF), BOKU Vienna, Konrad-Lorenz Straße 24, 3034 Tulln, Austria
3. Istituto di Ricerca sugli Ecosistemi Terrestri, Consiglio Nazionale delle Ricerche (IRET-CNR), Area della Ricerca di Roma 1, Via Salaria Km 29.300, 00015 Monterotondo Stazione (Roma), Italia
* Correspondence: forni@uniroma2.it

Received: 4 June 2019; Accepted: 7 August 2019; Published: 12 August 2019

Abstract: Soil salinity is considered one of the most severe abiotic stresses in plants; plant acclimation to salinity could be a tool to improve salt tolerance even in a sensitive genotype. In this work we investigated the physiological mechanisms underneath the response to gradual and prolonged exposure to sodium chloride in cultivars of *Brassica napus* L. Fifteen days old seedlings of the cultivars Dynastie (salt tolerant) and SY Saveo (salt sensitive) were progressively exposed to increasing soil salinity conditions for 60 days. Salt exposed plants of both cultivars showed reductions of biomass, size and number of leaves. However, after 60 days the relative reduction in biomass was lower in sensitive cultivar as compared to tolerant ones. An increase of chlorophylls content was detected in both cultivars; the values of the quantum efficiency of PSII photochemistry (ΦPSII) and those of the electron transport rate (ETR) indicated that the photochemical activity was only partially reduced by NaCl treatments in both cultivars. Ascorbate peroxidase (APX) activity was higher in treated samples with respect to the controls, indicating its activation following salt exposure, and confirming its involvement in salt stress response. A gradual exposure to salt could elicit different salt stress responses, thus preserving plant vitality and conferring a certain degree of tolerance, even though the genotype was salt sensitive at the seed germination stage. An improvement of salt tolerance in *B. napus* could be obtained by acclimation to saline conditions.

Keywords: acclimation; *Brassica napus*; salt stress; chlorophyll fluorescence; photosynthesis; anti-oxidant enzymes; polyamines; proline

---

1. Introduction

Soil salinity affects at least 20% of irrigated land worldwide, with a prevision of an increasing trend of arable lands loss up to 50% by the middle of the twenty-first century. Therefore, a devastating scenario is foreseen in the global scale both for the environmental resources and human health due to reduction of soil fertility and crop yield [1]. Italy is one of the salt-affected countries in Europe, and in the Mediterranean area this country is considered a hotspot for both land degradation and desertification, and salinization can be considered an important cause of soil degradation [2].

Salt stress is one of the most adverse abiotic stress impairing productivity of several important crops. The major consequences of plant exposure to saline conditions are cell dehydration and toxicity caused by salt, due to the inhibition of water uptake by the root hairs. Generally, soil salinity affects many aspects of plant development, such as seed germination, vegetative growth (through the reduction of the leaf area, the content of chlorophylls and the conductance of the stomata), and the reproduction by inducing the abortion of the ovules and senescence of fertilized embryos [3,4]. Sodium and chlorine are considered the ions mainly responsible for the salinization, reaching concentration that can inhibit or decrease the development of plants [1–4]. Leaves accumulate most of the excess of Na^+ and Cl^- absorbed at the roots level, and this accumulation causes an early senescence, associated with more or less marked chlorosis and/or necrosis phenomena, depending on the accumulation rate of these ions and the capacity to compartmentalize them. The effect of salinity on leaves can be either direct, due to the closure of the stomata and the consequent reduction in the rate of assimilation of CO_2 [5], or secondary; due to oxidative stress, which can seriously affect leaf photosynthetic machinery. Besides this, the intracellular accumulation of Na^+ changes the ratio of K^+/Na^+ and may lead to an osmotic imbalance and the production of Reactive Oxygen Species (ROS), which seems to affect the function of some enzymes, and the amounts of photosynthetic pigments [6]. Oxidative stress, caused by ROS overproduction, is the main cause of loss in productivity of the cultivated species [7].

The salt tolerance mechanism is a multigenic trait involving different biochemical pathways: i.e., control of ion uptake by roots and transport into leaves, synthesis of compatible solutes, change in photosynthetic pathway, induction of antioxidant enzymes as well as the synthesis of hormones. The stress response depends to a different extent on both genotypes and developmental stages of the plants; furthermore, the genetic plasticity is fundamental for conferring different degrees of tolerance, by depending on the effectiveness of stress response mechanisms. In non-tolerant species a gradual adaptation to saline conditions can ameliorate the plant response to stress, leading to a better performance [8].

The aim of this work was to reveal the role and relative contribution of plant acclimation to the amelioration of tolerance to salt stress in rapeseed (*Brassica napus* L.). This crop is mainly utilized for edible oil production, its use has been also considered for the production of renewable energy, in fact the large amount of oil present in the seeds makes them suitable as oilseed and fodder crop, and for the production of biodiesel [9]. Even though many cultivars of this species were described as 'salt-tolerant', yield and growth of this crop can decline with increasing soil salinity [10].

To assess the possibility to induce the tolerance to salt in cultivars of rapeseed, the effects of NaCl on plants were determined by a whole-plant assessment of agronomic parameters, used for salinity tolerance selection [11], and of physiological mechanisms underneath the response to gradual and prolonged exposure to sodium chloride.

2. Materials and Methods

2.1. Chemicals and Reagents

All reagents, analytical grade or equivalent, were purchased from Merck or Sigma-Aldrich, unless otherwise stated. In each set of experiments all working solutions were prepared immediately before use from stock reagents.

2.2. Plant Growth Conditions and Saline Treatments

The seeds of Italian cultivars of *B. napus* were kindly supplied by Dr. Montanari of the CREA-CIN (Centro di Ricerca per le Colture Industriali) Bologna, Italy. The cultivars were selected based on high-yield in oil production and low glucosinolates and erucic acid content. The seeds were stored at 4 °C until sowing.

Preliminary seed germination test in presence of salt were performed on seven cultivars cultivated in Italy (data not shown). Basing on the germination rates, we selected the most susceptible cultivar (SY Saveo 16% seed germinated at 72 h at 320 mM NaCl) and the most tolerant ones (Dynastie 82% seed germinated at 72 h at 320 mM NaCl) for the following experiments.

The seeds of these cultivars were first washed with running water and then soaked for 1 h in tap water. The seeds were then placed in Petri dishes (50 seeds each) containing two layers of filter paper soaked with 1:10 strength Hoagland solution [12]. The Petri dishes were incubated in the dark for 24 h at 10 °C, then they were placed at 24 °C for 48 h. Then, the germinated seeds were carefully sown in plastic pots (20 cm diameter), 8 seeds per pots. Each pot contained about 1.6 kg of water saturated soil material (total volume of 2.5 L) with the following chemical-physical characteristics: pH

6.2, EC 0.25 dS m^{-1}, dry bulk density 135 kg m^3, total porosity 91% v/v, components (neutral sphagnum peat, perlite (<5%), mineral compound fertilizer).

The experimental groups consisted of 3 plastic pots (24 cm diameter, 20 cm height) for each treatment, containing 8 germinated seeds. Seedlings were grown for 15 days before the beginning of salt treatments at 23 °C ± 2 °C (constant temperature) and 48% ± 2% of relative humidity, moved randomly every 4 days, with a photoperiod of 16 h, PAR between 20 (min.) and 30 (max.) μmoles photons m^{-2} s^{-1} (lamp: 2 × OSRAM, FLUORA t8 36.00 W and 2 × OSRAM, LUMILUX Cool Daylight t 8 36.00 W). During the first growing period of 15 days, each pot was irrigated once a week with 50 mL of 1/10 strength Hoagland solution each pot, ~20% GWC. Fifteen-days old seedlings were randomly assigned to three (different treatment groups and irrigated every 4 days with: 100 mL Hoagland (control) or 100 mL Hoagland containing 160 mM NaCl or 320 mM NaCl (saline treatments). Pots without plants irrigated and maintained in the same conditions were added and considered as blank. The experimental period was 60 days.

2.3. Growth and Morphological Parameters

Morphological parameters were monitored every 15 days starting from the beginning of the treatments to progressively evaluate the effects of NaCl on plant growth. Parameters taken in account were as follows: length of the stem, leaf area, number of leaves, and length of the longest root (for plants sampled at 30th and 60th day). The biomass was determined by evaluating the total weight of the plants (g fresh weight). The biomass of the control and of the treated plants was related to the number of plants that composed each group obtaining the average value of biomass per plant.

For subsequent determinations, samples (200 mg fresh weight) were frozen by dipping in liquid nitrogen and stored at −80 °C until further analyses.

2.4. Water Content of the Plants

The percentage of water content was determined by bringing the plants to dryness in a stove at a temperature of 70 °C for 48 h [13]. The average % of water was estimated as:

$$\text{Water content (\%)} = \frac{\text{f.w.} - \text{d.w.}}{\text{f.w.}} \times 100 \tag{1}$$

where: f.w. = Plant's fresh weight; d.w. = Plant's dry weight.

2.5. Membrane Injury Index (MII)

At the 60th day, the membrane injury index was calculated on fresh samples of shoots and roots by measuring the electrical conductivity according to the method of Blum and Ebercon [14]. Immediately after sampling, the samples were analysed by immersing in milli-Q water in a volume equal to 0.1 mL water mg f.w.$^{-1}$ The conductivity was measured at room temperature with the electrical conductivity (EC) meter (Hanna instrument 8333) after 30 min incubation at 40 °C and 10 min at 100 °C in a water bath (Gesellschaft für Labortechnik, GFL). The membrane injury index was calculated according to the formula:

$$\text{MII (\%)} = \frac{C_{40\,°C}}{C_{100\,°C}} \times 100 \tag{2}$$

where: $C_{40\,°C}$ = electrical conductivity measured after 30 min at 40 °C; $C_{100\,°C}$ = electrical conductivity measured after 10 min at 100 °C.

2.6. Soil Analysis

The chemical-physical parameters of the soils were determined in pots containing plants and in pots without plants (blank) in order to evaluate the absorption of minerals by the roots.

The gravimetric water content of soil was determined, at the end of the experiments. The percentage of water content was determined by oven-drying soils samples at a temperature of 70 °C for 48 h. The average % of water was estimated as:

$$\text{Gravimetric Water Content (\%)} = \frac{\text{f.w.} - \text{d.w.}}{\text{d.w.}} \times 100 \qquad (3)$$

where: f.w. = Soil fresh weight; d.w. = Soil dry weight.

The osmolality of the soil was determined according to Merchant et al. [15] with the cryogenic osmometer OSMOMAT 030 (Gonotec, Berlin, Germany).

Forty milligrams (dry weight) of soil samples, lacking of perlite, were homogenized and were crushed with pestles adding 1 mL of distilled water at 80 °C and incubated for 30 min at 80 °C in a water bath. The samples, were centrifuged for 5 min at 15,000× g at room temperature, and the supernatant was analysed for the osmolality.

The determination of the electrical conductivity (EC) was carried out according to Sairam et al. [16]. Soil samples (200 mg dry weight) were dipped in water milli-Q (0.1 mL H_2O mg d.w.$^{-1}$). They were subsequently incubated in a water bath at 100 °C for 10 min and then filtered with paper funnels. Electrical conductivity was measured on the filtrate at room temperature with an EC meter (HANNA Instrument 8333).

2.7. Imaging of Chlorophyll Fluorescence and Leaf Total Chlorophyll Content

At the end of the experiment (60th day), to assess the efficiency of the photosynthetic apparatus in plants of *B. napus* exposed to salt stress, imaging of chlorophyll fluorescence parameters were performed. In particular, the maximal quantum efficiency of PSII photochemistry (F_v/F_m), the quantum efficiency of PSII photochemistry (ΦPSII) and the non-photochemical quenching (NPQ) were measured on the last fully expanded leaf using a chlorophyll fluorescence imaging (MINI-Imaging-PAM, Walz, Germany). Leaves were dark adapted for at least 30 min before determining F_0 and F_m (minimum and maximum fluorescence, respectively). The F_v/F_m value was calculated as $(F_m - F_0)/F_m$. Subsequently, leaves were adapted to a photosynthetic photon flux density (PPFD) of 55 µmoles m^{-2} s^{-1} for at least 10 min to reach a steady-state condition. A saturation light pulse was then applied to determine the maximum fluorescence ($F_{m'}$) and steady-state fluorescence (F_s) during the actinic illumination. Saturation pulse images and values of the chlorophyll fluorescence parameters were captured. The ΦPSII value was calculated according to Genty et al. [17] using the formula $(F_{m'} - F_s)/F_{m'}$. Calculation of NPQ was determined as $(F_m - F_{m'})/F_{m'}$ [18]. Data of NPQ were divided by four to display values < 1.00. The apparent photosynthetic electron transport rate (ETR) was calculated as follows: ETR = ΦPSII × PPFD × 0.5 × Abs, where Abs is the apparent absorptivity of the leaf surface and 0.5 is the fraction of light absorbed by PSII antennae [19]. The Abs value was automatically calculated pixel by pixel from the R (red) and NI (near infrared) images using the formula: Abs = 1 − (R/NI). Furthermore, measurements of total chlorophyll content were performed by the chlorophyll meter readings (SPAD-502, Minolta Camera Co., Osaka, Japan) on the same leaves previously used for chlorophyll fluorescence determinations. The measure was taken from one fully developed leaf per plant. Four SPAD readings were taken from the widest portion of the leaf lamina, while avoiding major veins. The four SPAD readings were averaged to represent the SPAD value of each leaf. SPAD values were converted to chlorophyll content (µg cm^{-2}) using the following equation [20]:

$$\text{Chlorophyll content} = (99 \times \text{SPAD})/(144 - \text{SPAD}) \qquad (4)$$

2.8. Phenolic Compounds

The phenolic compounds were extracted from frozen samples according to Orzali et al. [21] with some modification. Briefly, samples were ground to fine powder in ceramic mortars with pestles at 4 °C, suspended in 3 mL of 0.1 M HCL (Scharlau), and incubated for 3 h at 4 °C. Extracts were centrifuged for 15 min at 8000× g. The supernatants were collected, and the pellets rinsed with an additional 2 mL of 0.1 M HCl to complete the extraction, and centrifuged again for 15 min at 8000× g. The supernatants were pooled and total phenolic amount was determined as described by Orzali et al. [21]. The concentration of the phenolic compounds was determined by a calibration curve of chlorogenic acid (CA) (Alfa Aesar) as standard (y = 0.0015x; R^2 = 0.9997). Results are expressed as micrograms of CA equivalents per g of plant fresh weight.

2.9. Determination of Enzymatic Activities

Enzymatic activities were determined in extracts of frozen samples (200 mg f.w.) homogenized with ceramics mortars and pestles. In each assay, the protein content of the extracts was estimated by the dye-binding method of Bradford [22] using bovine serum albumin (BSA) as a standard, calibration curve (y = 0.0315x; R^2 = 0.9913).

Polyphenol oxidase (PPO) (EC 1.14.18.1) activity was determined according to the method of Orzali et al. [21]. The kinetics of the enzymes were followed by spectrophotometer (VARIAN Cary 50 Bio) at the wavelength of 420 nm for 300 s. The activity was detected as enzymatic activity units (1 unit was defined as 1 µM of product produced minute^{-1} mL^{-1} reaction solution). The enzymatic activity of PPO was expressed as E.U. * mg protein^{-1}.

Superoxide dismutase (SOD) (EC 1.15.1.1) and ascorbate peroxidase (APX) (EC 1.11.1.11) activities were determined by NPAGE (native polyacrylamide gel electrophoresis). Two hundred milligrams (fresh weight) of frozen leaves were homogenized in liquid nitrogen and polyvinylpolypyrrolidone (PVPP). Homogenates were re-suspended in 1 mL of 0.2 M sodium phosphate buffer (pH 7.0) containing protease inhibitor cocktail for plant cell (Sigma). The extracts were centrifuged at 15,000× g at 4 °C for 30 min, the supernatants were recovered. Samples containing 40 µg proteins was added and separated on native polyacrilamide gel electrophoresis, and SOD activity was visualized of gel staining through the conversion of nitroblue tetrazolium to formazane, the gel was exposed to light to allow the development of the colour [23].

APX activity was detected according to the procedure by Mittler and Zilinskas [24]. PAGE was carried out under native conditions.

2.10. Polyamines

Polyamines (PA) were extracted from frozen leaves (200 mg f.w.) homogenized in liquid nitrogen using ceramic mortars and pestle. The homogenates were suspended in 1 mL of Dulbecco's Phosphate Buffered Saline (D-PBS, Biowest) and then centrifuged for 10 min at 17,000× g. After a second centrifugation the supernatants were collected, quantified and brought to 1200 µL volume with saline phosphate buffer (PBS). To precipitate the proteins, perchloric acid (PCA) (final concentration 5%) was added to the sample. The samples were kept on ice bath for 30 min. and then centrifuged at 17,000× g for 15 min.; 50 µL of 1 mM Diaminooctane, (DAO) was added as internal standard to quantify any losses during sample preparation. Free polyamines content of the supernatant was determined by HPLC according to the method of Beninati et al. [25]. The correspondence between the areas of the PA peaks and their concentration was obtained by comparison with the areas of the standards of the Spermine (SPM), Spermidine (SPD) and Putrescine (PUT). The concentration of PA was expressed in:

$$\text{nmoles} * \text{mg f.w.}^{-1} \tag{5}$$

2.11. Proline

Proline and total amino acids were extracted from homogenized samples (50 mg f.w.) using a cold extraction procedure according to Carillo and Gibon [26]. The homogenate was collected and resuspended in 10 mg/0.2 mL (*w/v*) of ethanol: Milli-Q water (40:60 *v/v*). The resulting mixture was left overnight at 4 °C, and then centrifuged for 5 min at 14,000× g. To 500 µL of ethanolic extract was added 1 mL of reaction mixture composed by 1% (*w/v*) Ninhydrin (2,2-dihydroxyindane-1,3-dione) dissolved in a mixture of 60% (*v/v*) acetic acid (Fluka) and 20% (*v/v*) ethanol (Fluka). The sample mixture was heated at 95 °C in a block heater for 20 min and then centrifuged for 1 min to 6000× g. Proline concentration was evaluated by detecting the absorbance at 520 nm wavelength at the spectrophotometer (VARIAN

Cary 50 Bio). Proline concentration was determined from a standard curve made with standard solutions of L-Proline (y = 0.0135 x; R^2 = 0.9946). The proline content is expressed in nmoles * mg f.w.$^{-1}$

2.12. Statistical Analysis

The experiment was set up in a completely randomized design with at least three replicates for each treatment and physiological determinations. Data are expressed as mean ± standard error (SE). The relationship between the measured morphological parameters and the saline treatment was assessed at 60th day, through the Kruskal Wallis test and the Mann–Whitney pairwise comparison test. Results of physiological analyses were evaluated by one-way analysis of variance (ANOVA). Tukey–Kramer method was used to evaluate the significance of the differences between the means of the parameters taken into consideration. Data regarding chlorophyll fluorescence parameters and leaf total chlorophyll were evaluated by two-way ANOVA, with cultivar and NaCl treatment as the main effects, using the Past3 software. All analyses were considered significant at $p \leq 0.05$. In the graphs, different letters are considered significantly different from each other according to the probabilistic value $p \leq 0.05$.

3. Results

In this work, we have focused the attention on the response of *B. napus* cultivars with different salt sensitivity exposed to a gradual increasing of salinity levels of soil medium, in order to verify if they can develop the ability to acclimatize to saline conditions. During the experiments, the irrigation with saline solution led to a gradually increasing salinity of the soil media with a significant enhancement of the electric conductivity (EC) at the end of the experiments (Figure 1). Moreover, in saline media, the water content was significantly higher respect to their control ($p \leq 0.05$) (Figure 2), suggesting that salinity hindered water absorption by the plants.

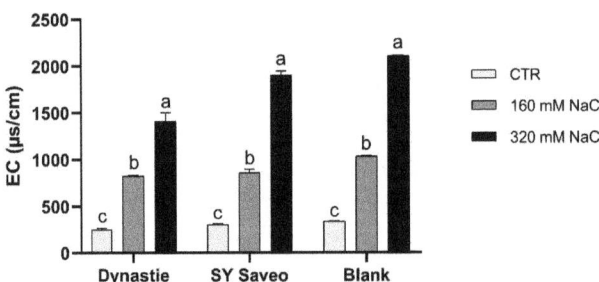

Figure 1. Electrical conductivity (EC) of the soils at the end of the experiments. Data are expressed as means ± standard error (SE, n = 3). Post hoc comparisons conducted with the Tukey–Kramer method. Means with different letters are statistically different ($p \leq 0.05$) within the same group.

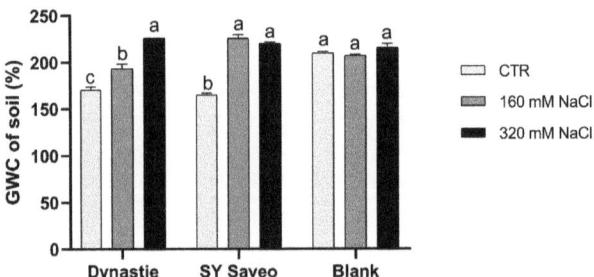

Figure 2. Gravimetric water content (GWC) of the soils at the end of experiments. Data are expressed as means ± SE (n = 3). Post hoc comparisons conducted with the Tukey—Kramer method. Means with different letters are statistically different ($p \leq 0.05$) within the same group.

3.1. Effects of Soil Salinization On Plants

Under the conditions reported above, the growth and consequently the production of biomass were reduced significantly with the increasing of soil salinity (Table 1; Figure 3) (In Supplementary Materials, photographs of the plants and statistical analysis of the height values are shown respectively in Figures S1 and S2, and Table S1). The relative reduction in biomass was lower after 60 days in sensitive cultivar as compared to tolerant. At the end of the experiments, significant differences in dry/fresh weight ratio of the treated plants compared to the controls were observed in both cultivars (Table 2). This was related to a decrease in water content of the plants exposed to NaCl (Table 3), which was caused by the reduced osmotic potential of the soil. The quantity of soluble proteins detected was generally higher in the treated plants than the controls, with the exception of Dynastie at 160 mM NaCl after 60 days (Table 2).

Table 1. Biomass production at 30 and 60 days of experiment. Data are expressed as means ± SE (n = 3). Post hoc comparisons conducted with the Tukey–Kramer method. Means with different letters are statistically different ($p \leq 0.05$) within the same period and cultivar. Photographs of the plants at the end of the experiments are shown in Supplementary Materials, Figures S1 and S2.

	Plant Biomass			
	Dynastie		SY Saveo	
NaCl	30 d	60 d	30 d	60 d
CTR	1.02 ± 0.03 a	4.31 ± 0.03 a	1.00 ± 0.05 a	3.27 ± 0.22 a
160 mM	0.94 ± 0.04 a	1.94 ± 0.08 b	0.76 ± 0.03 b	1.86 ± 0.07 b
320 mM	0.74 ± 0.04 b	1.05 ± 0.02 c	0.66 ± 0.02 c	0.96 ± 0.05 c

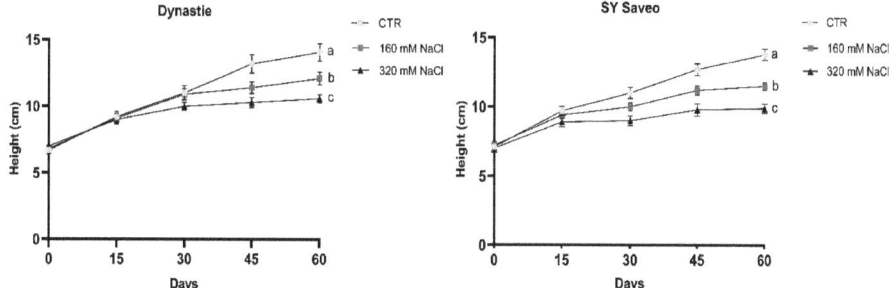

Figure 3. Effects of salt on the height of the cultivars. Data are expressed as means ± SE (n ≥ 8). Post hoc comparisons conducted with the Mann–Whitney pairwise comparison test. Means with different letters are statistically different ($p \leq 0.05$) within the same cultivar. All values with post hoc comparisons are shown in the Supplementary Materials, Table S1.

Table 2. Dry weight/fresh ratio and protein content. Data are expressed as means ± SE (n = 3). Post hoc comparisons conducted with the Tukey–Kramer method. Means with different letters are statistically different ($p \leq 0.05$) within the same cultivar.

	Dry Weight/Fresh Weight		Soluble Proteins (µg Protein/mg f.w.)			
	Dynastie	SY Saveo	Dynastie		SY Saveo	
NaCl	60 d	60 d	30 d	60 d	30 d	60 d
CTR	0.018 ± 0.0005 c	0.069 ± 0.001 b	4.62 ± 0.06 b	7.30 ± 0.12 a	3.21 ± 0.14 c	4.32 ± 0.14 b
160 mM	0.053 ± 0.003 b	0.077 ± 0.008 ab	4.62 ± 0.06 b	5.50 ± 0.09 b	4.40 ± 0.07 a	4.69 ± 0.07 b
320 mM	0.083 ± 0.0005 a	0.086 ± 0.005 a	7.66 ± 0.08 a	7.82 ± 0.16 a	3.75 ± 0.07 b	8.98 ± 0.15 a

Table 3. Water content (%) of the plants after 60 days of treatment. Data are expressed as means ± SE (n = 3). Post hoc comparisons conducted with the Tukey–Kramer method. Means with different letters are statistically different ($p \leq 0.05$) within the same cultivar.

NaCl	Plants Water Content (%)	
	Dynastie	SY Saveo
CTR	92.37 ± 0.12 a	93.10 ± 0.12 a
160 mM	91.67 ± 0.50 a	91.39 ± 0.17 b
320 mM	89.38 ± 0.07 b	90.82 ± 0.10 c

Leaf premature sensing of salt were observed in manner depending on NaCl concentration. Due to the small size, leaf area was measured after 15 days of treatments. The leaves were gradually affected by salt exposure (Figure 4); after 45 and 60 days both number of leaves per plants and leaf area were significantly reduced in the treated plants respect to the control (Figure 4). The length of the roots was not significantly affected by salt (Figure 5).

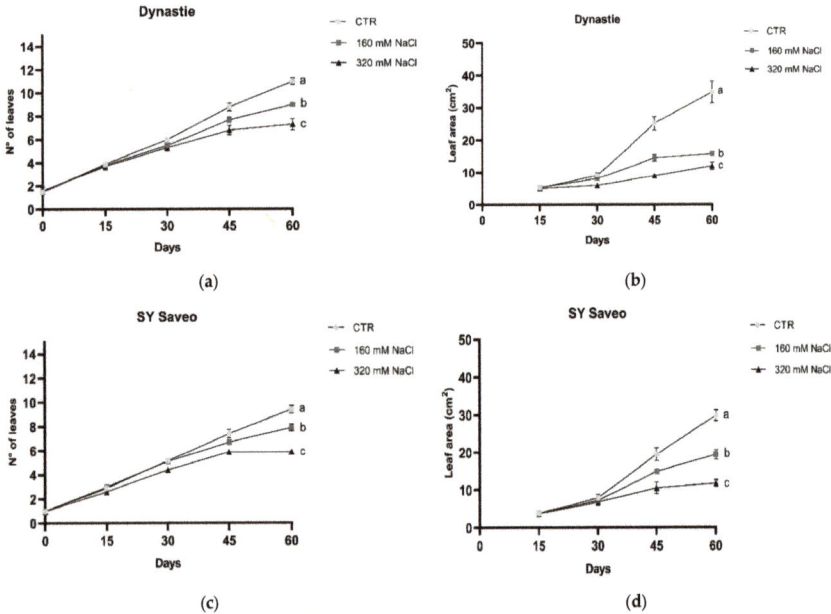

Figure 4. Number of leaves and leaf area in Dynastie (**a**,**b**) and SY Saveo (**c**,**d**). Post hoc comparisons conducted with the Mann–Whitney pairwise comparison test. Data are expressed as means ± SE (n ≥ 8). Means with different letters are statistically different ($p \leq 0.05$) within the same cultivar. All leaf area values with post hoc comparisons are shown in the Supplementary Materials, Table S2.

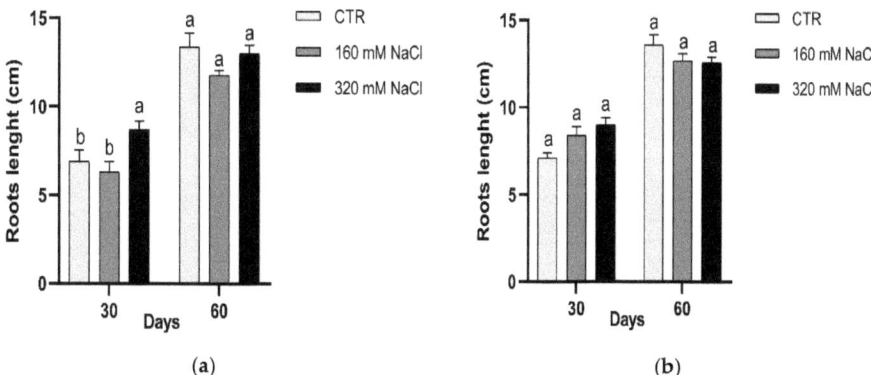

Figure 5. Effect of salt on the root growth: Dynastie (**a**) and SY Saveo (**b**). Data are expressed as means ± SE (n ≥ 8). Post hoc comparisons conducted with the Tukey–Kramer method. Means with different letters are statistically different ($p \leq 0.05$) within the same period and cultivar.

Membrane injury index (MII) increased significantly ($p \leq 0.05$) only in roots, while MII of leaves was similar in control and treated plants, with a little increase at higher salt concentration in Dynastie (Figure 6).

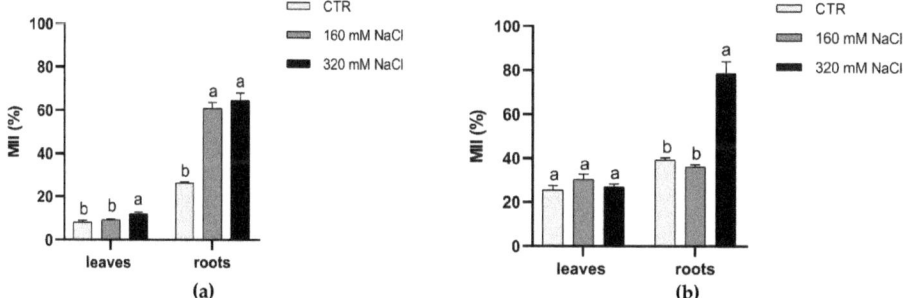

Figure 6. Membrane injury index of leaves and roots of Dynastie (**a**) and SY Saveo (**b**), determined at the end of the experiment. Data are expressed as means ± SE (n = 3). Post hoc comparisons conducted with the Tukey–Kramer method. Means with different letters are statistically different ($p \leq 0.05$) within the same organ of the cultivar.

To assess the capacity of *B. napus* to withstand NaCl treatments and to study the spatial heterogeneity of photosynthesis, measurements of chlorophyll fluorescence imaging and leaf chlorophyll content were performed (Table 4). A representative image of chlorophyll fluorescence parameters (F_v/F_m, ΦPSII and NPQ) in a single leaf of the two cultivars of *B. napus* treated and untreated is reported in Figure 7. At the end of the experimental period, our results showed that with respect to control, the maximal quantum efficiency of PSII (F_v/F_m) was unaffected by the treatment at both NaCl concentrations in the two cultivars (Table 4). Moreover, the data revealed that the response of the quantum efficiency of PSII photochemistry (ΦPSII) in the two treated cultivars, measured at the light intensity close to that experienced by plants during the growth conditions, was quite similar (Table 4). In this study, plant exposure to both NaCl concentrations caused a decrease of the ΦPSII, compared to the control, nevertheless the reduction of this parameter was just more evident in cv Dynastie. A quite similar trend was also observed in the electron transport rate (ETR) values (Table 4). The response of the non-photochemical quenching (NPQ) to NaCl exposure evidenced that the NPQ values were practically unchanged, compared to control, in plant exposed to higher NaCl concentration, and showed a reduction at lower salt concentration in Dynastie, whereas exhibited an increase in SY Saveo (Table 4). Finally, in both cultivars the leaf total chlorophyll content and the apparent absorptivity of the leaf surface (Abs) values were generally higher in salt treated plants respect to control (Table 4).

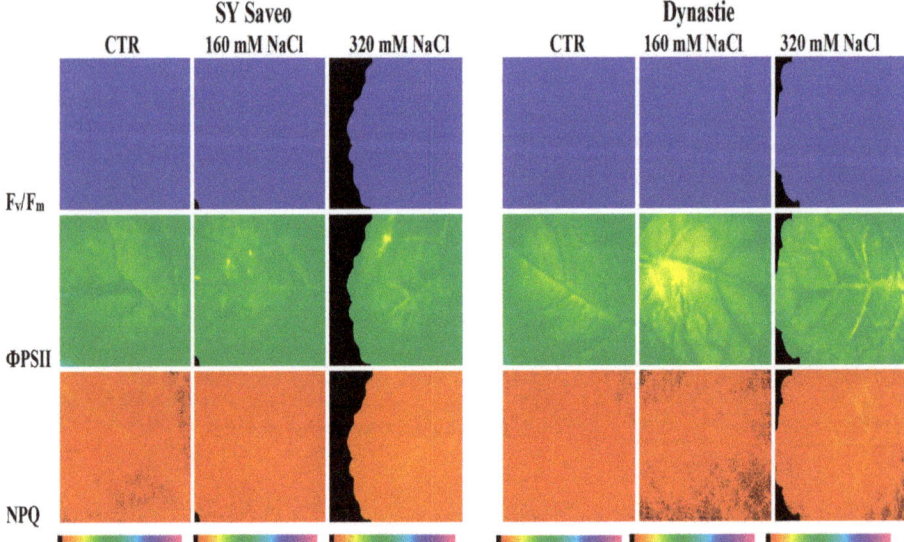

Figure 7. Chlorophyll fluorescence images of photochemistry (F_v/F_m) in a dark-adapted leaf and PSII photochemistry (ΦPSII) and non-photochemical quenching (NPQ) at steady-state with actinic illumination of 55 μmol photons m^{-2} s^{-1} measured at the end of the experiment (60th day) in two cultivars of *Brassica napus* (Saveo and Dynastie) exposed to different NaCl treatments (CTR, 160 mM NaCl, 320 mM NaCl). The false color code depicted at the bottom of the images ranges from 0.00 (black) to 1.00 (pink).

Table 4. Effects of NaCl on total chlorophyll content and chlorophyll fluorescence parameters, maximal quantum efficiency (Fv/Fm) measured in dark adapted leaves and quantum efficiency of PSII photochemistry (Φ PSII), non-photochemical quenching (NPQ), electron transport rate (ETR) and PAR-absorptivity (Abs) measured at steady state with actinic light illumination of 55 μmol photons m^{-2} s^{-1} at the end of the experiment (60th day) in two cultivars of *B. napus* (SY Saveo and Dynastie) exposed to different NaCl treatments (CTR, 160 mM NaCl, 320 mM NaCl). Data are expressed as means ± SE. Means with different letters are statistically different (LSD test, $p \leq 0.05$). ns = not significant. * $p < 0.05$. ** $p < 0.01$. *** $p < 0.001$.

Cvs.	NaCl	Fv/Fm (r.u.)	ΦPSII (r.u.)	NPQ (r.u.)	ETR (μmol Electrons m^{-2}s^{-1})	Abs (rel. un.)	Tot Chl Content (μg cm^{-2})
SY Saveo	CTR	0.810 ± 0.001 a	0.371 ± 0.013 ab	0.257 ± 0.005 b	8.10 ± 0.28 ab	0.804 ± 0.003 d	27.64 ± 0.63 a
	160 mM	0.814 ± 0.001 a	0.349 ± 0.009 bc	0.278 ± 0.010 b	7.98 ± 0.21 ab	0.846 ± 0.002 c	37.22 ± 0.06 b
	320 mM	0.812 ± 0.001 a	0.334 ± 0.002 cd	0.333 ± 0.010 a	7.74 ± 0.05 bc	0.859 ± 0.002 ab	41.26 ± 0.75 c
Dynastie	CTR	0.805 ± 0.001 b	0.385 ± 0.002 a	0.241 ± 0.010 b	8.35 ± 0.04 a	0.806 ± 0.004 d	28.15 ± 0.90 a
	160 mM	0.804 ± 0.003 b	0.316 ± 0.002 d	0.198 ± 0.019 c	7.33 ± 0.06 c	0.848 ± 0.005 bc	37.93 ± 1.66 b
	320 mM	0.805 ± 0.001 b	0.320 ± 0.005 d	0.277 ± 0.017 b	7.43 ± 0.12 c	0.860 ± 0.001 a	41.38 ± 0.30 c
p-Value (ANOVA)	Cv.	***	ns	***	ns	ns	ns
	Treat.	ns	***	***	**	***	***
	Cv.× Treat.	ns	*	ns	*	ns	ns

3.2. Plant Response to Salt Stress

3.2.1. Antioxidant Defence

In plants, the antioxidant activity depends on either an effective non-enzymatic antioxidant defence system or on the enzymatic antioxidant response or on both of them. In agreement with the literature, we observed a general increase in the amounts of phenolics, with significant changes in the production in treated plants of Dynastie at the end of the experiments (Figure 8).

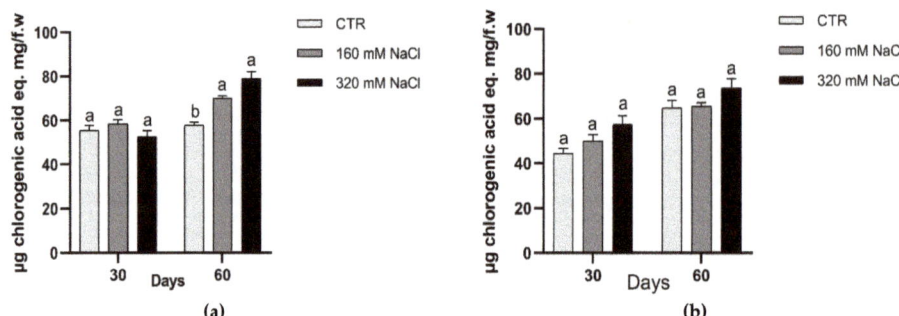

Figure 8. Amounts of phenolic compounds after 30 and 60 days of treatment: Dynastie (**a**) and SY Saveo (**b**). Data are expressed as means ± SE (n = 3). Post hoc comparisons conducted with the Tukey–Kramer method. Means with different letters are statistically different ($p \leq 0.05$) within the same period and cultivar.

Higher APX activity was detected in the treated samples respect to the controls (Figure 9), indicating the activation of this enzyme following salt exposure.

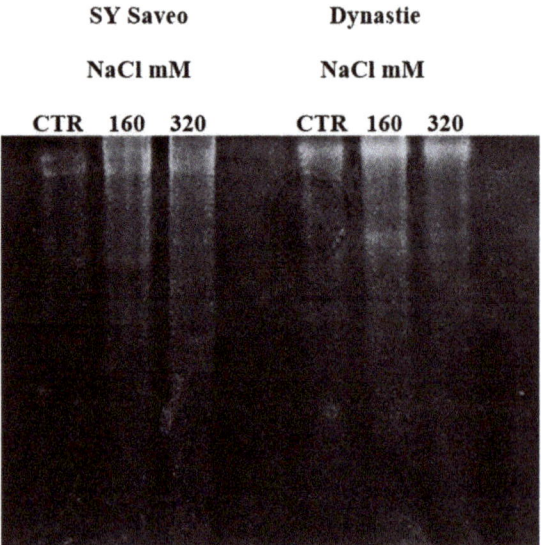

Figure 9. Activity of ascorbate peroxidase (APX) enzymes in plant exposed to different NaCl concentration after 60 days of treatment.

A critical role in the survival of plants under stressful environment has been ascribed to the activity of SOD, which upregulation can counteract the oxidative stress. In our experiment, after 60 days of treatment, SOD of both cultivars resulted to be less active in the plants treated with 320 mM NaCl compared to control and to 160 mM NaCl; while at 30 days the activity in SY Saveo was lower than that observed in Dynastie for both treated and control plants (Figure 10). These data suggest a different behaviour of SOD in the two genotypes.

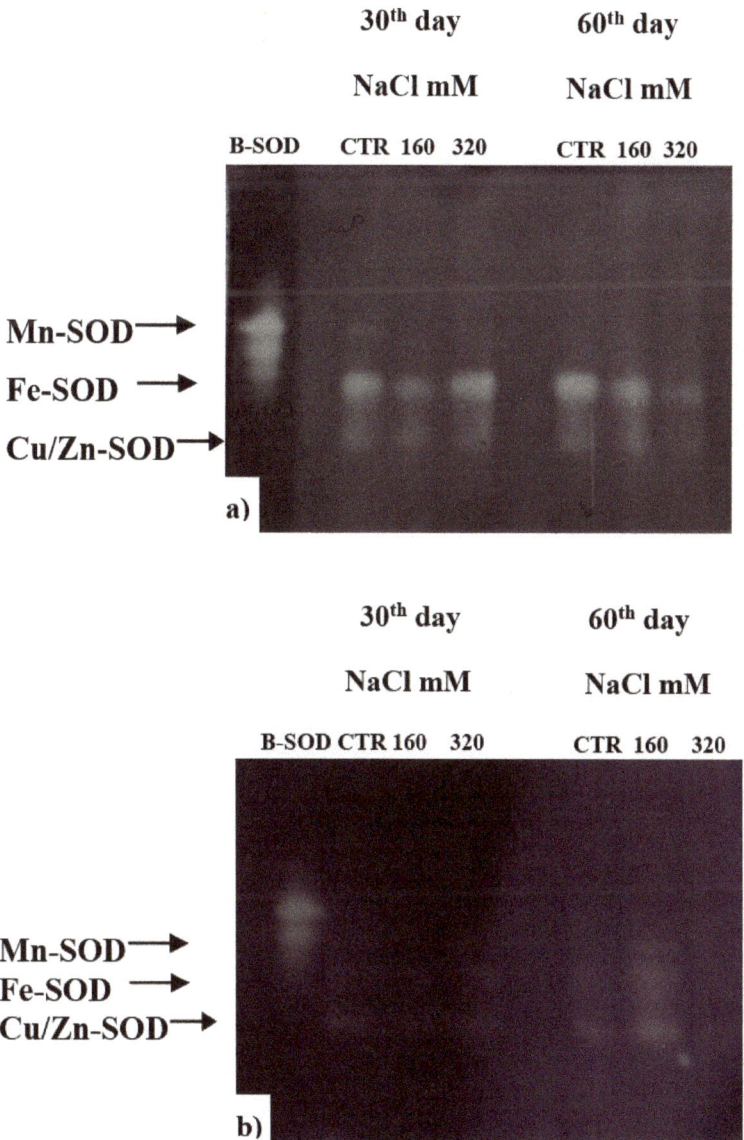

Figure 10. Activity of SOD enzymes in plant exposed to different NaCl concentration: (**a**) Dynastie, (**b**) SY Saveo.

In both cultivars, PPO activity was significantly higher in treated plants respect to the controls (Figure 11). A different timing in PPO response was detected, i.e., in Dynastie significantly enhanced activity was measured after longer exposure, while an opposite trend was observed in SY Saveo (Figure 11).

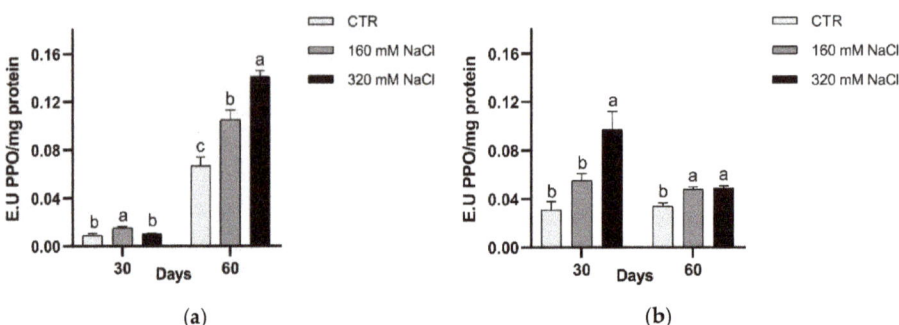

Figure 11. PPO activity after 30 and 60 days of treatment: Dynastie (a) and SY Saveo (b). Data are expressed as means ± SE (n = 3). Post hoc comparisons conducted with the Tukey–Kramer method. Means with different letters are statistically different ($p \leq 0.05$) within the same period and cultivar.

3.2.2. Osmotic Balancing

The accumulation of osmolytes, such as proline, glycine-betaine and polyamines, can help the plants to overcome osmotic unbalance. In this study we detected the concentrations of proline and PAs. The proline concentration increased significantly in both treatments and detection times (Figure 12). At the 60th day, Dynastie had the highest amount in the samples treated with 160 mM NaCl.

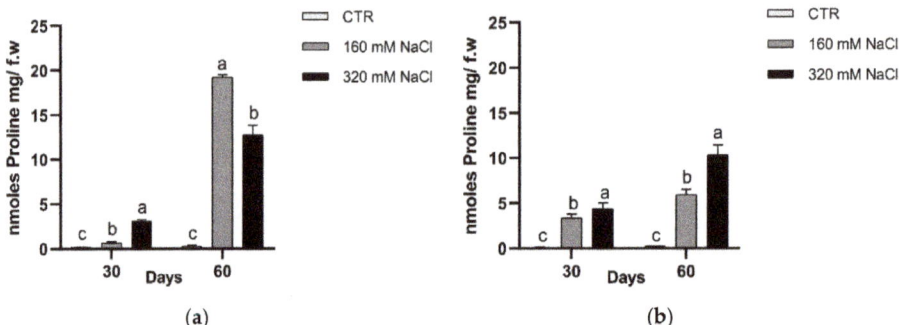

Figure 12. Proline concentration after 30 and 60 days of treatment: Dynastie (a) and SY Saveo (b). Data are expressed as means ± SE (n = 3). Post hoc comparisons conducted with the Tukey–Kramer method. Means with different letters are statistically different ($p \leq 0.05$) within the same period.

The presence of SPM, SPD and PUT was detected by HPLC (Tables 5 and 6). In SY Saveo, the amounts of PA changed within time, being higher at the end of exposure in the treated plants respect to the control (Table 5), especially the level of SPD. In Dynastie, significant changes respect to the control were detected in SPD and SPM content (Table 6).

Table 5. Polyamines in SY Saveo. Data are expressed as means ± SE. Post hoc comparisons conducted with the Tukey–Kramer method. Means with different letters are statistically different ($p \leq 0.05$) within the same period and PA.

	Polyamines (nmol/mg f.w.)					
	30 days			60 days		
NaCl	SPM	SPD	PUT	SPM	SPD	PUT
CTR	18.37 ± 1.92 b	100.71 ± 4.34 a	19.74 ± 1.31	27.29 ± 7.60 a	62.01 ± 4.61 b	14.08 ± 2.30 b
160 mM	29.20 ± 1.53 a	67.04 ± 1.83 b	N.D.*	24.83 ± 1.93 a	65.25 ± 3.69 b	39.66 ± 3.92 a
320 mM	22.76 ± 2.65 b	68.33 ± 2.31 b	N.D.*	30.76 ± 3.54 a	166.78 ± 12.18 a	35.78 ± 3.11 a

* N.D. = Not detected.

Table 6. Polyamines in Dynastie. Data are expressed as means ± SE. Post hoc comparisons conducted with the Tukey–Kramer method. Means with different letters are statistically different ($p \leq 0.05$) within the same period and PA.

	Polyamines (nmol/mg f.w.)					
	30 days			60 days		
NaCl	SPM	SPD	PUT	SPM	SPD	PUT
CTR	32.70 ± 1.98 b	97.83 ± 4.78 b	18.17 ± 1.54 a	13.89 ± 0.31 b	46.65 ± 0.34 c	12.58 ± 2.27 a
160 mM	44.70 ± 2.67 a	92.32 ± 4.76 a	12.63 ± 1.08 b	44.31 ± 7.29 a	99.83 ± 4.90 b	23.87 ± 4.65 a
320 mM	37.57 ± 2.71 ab	150.81 ± 9.97 a	22.76 ± 2.65 a	40.55 ± 7.81 a	151.08 ± 11.10 a	19.73 ± 5.62 a

4. Discussion

The deep changes in climate and soil conditions have led to a global warming with probable devastating effects on crops, such as phenology and/or yield. In the coming decades, the improvement of the crops ability to cope with both biotic and abiotic stresses, caused by the above reported changes, will likely play a fundamental role for adapting agriculture to the new environment.

Changes of soil conditions include the increase of soil salinity, which effects on plants are the results of complex interactions of different physiological processes depending on plant phenological stage.

For what salt stress is concern, the existence of salt-tolerant plants (halophytes) and the presence of different degree of tolerance within genotypes in the sensitive ones (glycophytes) indicate the presence of a genetic basis to salt tolerance [27]. Under this perspective, it is probable that a progressive acclimation to saline soil may be obtained, even in sensitive genotypes, by gradually exposing the plants to salt.

We have selected two cultivars that at seed germination stage were salt sensitive (Sy Saveo) and salt tolerant (Dynastie). To determine the possibility to obtain acclimation to saline conditions, we gradually exposed plants to increasing salt concentration and we detected the response in the sensitive cultivar and compared the data with those of tolerant ones, grown in the same conditions. At the end of the experiments, significant differences in dry/fresh weight ratio of the treated plants compared to the controls were observed in both cultivars. This was related to a lower water content in the plants exposed to NaCl, which was caused by the reduced osmotic potential of the soil, confirming the literature data [28]. Moreover, the detected differences in EC and osmolality (data not shown) between the soils where the plants were grown and the blank suggest that Na^+, present in the soil, was taken up and probably compartmentalized at root level to prevent toxic ionic effects, as proposed by Munns and Tester [29]. For what biomass production is concerned, we may say that in Sy Saveo some acclimation could be seen, since the relative reduction in its biomass was lower after 60 days as compared to tolerant cultivar. The quantity of soluble proteins detected was generally higher in the salt treated plants than controls. Several studies reported positive variations in the content of soluble proteins synthesized in response to the increase of NaCl in soil, suggesting a relation with salt stress tolerance [30].

Salinity effects were evident in the leaves that were the most affected organ. Generally, salt stress can cause leaf premature sensing together with reductions in photosynthesis, respiration and protein synthesis in sensitive species [3,29,31–33]. While, in comparison with the control, the length of the roots were not significantly affected by salt, even though this organ showed a significant increase of membrane injury index in the treated plants.

In this study, analyses of chlorophyll fluorescence parameters and images were used to investigate the effects of NaCl treatments on plants after the gradual exposure to saline conditions. Fluorescence image analysis has been already used to investigate the effects of salt stress on plants, providing information on the performance of the photosynthetic apparatus [34]. As previously described, our data showed that, with respect to control, F_v/F_m was unaffected by the treatment at both NaCl concentrations in the two cultivars, in agreement to the results reported by other authors [35,36]. Moreover, according to Maxwell and Johnson [19], the F_v/F_m ratio in the range of 0.79 to 0.84 is the approximate optimal value for many plant species, confirming that our cultivars were not affected by NaCl treatments. Lower values of this parameter, in fact, indicate that a proportion of the PSII reaction centres is damaged or inactivated, a phenomenon commonly observed in plants under stress [37]. Finally, some authors [38] suggested that F_v/F_m can be also considered an early indicator of salt stress. In order to study the possible changes in PSII photochemistry, the chlorophyll fluorescence characteristics under the steady-state of photosynthesis were investigated. Results showed that NaCl treatments caused a reduction of the ΦPSII values, with respect to control, in the two cultivars in both salt concentrations, nevertheless highlighting a just more pronounced decrease in cv Dynastie. As previously reported, a quite similar trend was also detectable for the electron transport rate (ETR) data. However, the slight reduction of ΦPSII and ETR values, with respect to control, measured at higher NaCl concentrations in plants of cv SY Saveo and cv Dynastie (less than 10 and 17% respectively), put in evidence the ability of the two cultivars to maintain a good photosynthetic activity even under NaCl treatments. In fact, ΦPSII and ETR are two widely used chlorophyll fluorescence parameters, which are employed to measure photochemistry and the overall photosynthetic capacity of plants [17,19]. Our data confirm, thus, that a relationship exists between the maintenance of photosynthesis (i.e., ΦPSII and ETR) as closer as possible to control rates and the effectiveness of protection or tolerance mechanisms towards different stress factors, as already reported by several authors [39–41]. As concerns the NPQ values, they showed, if compared to the control, a reduction in plants of cv Dynastie exposed to 160 mM NaCl, whereas highlighted an increase in plants of SY Saveo exposed to 320 mM NaCl. High values of NPQ in plants exposed to NaCl indicate the presence of a non-radiative energy dissipation mechanism [42], in which a higher proportion of absorbed photons is lost as thermal energy, instead of being used to drive photosynthesis [43]. Such thermal deactivation of the excess energy represents a protective mechanism, which maintains a high oxidative state of the primary electron acceptors of PSII and reduces the probability of photodamage [44]. On the other hand, Stepien and Johnson [45], in a study on *Arabidopsis* and *Thellungiella* under salt stress, concluded that increasing salinity resulted in a substantial increase in NPQ in salt-sensitive plant (*Arabidopsis*), while in salt-tolerant plant (*Thellungiella*) this parameter remained close to control levels at all salt concentrations. As shown by images of chlorophyll fluorescence, it is noteworthy that parameters measured in dark adapted leaves (F_v/F_m) for both cultivars revealed a homogeneous pattern of distribution of chlorophyll fluorescence, whereas showed an appreciable heterogeneous pattern of light utilization and photosynthetic activity in light adapted leaves (ΦPSII, and NPQ), especially in cultivar Dynastie. Regarding the response of the leaf chlorophyll content in controls in comparison to salt treated plants, this parameter showed higher values in the latter in both cultivars. Leaf chlorophyll content is considered one of the most important factors in determining photosynthetic potential and primary production [46]. In the literature, opposite plant behaviours have been reported [31,32]: i.e., some authors suggest that salinity reduces the chlorophyll content in salt susceptible plants and increases it in tolerant ones [47,48]. In our experiments, the analysis of apparent absorptivity of the leaf surface (Abs) showed a trend similar to chlorophyll content, in fact a strong correlation between the leaf absorbance and the total chlorophyll content has been described in the literature [49]. Therefore, basing on our results, we may hypothesize that in Sy Saveo a gradual exposure to salt enhanced the level of tolerance to saline growth conditions.

To counteract oxidative damage caused by the unbalance between ROS production and quenching, plant cell elicits antioxidant activity, which can depend on an effective non-enzymatic antioxidant defence system as well as on simultaneous antioxidant enzymatic response. In *Brassica* species the phenolic compounds have shown a higher antioxidant action compared to vitamins and carotenoids; as reported by Cartea et al. [50] the winter *B. napus* cultivars are rich in phenolic compounds. The biosynthesis of the phenolic compounds depends on both genetic and environmental factors. In *Brassica* many studies have shown that the intra-interspecific variation of these molecules is genotype dependent. In agreement with the literature, we observed a general increase in the amounts of phenolics, with significant changes in the production in treated plants of Dynastie at the end of the experiments.

SOD and APX constitute first line of enzymatic defence against ROS, thus protecting cells against superoxide radicals. A dual role has been ascribed to H_2O_2, which depends on the concentration. In fact, a lower amount of H_2O_2 has a beneficial role in both signalling pathways and cellular responses to stress, including the synthesis of stress related proteins, upregulation of antioxidant molecules and enzymes and in the accumulation of compatible solutes [51]. Vice versa, the effects of higher concentration of H_2O_2 can be deleterious, because it oxidizes cellular components like lipids, proteins and nucleic acids, cause inhibition of photosynthetic apparatus and can initiate chain reactions triggering cellular apoptosis. Therefore, the overexpression of SOD and APX and other peroxidases, and their coordinated action allows the maintenance of the balance between production and removal of ROS [29,52]. The observed higher APX activity of the treated samples respect to the controls confirmed the involvement of this enzyme in the response to salt stress, thus conferring tolerance to salinity [52]. In fact, several studies demonstrated that plant species with APX deficiency are more susceptible to oxidative damage with respect to those that over-express this enzyme, such as *Nicotiana tabacum*, where a higher expression of APX confers an increase in salinity tolerance [7].

The upregulation of SODs can counteract the oxidative stress, and plays a critical role in the plant survival under stressful environment [7]. Significant increase in the activities of Cu/Zn-SOD and Mn-SOD isozymes under salt stress was observed by Eyidogan and Oz [53]. In our experiment, we observed a lower SOD activity after a prolonged exposure to NaCl; this may be compensated by the activation of other antioxidant enzymes or by the production of antioxidant molecules, suggesting that in rapeseed SOD has a role in the first phases of the exposure to salt stress rather than during the acclimation phase. For what the latter is concern, a possible role might be ascribed to PPO activity, but in a manner that could be genotype dependent; in fact, in Dynastie a significant enhancement of the activity was measured after longer exposure, while an opposite trend was observed in SY Saveo.

Besides antioxidant response, plants possess mechanisms by which they re-establish osmotic and ionic homeostasis after salt stress imposition, and then a maintain steady state that is fundamental for their growth in the saline environment. To counteract the osmotic stress, plants accumulate osmolytes, such as proline, glycine-betaine and polyamines, allowing an intracellular osmotic balance [4], especially in the halophytes. Many studies on genes controlling the synthesis or metabolism of these compounds have indicated their essential role in tolerance to abiotic stress. However, a compatible solute can lead to potential growth reduction due to the high energy cost of their synthesis [29]. The protective role of osmolytes is species specific and depends on factors, such as growth conditions and phenological status of the plant [4]. An important role in stress response has been attributed to proline, being involved in redox balance, osmoprotection, and stress signalling. In this study, the proline concentration increased significantly in both treatments and detection times. Even though controversy exists about the role of this amino acid in counteracting stress in different species, in rapeseed, proline may be involved in the response and tolerance to salt stress [10].

The increase of proline levels in response to salt stress has been related to PA catabolism [54]; this can be due to the sharing of ornithine as common precursor. Polyamines have been reported to have regulatory functions in plant abiotic stress tolerance [4]. Changes in PA content and catabolism can occur in the interaction between plants and stressful environments. In plants, PUT is required for stress tolerance; SPD is essential for the maintenance of plant growth, while SPM has a crucial role in signal transduction [55]. Moreover, there is an increasing evidence that the stress-induced accumulation of PAs in several plant species can be related to the response to salinity [56]. However, the amount of PAs in plants can change depending on different factors, such as concentration of NaCl, time of exposure, or plant tolerance to salinity [56,57]. In SY Saveo, the amounts of PA changed within time, being higher at the end of exposure in the treated plants respect to the control, especially the level of SPD. In Dynastie, significant changes respect to the control were detected in SPD and SPM content. In rice, a higher level of cellular SPD was reported in salt-resistant cultivars, but not in salt sensitive ones [58]. Osmotic stress induced an increase in the level of PUT with a decrease in the level of SPM in *B. napus* plants [54]. Moreover, the increasing of spermidine + spermine/putrescine ratio has been reported in different plant species, such as spinach, lettuce, melon, pepper, broccoli, beetroot and tomato, and related to the enhancement of salinity tolerance [57].

5. Conclusions

Our study has shown that in *B. napus* a gradual exposure to salt can elicit several mechanisms of response to stress that preserve plant vitality and then maintain it, even though the genotype can be salt sensitive at the seed germination stage. Acclimation to saline conditions may represent a metabolic imprint in the struggle against soil salinization, where the adoption of different defense mechanisms can improve plant performance, leading to a more efficient stress response even in the more sensitive genotypes. Further exploitation of acclimation to salt conditions can open new alternative for both basic and applied research.

Supplementary Materials: The following are available online at http://www.mdpi.com/2073-4441/11/8/1667/s1, Figures S1 and S2: plants of Dynastie and SY Saveo after 60 days of treatments; Table S1: statistical analysis of plant height data, Table S2: statistical analysis of leaf area data.

Author Contributions: M.S. performed the experiments and data analyses; C.F. conceived and designed the experiments; F.P. performed photosynthesis experiments and analysed the data; C.C. and S.B. contributed to biochemical data analysis and interpretation; M.S., F.P., S.B., C.F. wrote the original draft of the paper; C.F. reviewed, edited and supervised the manuscript. All authors read and approved the final manuscript.

Funding: This research received no external funding.

Acknowledgments: This research did not receive any specific grant from funding agencies in the public, commercial, or not-for-profit sectors. We thank Montanari (CREA-CIN) for providing seeds and the personnels of Micropropagation Laboratories of the CREA-OFA (Centro di Ricerca per l'Olivicoltura, Frutticoltura, Agrumicoltura, Ciampino, Rome) for allowing the use of some equipment.

Conflicts of Interest: The authors declare no conflicts of interest.

Abbreviations

APX	ascorbate peroxidase
PA	polyamines
SOD	superoxide dismutase
ROS	reactive oxygen species
SPM	spermine
SPD	spermidine
PUT	putrescine

References

1. Mahajan, S.; Tuteja, N. Cold, salinity and drought stresses: An overview. *Arch. Biochem. Biophys.* **2005**, *444*, 139–158. [CrossRef]
2. Canfora, L.; Salvati, L.; Benedetti, A.; Dazzi, C.; Lo Papa, G. Saline soils in Italy: Distribution, ecological processes and socioeconomic issues. *Riv. Econ. Agrar.* **2017**, *1*, 63–77.

3. Shrivastava, P.; Kumar, R. Soil salinity: A serious environmental issue and plant growth promoting bacteria as one of the tools for its alleviation. *Saudi J. Biol. Sci.* **2015**, *22*, 123–131. [CrossRef]
4. Forni, C.; Duca, D.; Glick, B.R. Mechanisms of plant response to salt and drought stress and their alteration by rhizobacteria. *Plant Soil* **2017**, *410*, 335–356. [CrossRef]
5. Flexas, J.; Diaz-Espejo, A.; Galmés, J.; Kaldenhoff, R.; Medrano, H.; Ribas-Carbo, M. Rapid variations of mesophyll conductance in response to changes in CO_2 concentration around leaves. *Plant Cell Environ.* **2007**, *30*, 1284–1298. [CrossRef]
6. Chaves, M.M.; Flexas, J.; Pinheiro, C. Photosynthesis under drought and salt stress: Regulation mechanisms from whole plant to cell. *Ann. Bot.* **2009**, *103*, 551–560. [CrossRef]
7. Gill, S.S.; Tuteja, N. Reactive oxygen species and antioxidant machinery in abiotic stress tolerance in crop plants. *Plant Physiol. Biochem.* **2010**, *48*, 909–930. [CrossRef]
8. Bartels, D.; Sunkar, R. Drought and salt tolerance in plants. *Crit. Rev. Plant Sci.* **2005**, *24*, 23–28. [CrossRef]
9. Wanasundara, J.P.D. Proteins of Brassicaceae oilseeds and their potential as a plant protein source. *Crit. Rev. Food Sci. Nutr.* **2011**, *51*, 635–677. [CrossRef]
10. Shokri-Gharelo, R.; Noparvar, P.M. Molecular response of canola to salt stress: Insights on tolerance mechanisms. *Peer J.* **2018**, *6*, e4822. [CrossRef]
11. Ashraf, M.; Harris, P.J.C. Potential biochemical indicators of salinity tolerance in plants. *Plant Sci.* **2004**, *166*, 3–16. [CrossRef]
12. Hoagland, D.R.; Arnon, D.I. The water-culture method for growing plants without soil. *Calif. Agric. Exp. Stn. Circ.* **1950**, *347*, 1–32.
13. Zeng, F.; Shabala, L.; Zhou, M.; Zhang, G.; Shabala, S. Barley responses to combined waterlogging and salinity stress: Separating effects of oxygen deprivation and elemental toxicity. *Front. Plant Sci.* **2013**, *4*, 313. [CrossRef]
14. Blum, A.; Ebercon, A. Cell membrane stability as a measure of drought and heat tolerance in wheat. *Crop Sci.* **1981**, *21*, 43–47. [CrossRef]
15. Merchant, A.; Tausz, M.; Arndt, S.K.; Adams, M.A. Cyclitols and carbohydrates in leaves and roots of 13 Eucalyptus species suggest contrasting physiological responses to water deficit. *Plant Cell Environ.* **2006**, *29*, 2017–2029. [CrossRef]
16. Sairam, R.K.; Deshmukh, P.S.; Shukla, D.S. Tolerance of drought and temperature stress in relation to increased antioxidant enzyme activity in wheat. *J. Agron. Crop Sci.* **1997**, *178*, 171–178. [CrossRef]
17. Genty, B.; Briantais, J.M.; Baker, N.R. The relationship between the quantum yield of photosynthetic electron transport and quenching of chlorophyll fluorescence. *Biochim. Biophys. Acta* **1989**, *990*, 87–92. [CrossRef]
18. Bilger, W.; Björkman, O. Temperature dependence of violaxanthin de-epoxidation and non-photochemical fluorescence quenching in intact leaves of *Gossypium hirsutum* L. and *Malva parviflora* L. *Planta* **1991**, *184*, 226–234. [CrossRef]
19. Maxwell, K.; Johnson, G.N. Chlorophyll fluorescence-a practical guide. *J. Exp. Bot.* **2000**, *51*, 659–668. [CrossRef]
20. Cerovic, Z.G.; Masdoumier, G.; Ghozlen, N.B.; Latouche, G. A new optical leaf-clip meter for simultaneous non-destructive assessment of leaf chlorophyll and epidermal flavonoids. *Physiol. Plant* **2012**, *146*, 251–260. [CrossRef]
21. Orzali, L.; Forni, C.; Riccioni, L. Effect of chitosan seed treatment as elicitor of resistance to *Fusarium graminearum* in wheat. *Seed Sci. Technol.* **2014**, *42*, 132–149. [CrossRef]
22. Bradford, M.M. A rapid and sensitive method for the quantitation of microgram quantities of protein utilizing the principle of protein-dye binding. *Anal. Biochem.* **1976**, *72*, 248–254. [CrossRef]
23. Beauchamp, C.; Fridovich, I. Superoxide dismutase: Improved assays and an assay applicable to acrylamide gels. *Anal. Biochem.* **1971**, *44*, 276–287. [CrossRef]
24. Mittler, R.; Zilinskas, B.A. Detection of ascorbate peroxidase activity in Native Gels by inhibition of the ascorbate-dependent reduction of nitroblue tetrazolium. *Anal. Biochem.* **1993**, *212*, 540–546. [CrossRef]
25. Beninati, S.; Martinet, N.; Folk, J.E. High-performance liquid chromatographic method for the determination of ε-(γ-glutamyl) lysine and mono- and bis-γ-glutamyl derivatives of putrescine and spermidine. *J. Chromatogr. A* **1988**, *443*, 329–335. [CrossRef]
26. Carillo, P.; Gibon, Y. PROTOCOL: Extraction and Determination of Proline. PrometheusWiki 2011. Available online: http://prometheuswiki.publish.csiro.au/tiki (accessed on 10 January 2018).

27. Yamaguchi, T.; Blumwald, E. Developing salt-tolerant crop plants: Challenges and opportunities. *Trends Plant Sci.* **2005**, *10*, 615–620. [CrossRef]
28. Khan, M.H.; Panda, S.K. Alterations in root lipid peroxidation and antioxidative responses in two rice cultivars under NaCl-salinity stress. *Acta Physiol. Plant.* **2008**, *30*, 81–89. [CrossRef]
29. Munns, R.; Tester, M. Mechanisms of salinity tolerance. *Ann. Rev. Plant Biol.* **2008**, *59*, 651–681. [CrossRef]
30. Kapoor, K.; Srivastava, A. Assessment of salinity tolerance of *Vinga mungo* var. Pu-19 using *ex vitro* and in vitro methods. *Asian J. Biotechnol.* **2010**, *2*, 73–85.
31. Ashraf, M.Y.; Akhtar, K.; Sarwar, G.; Ashraf, M. Role of rooting system in salt tolerance potential of different guar accessions. *Agron. Sustain. Dev.* **2005**, *25*, 243–249. [CrossRef]
32. Khan, M.A.; Shirazi, M.U.; Khan, M.A.; Mujtaba, S.M.; Islam, E.; Mumtaz, S.; Shereen, A.; Ansari, R.U.; Ashraf, M.Y. Role of proline, K^+/Na^+ ratio and chlorophyll content in salt tolerance of wheat (*Triticum aestivum* L.). *Pak. J. Bot.* **2009**, *41*, 633–638.
33. Panda, D.; Dash, P.K.; Dhal, N.K.; Rout, N.C. Chlorophyll fluorescence parameters and chlorophyll content in mangrove species grown in different salinity. *Gen. Appl. Plant Physiol.* **2006**, *32*, 175–180.
34. Guidi, L.; Landi, M.; Penella, C.; Calatayud, A. Application of modulated chlorophyll fluorescence and modulated chlorophyll fluorescence imaging to study the environmental stresses effect. *Ann. Bot.* **2016**, *6*, 39–56.
35. Kalaji, H.M.; Bosa, K.; Kościelniak, J.; Żuk-Gołaszewska, K. Effects of salt stress on photosystem II efficiency and CO_2 assimilation of two Syrian barley landraces. *Environ. Exp. Bot.* **2011**, *73*, 64–72. [CrossRef]
36. Singh, D.P.; Sarkar, R.K. Distinction and characterisation of salinity tolerant and sensitive rice cultivars as probed by the chlorophyll fluorescence characteristics and growth parameters. *Funct. Plant Biol.* **2014**, *41*, 727–736. [CrossRef]
37. Baker, N.R.; Rosenqvist, E. Applications of chlorophyll fluorescence can improve crop production strategies: An examination of future possibilities. *J. Exp. Bot.* **2004**, *55*, 1607–1621. [CrossRef]
38. Bongi, G.; Loreto, F. Gas-exchange properties of salt-stressed olive (*Olea europea* L.) leaves. *Plant Physiol.* **1989**, *90*, 1408–1416. [CrossRef]
39. Pietrini, F.; Zacchini, M.; Iori, V.; Pietrosanti, L.; Bianconi, D.; Massacci, A. Screening of poplar clones for cadmium phytoremediation using photosynthesis, biomass and cadmium content analyses. *Int. J. Phytorem.* **2010**, *12*, 105–120. [CrossRef]
40. Mahlooji, M.; Sharifi, R.S.; Razmjoo, J.; Sabzalian, M.R.; Sedghi, M. Effect of salt stress on photosynthesis and physiological parameters of three contrasting barley genotypes. *Photosynthetica* **2018**, *56*, 549–556. [CrossRef]
41. Acosta-Motos, J.; Ortuño, M.; Bernal-Vicente, A.; Diaz-Vivancos, P.; Sanchez-Blanco, M.; Hernandez, J. Plant responses to salt stress: Adaptive mechanisms. *Agronomy* **2017**, *7*, 18. [CrossRef]
42. Schreiber, U.; Bilger, W.; Neubauer, C. Chlorophyll fluorescence as a nonintrusive indicator for rapid assessment of in vivo photosynthesis. In *Ecophysiology of Photosynthesis*; Schulze, E.D., Caldwell, M.M., Eds.; Springer: Berlin, Germany, 1994; pp. 49–70.
43. Shangguan, Z.; Shao, M.; Dyckmans, J. Effects of nitrogen nutrition and water deficit on net photosynthetic rate and chlorophyll fluorescence in winter wheat. *J. Plant Physiol.* **2000**, *156*, 46–51. [CrossRef]
44. Epron, D.; Dreyer, E.; Bréda, N. Photosynthesis of oak trees [*Quercus petraea* (Matt.) Liebl.] during drought under field conditions: Diurnal courses of net CO2 assimilation and photochemical efficiency of photosystem II. *Plant Cell Environ.* **1992**, *15*, 809–820. [CrossRef]
45. Stepien, P.; Johnson, G.N. Contrasting responses of photosynthesis to salt stress in the glycophyte *Arabidopsis* and the halophyte *Thellungiella*: Role of the plastid terminal oxidase as an alternative electron sink. *Plant Physiol.* **2009**, *149*, 1154–1165. [CrossRef]
46. Dai, Y.J.; Shen, Z.G.; Liu, Y.; Wang, L.L.; Hannaway, D.; Lu, H.F. Effects of shade treatments on the photosynthetic capacity, chlorophyll fluorescence, and chlorophyll content of *Tetrastigma hemsleyanum* Diels et Gilg. *Environ. Exp. Bot.* **2009**, *65*, 177–182. [CrossRef]
47. Jamil, M.; Lee, K.J.; Kim, J.M.; Kim, H.S.; Rha, E.S. Salinity reduced growth PS2 photochemistry and chlorophyll content in radish. *Sci. Agric.* **2007**, *64*, 111–118. [CrossRef]
48. Ayyub, C.; Rashid Shaheen, M.; Raza, S.; Sarwar Yaqoob, M.; Khan Qadri, R.; Azam, M.; Ghani, M.; Khan, I.; Akhtar, N. Evaluation of different radish (*Raphanus sativus*) genotypes under different saline regimes. *Am. J. Plant Sci.* **2016**, *7*, 894–898. [CrossRef]

49. Lin, Z.F.; Ehleringer, J.R. Effects of leaf age on photosynthesis and water use efficiency of papaya. *Photosynthetica* **1982**, *16*, 514–519.
50. Cartea, M.E.; Francisco, M.; Soengas, P.; Velasco, P. Phenolic compounds in *Brassica* vegetables. *Molecules* **2011**, *16*, 251–280. [CrossRef]
51. Miller, G.; Suzuki, N.; Ciftci-Yilmaz, S.; Mittler, R. Reactive oxygen species homeostasis and signalling during drought and salinity stresses. *Plant Cell Environ.* **2010**, *33*, 453–467. [CrossRef]
52. Sofo, A.; Scopa, A.; Nuzzaci, M.; Vitti, A. Ascorbate peroxidase and catalase activities and their genetic regulation in plants subjected to drought and salinity stresses. *Int. J. Mol. Sci.* **2015**, *16*, 13561–13578. [CrossRef]
53. Eyidogan, F.; Oz, M.T. Effect of salinity on antioxidant responses of chickpea seedlings. *Acta Physiol. Plant.* **2005**, *29*, 485–493. [CrossRef]
54. Aziz, A.; Martin-Tanguy, J.; Larher, F. Plasticity of polyamine metabolism associated with high osmotic stress in rape leaf discs and with ethylene treatment. *Plant Growth Regul.* **1997**, *21*, 153–163. [CrossRef]
55. Takahashi, T.; Kakehi, J.I. Polyamines: Ubiquitous polycations with unique roles in growth and stress responses. *Ann. Bot.* **2010**, *105*, 1–6. [CrossRef]
56. Janicka-Russak, M.; Kabała, K.; Młodzińska, E.; Kłobus, G. The role of polyamines in the regulation of the plasma membrane and the tonoplast proton pumps under salt stress. *J. Plant Physiol.* **2010**, *167*, 261–269. [CrossRef]
57. Zapata, P.J.; Serrano, M.; Pretel, M.T.; Amoros, A.; Botella, M.A. Polyamines and ethylene changes during germination of different plant species under salinity. *Plant Sci.* **2004**, *167*, 781–788. [CrossRef]
58. Quinet, M.; Ndayiragije, A.; Lefèvre, I.; Lambillotte, B.; Dupont-Gillain, C.C.; Lutts, S. Putrescine differently influences the effect of salt stress on polyamine metabolism and ethylene synthesis in rice cultivars differing in salt resistance. *J. Exp. Bot.* **2010**, *61*, 2719–2733. [CrossRef]

© 2019 by the authors. Licensee MDPI, Basel, Switzerland. This article is an open access article distributed under the terms and conditions of the Creative Commons Attribution (CC BY) license (http://creativecommons.org/licenses/by/4.0/).

Article

Application of *Arthrospira (Spirulina) platensis* against Chemical Pollution of Water

Inga Tabagari [1], Maritsa Kurashvili [1], Tamar Varazi [1], George Adamia [1], George Gigolashvili [1], Marina Pruidze [1], Liana Chokheli [1], Gia Khatisashvili [1] and Peter von Fragstein und Niemsdorff [2,*]

1. Durmishidze Institute of Biochemistry and Biotechnology of Agricultural University of Georgia, Davit 240 David Agmashenebeli Alley, Tbilisi 0131, Georgia
2. Department of Organic Vegetable Production (before Retirement), Faculty of Organic Agricultural Sciences, University of Kassel, Steinstraße 19, 37213 Witzenhausen, Germany
* Correspondence: pvf@uni-kassel.de; Tel.: +49-5542-911-882

Received: 17 July 2019; Accepted: 20 August 2019; Published: 23 August 2019

Abstract: The basis of phytoremediation technology for cleaning chemically polluted water was developed in the framework of the presented work. This technology is based on the ability of blue-green alga *Arthrospira platensis* to eliminate different environmental toxicants from water. This technological approach was conducted for the following pollutants: 1,1,1-trichloro-2,2-bis(4-chlorophenyl)ethane (DDT), 2,4,6-trinitrotoluene (TNT), and cesium ions. The effectiveness of the technology was tested in model experiments, which were carried out in glass containers (volume 40 L). In particular, the different concentrations of alga biomass with the aforementioned pollutants were incubated with permanent illumination conditions and air barbotage, at a temperature of 25 °C. The results of the model experiments showed that after two weeks from the start of remediation *Arthrospira* effectively cleaned artificially polluted waters. Particularly in the case of TNT 56 mg/L concentration, the effect of water remediation was 97%. In the case of DDT 10 mg/L concentration, the degree of cleaning was 90%. Similar results were obtained in the case of 100 mg/L concentration of cesium ions. Thus, the model experiments confirmed that the alga *Arthrospira* effectively removed tested pollutants from water. That is the basis of phytoremediation technology.

Keywords: phytoremediation; water pollution; DDT; TNT; heavy metals; cesium ions

1. Introduction

For the last few decades, distribution of chemical pollutants from urban, agricultural, and industrial sources caused the accumulation of toxic substances potentially harmful for plants and animals, becoming a serious concern for the ecosystem survival. In this context, the development of soil and water remediation methods for the removal of chemical contaminants is a widely recognized, challenging problem. Nowadays, the technological implementation of methodologies devoted to the removal of pharmaceutical and waste residues, pesticides, and heavy metals from natural and agricultural ecosystems is urgent for all countries.

Currently, urban environmental pollution comprises hundreds of substances, including aliphatic, aromatic and polycyclic hydrocarbons, phenols, pesticides, organochlorine compounds, explosives based on nitro compounds, heavy metals, radionuclides, etc. [1]. Despite the major efforts that have been made over recent years to clean up the environment, pollution remains a major problem and poses continuous risks to health.

At present, the freshwaters from rivers and lakes of industrialized countries are mainly used to produce tap water for large cities. Proceeding from the information mentioned above, the creation of quick-response, strategic approaches against chemical pollution of numerous precipitates of sewage sludge will resolve the important environmental problem.

The given work is focused on the setup of the technological approach on the bioaccumulation and removal of pollutants from water, based on the integrated assimilation of target toxicants by using modern phytoremediation technology [1–3].

Modern phytoremediation technologies may include different methodological approaches, depending on their purposes. One of the methods of cleaning the chemically polluted waters is phycoremediation, based on the application of algae [4]. There are some examples of using algae for cleaning water polluted with heavy metals, petroleum hydrocarbons, pesticides, etc. [4–11].

The main idea of the presented work is to create the basis for an effective cleaning technology of water that is artificially polluted with cesium ions, 1,1,1-trichloro-2,2-bis(4-chlorophenyl)ethane (or 1,1'-(2,2,2-trichloroethane-1,1-diyl)bis(4-chlorobenzene)—DDT) and 2,4,6-trinitrotoluene (TNT) by using blue-green alga *Arthrospira platensis* (formerly *Spirulina platensis*) as a phytoremediator.

Arthrospira, which is currently widely used in the food industry and in pharmacology [9,10], can be applied also as a phytoremediation agent for cleaning the chemically polluted waters. The cells of *Arthrospira* contain large amounts of compounds (enzymes, peptides, amino acids, etc.) capable of binding as ions of heavy metals to organic toxic compounds. The structures inside the cells of *Arthrospira* swell with air bubbles, which promotes their emerging on the water surface. Thus, the algae biomass is easily removed from the testing reservoirs.

It is known that *Arthrospira* is applied for remediation of water that is contaminated mainly with heavy metals [5,11–17]. Additionally, it is known for its use in cases of water polluted with petroleum hydrocarbons [12], pesticides [13], radioactive elements [18], fluoride ions [19], some estrogens [20], etc.

Water pollution caused by hazardous heavy metals and radionuclides is one of the most important ecological problems. Among them, ^{137}Cs is very dangerous. Nonradioactive cesium ions that occur in nature and are released into the environment through mining and milling of ores are only mildly toxic [21]. Both radioactive and stable cesium, when they occur in a human or animal body, act the same way [22]. There are only limited data concerning the use of *Arthrospira* as a phytoremediation agent in cases of water polluted with Cs^+ ions.

TNT that is used as a military high explosive is one of the most toxic compounds. It can accumulate in human organs, stimulating some chronic diseases, and it is considered a hazardous contaminant of the environment [23].

DDT was a widely used insecticide in the 20th century because it was applied to fight malaria-carrying mosquitoes [24]. Despite banning DDT application, this pesticide, as other organochlorine pollutants, stays in the environment in undestroyed form for a long time due to its chemical stability. Its high hydrophobicity allows the insertion of DDT into the food chain. DDT easily penetrates human organisms through the digestive tract or the skin and damages nerve tissue [1].

In our previous works it we showed *Arthrospira* has a high ability to eliminate Cs^+ ions, TNT, and DDT from artificially polluted water via adsorption on the surface of cellular lipopolysaccharides and then by their moving into cells [25–27]. The presented article gives the results of work that was performed to test the phytoremediation potential of *Arthrospira* in the model experiments.

Thus, the goal of the presented work is to investigate opportunities to use *Arthrospira platensis* as a tool for removing pollutants with different chemical natures from water. For this aim, model experiments were conducted to establish the optimal concentration of biomass of *Arthrospira* that allows elimination of Cs^+ ions, TNT, and DDT from artificially contaminated water with maximal effectiveness.

2. Materials and Methods

In the experiments we used the biomass *Arthrospira platensis* obtained via cultivation in standard Zarrouk's medium (pH—8.7; content in g/L: $NaHCO_3$—16.8, K_2HPO_4—0.5, $NaNO_3$—2.5, K_2SO_4—1.0, NaCl—1.0, $MgSO_4 \cdot 7H_2O$—0.2, $CaCl_2 \cdot 2H_2O$—0.04, $FeSO_4 \cdot 7H_2O$—0.01, EDTA—0.08; and microelements kit A5—1 mL). The cultivation was conducted in the following conditions: Permanent air barbotage (rate of air flow 2 L/min); temperature of 25 °C; a photoperiod of lighting of 16 L/8 D (16 h of light:8 h of dark); a total photosynthetic photon flux density (PPFD) of ≈15 $\mu mol \cdot m^{-2} \cdot s^{-1}$.

The biomass of *Arthrospira* in the incubation medium was measured spectrophotometrically at 750 nm [28].

For determination of DDT, the incubation medium was centrifuged at 1000 g for 20 min and DDT was extracted from the obtained supernatant by hexane. The extract was concentrated by evaporation, and DDT content was determined by gas chromatographic analysis (Instrument—Agilent 7890A (Beijing, China)) [29]. Limit of detection—0.01 mg/L. Detailed conditions for analysis are described in our previous work [26].

Concentration of residual TNT in the incubation area was measured spectrophotometrically at 447 nm, in a highly alkaline (pH > 12.2) area (spectrophotometer—Shimadzu UV1900 (Kyoto, Japan)) [30]. Limit of detection—0.2 mg/L.

The content of Cs^+ ions in the samples was determined by the method of atomic absorption (flame emission) analysis [31]. Conditions for analysis are the following: Instrument—PerkinElmer HGA900 Graphite Furnace (Waltham, MA, USA); wavelength—852.1 nm; slit—0.2/0.4 nm; flame—air-acetylene; stock standard solution—CESIUM 1000 mg/L; light source—EDL; interface—ionization controlled by addition of 0.1% KCl. Limit of detection—0.5 mg/L.

The model experiments for cleaning water polluted by tested toxicants were carried out in the following conditions: (1) *Arthrospira* was cultivated in Zarrouk's medium (volume 20 L) for 7 days; (2) thereafter, the solution of DDT, TNT, or Cs^+ was added (volume 20 L); (3) the incubation period was finished after 15 days. According to our previous experiments [21–23], the concentrations of pollutants were as follows: 10 mg/L in case of DDT, 22.5 or 56 mg/L in case of TNT, and 100 mg/L in case of Cs^+ ions. The biomass of *Arthrospira* varied from 3.0 to 4.5 g/L (see tables in Section 3). The incubation was carried out in a glass container with sizes 60 × 21 × 40 (in cm, length × width × height), with permanent air barbotage (rate of air flow 2 L/min), at a temperature of 25 °C, under the following illumination conditions: 24 L/0 D, PPFD ≈ 15 $\mu mol \cdot m^{-2} \cdot s^{-1}$. In the control variant, instead of the tested contaminant solution, we added 20 L of water. Model experiments were carried out in periods: April–June, 2017; September–October, 2017; April–June, 2018; September–October, 2018.

Presented data are the mean of three replicates ± standard deviation (SD). The replicates represent an assay of a sample from the same source multiple times. The statistical analysis of obtained data was performed using the method of Descriptive Statistics of Excel.

3. Results and Discussion

The model experiments for development of the phytoremediation technology based on the application of *Arthrospira* for cleaning chemically polluted waters were carried out. In the model experiments the initial biomass of *Arthrospira* was changed until we got maximal absorption of the toxicant.

In the case of DDT, *Arthrospira* was cultivated in artificially polluted waters containing different concentrations of the pollutant and *Arthrospira* biomass. The initial biomass of *Arthrospira* was changed in the model experiments to reach maximal elimination of the toxicant from the polluted water. The obtained results are given in Table 1.

The obtained data show that the remediation process was optimal when the biomass content of *Arthrospira* in the incubation medium was 4 g/L (Table 1, Experiment No. DDT-3). In this case the mass ratio of the DDT and the raw biomass of the alga was approximately 1:400. In this experiment alga removes about 90% of the tested pollutant, indicating the effectiveness of the presented method. At an increase of the alga biomass up to 4.5 g/L, a significant decrease of the DDT amount in the polluted water was not observed. On higher concentrations of DDT (15 mg/L), the effectiveness of elimination out of the polluted water decreased up to approximately 77% in spite of an increase of the initial content of the *Arthrospira* biomass (Table 1, Experiment No. DDT-5).

Table 1. Results of model experiments for cleaning water polluted with DDT by using *Arthrospira*. Mean values (n = 3) and ±SD are given.

Number of Model Experiment	Initial Content of *Arthrospira* in Polluted Water, g/L	Content of DDT in Polluted Water, mg/L	
		Initial (before Incubation)	Final (after Incubation)
DDT-1	3.0	10	2.56 ± 0.15
DDT-2	3.5	10	1.42 ± 0.08
DDT-3	4.0	10	0.98 ± 0.06
DDT-4	4.5	10	0.97 ± 0.06
DDT-5	4.5	15	3.49 ± 0.21

The results of the model experiments for the development of phytoremediation technology for cleaning waters polluted by TNT are given in Table 2. As seen from the obtained results, the most efficient cleaning rate could be achieved by a ratio of 1:60 from the contaminant and algal concentration (Table 2, Experiment No. TNT-3). In this case, alga removes about 98% of TNT from the polluted water. These data indicate the high effectiveness of the presented method for cleaning water polluted by TNT.

Table 2. Results of model experiments for cleaning water polluted with TNT by using *Arthrospira*. Mean values (n = 3) and ±SD are given.

Number of Model Experiment	Initial Content of *Arthrospira* in Polluted Water, g/L	Content of TNT in Polluted Water, mg/L	
		Initial (before Incubation)	Final (after Incubation)
TNT-1	3.0	22.5	2.61 ± 0.13
TNT-2	3.3	56.0	4.50 ± 0.25
TNT-3	3.5	56.0	1.80 ± 0.11

The model experiments for testing of *Arthrospira* as a remediator for cleaning water polluted with 100 mg/L of Cs^+ ions show that the best results were obtained in cases of using 3.5 and 4.0 g/L of the *Arthrospira* biomass (Table 3, Experiments Nos. CS-2 and CS-3). As can be seen from the obtained data, in these cases the efficiency of remediation was achieved up to 90%. When both the Cs^+ ions concentration and the *Arthrospira* initial biomass were increased, the effectiveness of uptake was only about 83%. These results indicate that the most effective ratio of heavy metal and raw biomass of alga was approximately 1:35.

Table 3. Results of model experiments for cleaning water polluted with Cs^+ ions by using *Arthrospira*. Mean values (n = 3) and ±SD are given.

Number of Model Experiment	Initial Content of *Arthrospira* in Polluted Water, g/L	Content of Cs^+ Ions in Polluted Water, mg/L	
		Initial (before Incubation)	Final (after Incubation)
CS-1	3.0	100	18.3 ± 0.91
CS-2	3.5	100	10.3 ± 0.06
CS-3	4.0	100	9.90 ± 0.05
CS-4	4.5	150	25.1 ± 2.26

Comparison of the obtained results with the results of the previous publications will corroborate that the cleaning rate was improved by choosing the proper correlation of the biomass of *Arthrospira* and the organic pollutants. In particular, in the case of DDT, cleaning efficiency was improved from 70% [26] to 90%, and in the case of TNT, from 87% [25] to 97%. As for cesium ions, nothing different from the previous experiments [27] was received.

The ability of *Arthrospira* to remove Cs^+ ions from polluted water supposedly might be related to the presence of heavy metal chelating compounds in this alga. This mechanism is still unknown and should be clarified in further studies. As for the organic pollutants, supposedly their uptake should occur via metabolizing by alga, which implies the primary transformations of compounds penetrated in the cells and then the conjugation of formed intermediates with intracellular compounds. In the case of

TNT, the primary transformations should be performed via reduction of nitro groups to amine groups. Such transformations must be catalyzed by enzymes similar to nitroreductases, which participate in TNT transformation in plants and microorganisms [1]. DDT may be undergoing dehalogenation and/or hydroxylation analogous to plants, by the enzymes such as monooxygenases, peroxidases, phenoloxidases, and dehalogenases [1,32]. Intermediates of transformation of DDT and TNT probably conjugate with the following intracellular compounds: Saccharides, amino acids, peptides, proteins, etc., which are in *Arthrospira* in large amounts. Therefore, to determine the mechanism of removing organic pollutants by *Arthrospira*, it is necessary to study the above-mentioned enzymes in this alga.

The above results show that the application of *Arthrospira* is indicated for cleaning chemically polluted waters. Until now, *Arthrospira* was considered effective mainly in cases of pollution with heavy metals [5,7,14–18]; but the presented results corroborate universal phytoremediation features and the high cleaning potential of *Arthrospira*. Further, it is desirable to continue investigations in the following directions:

- Optimization of incubation conditions of *Arthrospira*;
- Investigation of the processes of organic toxicant metabolism and binding heavy metals in cells of *Arthrospira*;
- Revelation of enzymes and enzymatic systems of *Arthrospira*, participating in the transformation of organic toxicants and chelating of heavy metals.

Moreover, it will be possible to refine and improve *Arthrospira*-based phytoremediation technology, due to the development of proper equipment providing, in automatic mode, cultivation of *Arthrospira*, a polluted water supply, and purified water removal.

Although the information and works are often found on remediation of sewage from heavy metals and microorganisms where *Arthrospira* is the key bio-tool, the presented work contains specific novelty. Removal from water and biodegradation of such persistent and toxic organic pollutants as TNT and DDT via the application of ecological potential *Arthrospira* is the first attempt and has no direct analogy.

4. Conclusions

The overall results of completed investigations can become a basis for the development of phytoremediation technology based on the application of *Arthrospira* for cleaning waters polluted with research toxicants.

The main idea of these technologies is to add the effective amount of *Arthrospira* biomass to polluted water and remove it from the artificially contaminated areas consecutive times.

The effectiveness of the technological approach tested in the model experiments was carried out in glass containers (volume 40 L). The results of the model experiments showed that after 15 days from the start of the experiments using *Arthrospira* the following results are obtained:

- Water remediation was occurring by 97%, in the case of 56 mg/L concentration of TNT and 3.5 g/L initial content of *Arthrospira*;
- The degree of cleaning was 90%, in the case of 10 mg/L concentration of DDT and 4.0 g/L initial content of *Arthrospira*;
- The degree of cleaning was 90%, in the case of 100 mg/L concentration of cesium ions and 3.5 g/L initial content of *Arthrospira*.

From the above-mentioned data, it can be supposed that *Arthrospira platensis* has high phytoremediation potential to pollutants with different chemical natures.

Author Contributions: Conceptualization: G.K., G.G. and G.A.; investigation: I.T., L.C., T.V. and G.A.; methodology: G.K., G.G., M.K., M.P. and G.A.; formal analysis: M.K., G.K., T.V. and M.P.; writing: M.K., G.K., P.v.F.u.N. and M.P. All the authors have approved of the submission of this manuscript.

Funding: The research study was funded by Shota Rustaveli National Science Foundation (SRNSF) and the National Research Council of Italy (CNR) grant #04/51 and the Volkswagenstiftung under the collaborative project,

Sustainable Agricultural and Food Systems (SAFS), between University of Kassel, Germany, and the Agricultural University of Georgia. This research paper is published as an open access journal article with the financial support from the Open-Access-Publikationsfonds from the University of Kassel, financed by the German Research Foundation (DFG) and the Library of University of Kassel, Germany.

Conflicts of Interest: The authors declare no conflict of interest.

References

1. Kvesitadze, G.; Khatisashvili, G.; Sadunishvili, T.; Ramsden, J.J. *Biochemical Mechanisms of Detoxification in Higher Plants. Basis of Phytoremediation*; Springer: Berlin/Heidelberg, Germany; New York, NY, USA, 2006.
2. Salt, D.E.; Smith, R.D.; Raskin, I. Phytoremediation. *Annu. Rev. Plants Physiol. Mol. Biol.* **1998**, *49*, 643–668. [CrossRef] [PubMed]
3. Tsao, D.T. *Phytoremediation*; Advances in Biochemical Engineering and Biotechnology; Springer: Berlin/Heidelberg, Germany; New York, NY, USA, 2003.
4. Phang, S.M.; Chu, W.L.; Rabiei, R. Phycoremediation. In *The Algae World*; Sahoo, D., Seckbach, J., Eds.; Springer: Dordrecht, The Netherlands, 2015; pp. 357–389.
5. Rangsayatorn, N.; Upatham, E.S.; Kruatrachue, M.; Pokethitiyook, P.; Lanza, G.R. Phytoremediation potential of *Spirulina Arthrospira platensis*: Biosorption and toxicity studies of cadmium. *Environ. Pollut.* **2002**, *119*, 45–53. [CrossRef]
6. Jacques, N.R.; McMartin, D.W. Evaluation of algal phytoremediation of light extractable petroleum hydrocarbons in subarctic climates. *Remediat. J.* **2009**, *20*, 119–132. [CrossRef]
7. Ahmad, A.; Ghufran, R.; Wahid, Z.A. Cd, As, Cu and Zn transfer through dry to rehydrated biomass of Spirulina Platensis from wastewater. *Polish J. Environ. Stud.* **2010**, *19*, 887–893.
8. Arif, I.A.; Bakir, M.A.; Khan, H.A. Microbial Remediation of Pesticides. In *Pesticides: Evaluation of Environmental Pollution*; Rathore, H.S., Nollet, L.M.L., Eds.; CRC Press: Boca Raton, FL, USA, 2012; pp. 131–144.
9. Matamoros, V.; Gutiérrez, R.; Ferrer, I.; García, J.; Bayona, J.M. Capability of microalgae-based wastewater treatment systems to remove emerging organic contaminants: A pilot-scale study. *J. Hazard. Mater.* **2015**, *288*, 34–42. [CrossRef]
10. Iasimone, F.; Panico, A.; De Felice, V.; Fantasma, F.; Iorizzi, M.; Pirozzi, F. Effect of light intensity and nutrients supply on microalgae cultivated in urban wastewater: Biomass production, lipids accumulation and settleability characteristics. *J. Environ. Manag.* **2018**, *223*, 1078–1085. [CrossRef]
11. Markou, G.; Depraetere, O.; Vandamme, D.; Muylaert, K. Cultivation of *Chlorella vulgaris* and *Arthrospira platensis* with Recovered Phosphorus from Wastewater by Means of Zeolite Sorption. *J. Mol. Sci.* **2015**, *16*, 4250–4264. [CrossRef]
12. Ciferri, O. Spirulina, the edible microorganism. *Microbiol. Rev.* **1983**, *47*, 551–578.
13. Khan, Z.; Bhadouria, P.; Bisen, P.S. Nutritional and therapeutic potential of Spirulina. *Curr. Pharm. Biotechnol.* **2005**, *6*, 373–379. [CrossRef]
14. Chojnacka, K.; Chojnacki, A.; Górecka, H. Biosorption of Cr^{3+}, Cd^{2+} and Cu^{2+} ions by blue–green *algae Spirulina sp.*: Kinetics, equilibrium and the mechanism of the process. *Chemosphere* **2005**, *59*, 75–84. [CrossRef]
15. Zinicovscaia, I.; Cepoi, L.; Chiriac, T.; Mitina, T.; Grozdov, D.; Yushin, N.; Culicov, O. Application of Arthrospira (Spirulina) platensis biomass for silver removal from aqueous solutions. *Int. J. Phytoremediat.* **2017**, *19*, 1053–1058. [CrossRef] [PubMed]
16. Chen, H.; Pan, S. Bioremediation potential of Spirulina: Toxicity and biosorption studies of lead. *J. Zhejiang Univ. Sci. B* **2005**, *6*, 171–174. [CrossRef] [PubMed]
17. Murali, O.; Mehar, S. Bioremediation of heavy metals using Spirulina. *Int. J. Geol. Earth Environ. Sci.* **2014**, *4*, 244–249.
18. Fukuda, S.; Lwamoto, K.; Asumi, M.; Yokoyama, A.; Nakayama, T.; Ishida, K.; Inouye, I.; Shraiwa, Y. Global searches for microalgae and aquatic plants that can eliminate radioactive cesium, iodine and strontium from the radio-polluted aquatic environment. *J. Plant Res.* **2014**, *127*, 79–89. [CrossRef] [PubMed]
19. Zazouli, M.A.; Mahvi, A.H.; Dobaradaran, S.; Barafrashtehpour, M.; Mahdavi, Y.; Balarak, D. Adsorption of fluoride from aqueous solution by modified *Azolla filiculoides*. *Fluoride* **2014**, *47*, 349–358.

20. Shi, W.; Wang, L.; Rousseau, D.P.L.; Lens, P.N.L. Removal of estrone, 17α-ethinylestradiol, and 17β-estradiol in algae and duckweed-based wastewater treatment systems. *Environ. Sci. Pollut. Res.* **2010**, *17*, 824–833. [CrossRef] [PubMed]
21. Pinsky, C.; Bose, R.; Taylor, J.R.; McKee, J.; Lapointe, C.; Birchall, J. Cesium in mammals: Acute toxicity, organ changes and tissue accumulation. *J. Environ. Sci. Health Part A* **1981**, *16*, 549–567. [CrossRef]
22. Rundo, J. A Survey of the Metabolism of Caesium in Man. *Br. J. Radiol.* **1964**, *37*, 108–114. [CrossRef] [PubMed]
23. Juhasz, A.L.; Naidu, R. Explosives: Fate, dynamics, and ecological impact in terrestrial and marine environments. *Rev. Environ. Contam. Toxicol.* **2007**, *191*, 163–215.
24. Konradsen, F.; Van der Hoek, W.; Amerasinghe, F.P.; Mutero, C.; Boelee, E. Engineering and malaria control: Learning from the past 100 years. *Acta Trop.* **2004**, *89*, 99–108. [CrossRef]
25. Adamia, G.; Chogovadze, M.; Chokheli, L.; Gigolashvili, G.; Gordeziani, M.; Khatisashvili, G.; Kurashvili, M.; Pruidze, M.; Varazi, T. About possibility of alga Spirulina application for phytoremediation of water polluted with 2,4,6-trinitrotoluene. *Ann. Agrar. Sci.* **2018**, *16*, 348–351. [CrossRef]
26. Kurashvili, M.; Varazi, T.; Khatisashvili, G.; Gigolashvili, G.; Adamia, G.; Pruidze, M.; Gordeziani, M.; Chokheli, L.; Japharashvili, S.; Khuskivadze, N. Blue-green Alga Spirulina as a Tool Against Water pollution by 1,1′-(2,2,2-Trichloroethane-1,1-diyl)bis(4-chlorobenzene) (DDT). *Ann. Agrar. Sci.* **2018**, *16*, 405–409. [CrossRef]
27. Kurashvili, M.; Adamia, G.; Varazi, T.; Khatisashvili, G.; Gigolashvili, G.; Pruidze, M.; Chokheli, L.; Japharashvili, S. Application of Blue-green Alga Spirulina for removing Caesium ions from polluted water. *Ann. Agrar. Sci.* **2019**, in press.
28. Butterwick, C.; Heaney, S.I.; Talling, J.F. A comparison of eight methods for estimating the biomass and growth of planktonic algae. *Br. Phycol. J.* **1982**, *17*, 69–79. [CrossRef]
29. Garrido Frenich, A.; Martínez Vidal, J.L.; Moreno Frías, M.; Olea-Serrano, F.; Olea, N.; Cuadros Rodriguez, L. Determination of organochlorine pesticides by GC-ECD and GC-MS-MS techniques including an evaluation of the uncertainty associated with the results. *Chromatographia* **2003**, *57*, 213–220. [CrossRef]
30. Oh, B.; Sarath, G.; Drijber, R.A.; Comfort, S.D. Rapid spectrophotometric determination of 2,4,6-trinitrotoluene in a Pseudomonas enzyme assay. *Microbiol. Methods* **2000**, *42*, 149–158. [CrossRef]
31. Guba, L.V.; Dovgyy, I.I.; Rizhkova, M.A. Method of measuring of cesium by flame emission photometry method. *Sci. Notes Taurida V. Vernadsky Natl. Univ. Ser. Biol. Chem.* **2012**, *25*, 284–288.
32. Mondal, M.; Halder, G.; Oinam, G.; Indrama, T.; Tiwari, O.N. Bioremediation of Organic and Inorganic Pollutants Using Microalgae. In *New and Future Developments in Microbial Biotechnology and Bioengineering*; Gupta, V.K., Pandey, A., Eds.; Elsevier: Amsterdam, The Netherlands, 2019; pp. 223–235.

© 2019 by the authors. Licensee MDPI, Basel, Switzerland. This article is an open access article distributed under the terms and conditions of the Creative Commons Attribution (CC BY) license (http://creativecommons.org/licenses/by/4.0/).

Article

The Influence of Bottom Sediments and Inoculation with Rhizobacterial Inoculants on the Physiological State of Plants Used in Urban Plantings

Anna Wyrwicka [1,*], Magdalena Urbaniak [2], Grzegorz Siebielec [3], Sylwia Siebielec [4], Joanna Chojak-Koźniewska [5], Mirosław Przybylski [6], Aleksandra Witusińska [1] and Petra Susan Kidd [7]

1. Department of Plant Physiology and Biochemistry, Faculty of Biology and Environmental Protection, University of Lodz, 90-237 Lodz, Poland
2. European Regional Centre for Ecohydrology of the Polish Academy of Sciences, 90-364 Lodz, Poland
3. Department of Soil Science Erosion and Land Protection, Institute of Soil Science and Plant Cultivation, State Research Institute, 24-100 Pulawy, Poland
4. Department of Agricultural Microbiology, Institute of Soil Science and Plant Cultivation, State Research Institute, 24-100 Pulawy, Poland
5. Genetically Modified Organisms Controlling Laboratory, Plant Breeding and Acclimatization Institute—National Research Institute, Radzikow, 05-870 Blonie, Poland
6. Department of Ecology and Vertebrate Zoology, Faculty of Biology and Environmental Protection; University of Lodz, 90-237 Lodz, Poland
7. Instituto de Investigaciones Agrobiológicas de Galicia (IIAG), Consejo Superior de Investigaciones Científicas (CSIC), 15780 Santiago de Compostela, Spain
* Correspondence: anna.wyrwicka@biol.uni.lodz.pl; Tel.: +48-42-635-44-16

Received: 15 July 2019; Accepted: 22 August 2019; Published: 28 August 2019

Abstract: Bottom sediments accumulate rapidly in urban reservoirs and should be periodically removed. Their high organic matter content makes them valuable fertilizers, but they often contain toxic substances. The present study compares the responses of the dicotyledonous *Tagetes patula* and monocotyledon *Festuca arundinacea* to the presence of such sediments in soil and to soil inoculation with two rhizobacterial strains (*Massilia niastensis* p87 and *Streptomyces costaricanus* RP92) isolated from contaminated soil. Total soluble protein, total chlorophyll content, as well as chlorophyll a/b ratio, degree of lipid peroxidation (TBARS), α-tocopherol content, total phenolic compounds (TPC) content and anthocyanins content were examined in the leaves of investigated plants. *T. patula* was more sensitive to the toxic substances in the sediments than *F. arundinacea*. Rhizobacterial inoculation reduced the toxic effect of the sediment. RP92 has a more favorable effect on the condition of *T. patula* than p87. *F. arundinacea* was not adversely affected by the addition of sediments or inoculation with the p87 or RP92 strains. Both tested plant species are suitable for planting on soils enriched with urban sediments, and the addition of bacterial inoculums promote plant growth and reduce the damage caused by the xenobiotics contained in the sediments.

Keywords: *Tagetes patula*; *Festuca arundinacea*; bottom urban sediment; phytoremediation; plant growth promoting bacteria; oxidative stress; plant stress reactions

1. Introduction

The reasonable use of natural resources requires a balance to be maintained between the improvement of the quality of life and fast economic growth on the one hand, and the improvement of the natural environment and the protection of its resources on the other. Overuse of environmental resources through intensive agriculture, industry, spatial development of cities and villages, architecture, transport, infrastructure development and inappropriate land management can contribute greatly towards the destabilization of the environment, especially in urbanized ecosystems [1].

Urban bottom sediments are naturally formed in water reservoirs during the sedimentation of mineral and organic suspensions, as well as components precipitated from water. Due to the rapid rate of accumulation of bottom sediments, they should be periodically removed in order to enhance reservoir capacity and to improve its functional value [2,3]. One example of a reservoir that is also an urban area sedimentation pond is the Sokołówka sequential biofiltration system (SSBS). The SSBS was constructed in the upper section of the Sokołówka River with the aim of removing various pollutants, such as sediments, suspended solids, particulate pollutants, petroleum hydrocarbons, heavy metals and nutrients from stormwater runoff. To do so, it employs a system of sedimentation and filtration mechanisms. The system comprises of three different zones to improve efficiency: A hydrodynamically intensified sedimentation zone, an intensive biogeochemical zone and an intensive biofiltration zone [4].

The composition of the bottom sediments depends on many factors, such as the way the catchment is used, its natural conditions and on the type and amount of contaminants reaching the surface waters. In many locations, bottom sediments include the material brought along with industrial and municipal sewage, as well as surface runoff from urbanized, industrial and agricultural areas [5]. The most common contaminants in bottom sediments include heavy metals and inorganic compounds such as silicates and aluminates, as well as pathogenic and other microbial pollutants; these are accompanied by a number of toxic organic pollutants, such as persistent organic pollutants (POPs), which are characterized by high persistence and great potential for bioaccumulation, biomagnification and toxicity; e.g., polychlorinated biphenyls (PCBs), polychlorinated dibenzo-*p*-dioxin (PCDD), polychlorinated dibenzofurans (PCDFs) and polycyclic aromatic hydrocarbons (PAH) [6–10].

On the other hand, bottom sediments may also act as fertilizers: Following decomposition, they serve as a rich source of essential organic and mineral matter for plants, including nitrogen and phosphorus compounds, that have important agricultural applications [9,11,12]. In addition, organic fertilization supplements the organic carbon and nitrogen stocks in the soil to a greater degree than mineral fertilization [13,14]. Such supplementation improves soil carbon sequestration and is particularly important in the face of climate change. Soils with higher levels of organic matter also demonstrate better structure, greater water holding capacity and increased soil nutrient availability; they also have greater microbial biomass, as well as a more diverse microbial community structure and greater biodiversity in general [15,16]. Organic fertilization can be particularly useful in cities, with urban bottom sediments offering the potential for the reclamation of degraded land, communication areas and recreational areas, such as flower beds and greenery, where ornamental plants and lawns are mainly grown.

It is possible to enhance the fertilizer properties of urban sediments while also negating their toxicity through the use of remediation techniques. One such technique is phytoremediation. It uses the ability and metabolic activity of selected plant species and their symbiotic microorganisms to reclaim a contaminated environment [17–21]. The very presence of a root system allows the immobilization of the soil and pollutants in a contaminated substrate, inhibiting erosion by air and water while also preventing the movement of contaminants to deeper layers of the soil profile. At this stage, such immobilization is particularly important in the presence of contaminants characterized by poor water solubility and low biodegradability, including POPs such as PCBs, PCDD, PCDFs or PAH. Although such remediation typically takes place via uptake by plant roots, it also occurs outside the roots through the participation of enzymes secreted by the plants into the rhizosphere, which stimulate the development of microorganisms responsible for detoxification. This use of plant roots to maintain a favorable growth environment for the contaminant-degrading rhizosphere microorganisms in soil is named rhizodegradation [22,23]. The rhizosphere, the area comprising the soil surrounding the root and the root surface itself, is characterized by intense microbiological activity. The biological activity taking place in this zone is one of the factors conditioning the growth of plants and their resistance to pollution [24]. However, the development of the root zone not only increases the total number and diversity of microorganisms, the plants also change the physicochemical properties of the local soil, improve the water and air ratios and maintain the correct, crumbly soil structure; they also modulate the pH, organic carbon content and mineral content of the soil. These factors contribute to the creation of favorable conditions for the course of biological processes, thereby stimulating bioremediation.

Many examples of relationships between plants and microorganisms that show potential applications in phytoremediation techniques have been observed, some of which use selected plant growth-promoting bacteria (PGPB). However, despite being a potentially valuable method for managing sediment-contaminated soils, there is currently a lack of evidence surrounding the use of bacterial inoculants in remediation, with most previous studies on the topic focusing on contaminated soils [25]. One of the bacterial strains used in recent research is *Massilia niastensis* p87, which was isolated from the rhizosphere of *Festuca rubra* growing on mine tailings with elevated concentrations of Cd, Pb and Zn. Strain p87 is a Cd/Zn-resistant bacterium that is an indole-3-acetic acid-producer, and was previously found to increase the biomass of both *Salix viminalis* and *Festuca arundinacea* growing in Cd/Zn-enriched Hoagland solutions [26]. Another bacterial strain is *Streptomyces costaricanus* RP92, which was isolated from the rhizosphere of *Cytisus striatus* (Hill) Rothm. growing in hexachlorocyclohexane (HCH)-contaminated soil; it has been characterized as an IAA and siderophore-producer [27]. Strain RP92 was previously shown to stimulate the growth of *Lupinus luteus* in diesel-contaminated soils, as well as to improve the dissipation of diesel range organics [28].

Tagetes patula L., commonly called French marigold, is a dicotyledonous annual plant species from the Asteraceae family, native from Mexico to Argentina. It is a well-known ornamental and medicinal plant, known to be a source of various secondary compounds including triterpene, thiophene and steroids. These compounds may be variably excreted by the plant in response to altering environmental conditions and have been shown to have antimicrobial, insecticidal and nematicidal effects [29,30]. It has been reported that *T. patula* is a very effective plant in benzo[*a*]pyrene remediation and offers substantial potential for soil metal remediation [31,32].

Festuca arundinacea (Schreb.), popularly known as tall fescue, is a monocotyledon species of perennial grass belonging to the Poaceae family. It is very resistant to adverse climatic conditions. Due to its low habitat requirements, it is used on areas devastated by industry and is a popular choice for phytoremediation [33,34].

In addition to certain physical characteristics such as fast growth with high biomass production, and a deep rooting system that is easily harvestable, the plants used for phytoremediation should also possess many features that will be conducive to survival in the difficult conditions of a contaminated environment, e.g., large tolerance for the uptake and accumulation of pollutants in their tissues. The potential presence of toxic substances in bottom sediments may act as an environmental stress factor and may have an adverse effect on plant growth and development by disturbing plant internal homeostasis. On the other hand, inoculation of the soil with selected plant growth-promoting bacterial strains may not only increase the effectiveness of the phytoremediation process but also contribute to alleviating the effects of stress factors on plant organisms.

The purpose of the present work was (1) to examine the response of selected plant species to soil application of bottom sediments from the urban reservoir, and to (2) determine whether the addition of selected bacterial strains supports the development and condition of plants grown on the soil without the addition of sediments; in addition, it examines (3) whether the addition of these bacterial strains influences the reaction of plants to the presence of bottom sediments in the substrate. The yield of the aboveground biomass and physiological status of the two species of investigated plants is determined by measuring total soluble protein and total chlorophyll content and chlorophyll a/b ratio. In addition, to determine the influence of the contaminants on the redox equilibrium of the plant, lipid peroxide content was measured as thiobarbituric acid reactive substances (TBARS), as well as non-enzymatic antioxidants such as α-tocopherol and total phenolic compound content (TPC). The anthocyanins content was also analyzed as a marker for the occurrence of environmental stress.

2. Materials and Methods

2.1. Substrate Preparation

Bottom sediment was collected from the Sokołówka sequential biofiltration system (SSBS) located in Lodz (Poland) on the Sokołówka River (51°48′27″ N; 19°27′4.5″ E). The sediment samples used in the experiment were collected from the first zone of the SSBS, intended for accelerated sedimentation of suspended matter and associated pollutants. The pH of the sediment was 7.15; organic carbon (OC) content and total N content was 108 g·kg^{-1} and 8.3 g·kg^{-1}, respectively. Fresh sediments (35% dry matter) were mixed with an uncontaminated agricultural soil in a proportion of 1:10 w/w and transferred into 2 kg pots. The uncontaminated soil was collected from Osiny/Puławy and sieved to 2 mm. The agricultural soil used in this experiment had loamy sand structure and pH 6.7; soil organic carbon (SOC) content was 11.0 g·kg^{-1}. Soil with bottom sediment mixtures was left for four weeks to allow for equilibrium. The detailed geochemical characterization of the soil and sediments is given in our previous article [35].

2.2. Plant Material and Growth Conditions

French marigold (*Tagetes patula* L. var. Bolero) and tall fescue (*Festuca arundinacea* Schreb.) were sown to two groups of previously-prepared pots: (1) Contained soil without the addition of sediment and (2) soil mixed with bottom sediment (for *T. patula* it was in a rate of 15 seeds per pot and for *F. arundinacea* at rate of 2 g seeds per pot). After germination, the number of plants per pot was reduced to 10.

The experiment was conducted for 10 weeks in a greenhouse under a day/night cycle of 16/8 h, temperature 27/20 °C and 65% relative humidity. A photosynthetic photon flux density of 400 mmol m^{-2} s^{-1} was applied as supplementary light.

After 10 weeks of growth the plants were harvested. Half of the harvested material was dried and weighed to determine the dry shoot weight yield, while the other half of the harvested material was used to analyze the biochemical parameters of the plants.

2.3. Bacteria Strains and Soil Inoculation

Two rhizobacterial strains previously isolated from contaminated soils were used as inoculants in this experiment: *Massilia niastensis* p87 (p87) and *Streptomyces costaricanus* RP92 (RP92). Fresh cultures of the bacterial strains were grown in 869 liquid medium [36] for 24 h. Five milliliters of this pre-culture were then transferred into fresh 869 liquid medium and grown for 12 h. The obtained bacterial biomass was harvested by centrifugation (6000 rpm, 15 min.), washed once with sterile 10 mM MgSO$_4$, and resuspended in 10 mM MgSO$_4$ to an OD$_{660}$ of 1.0 (about 10^7 cells per mL).

2.4. Experimental Design

Four weeks after seedling pots with plants growing on the soil, both with and without sediment, were divided into three groups (Table 1):

(1) In the first group, sterile 10 mM MgSO$_4$ was added in a volume of a 100 mL per pot (control, non-inoculated),
(2) In the second group, 100 mL per pot of p87 bacterial suspension was added (p87 inoculated) and,
(3) In the third group, 100 mL per pot of RP92 bacterial suspension was added (RP92 inoculated).

Table 1. Experimental model (TP—*Tagetes patula*; FA—*Festuca arundinacea*; p87—*Massilia niastensis* p87 strain; RP92—*Streptomyces costaricanus* RP92 strain).

	Soil without Inoculation	Soil with p87 Inoculation	Soil with RP92 Inoculation
Soil with sediment	TP 10 plants per pot × 4 FA 10 plants per pot × 4	TP 10 plants per pot × 4 FA 10 plants per pot × 4	TP 10 plants per pot × 4 FA 10 plants per pot × 4
Soil without sediment	TP 10 plants per pot × 4 FA 10 plants per pot × 4	TP 10 plants per pot × 4 FA 10 plants per pot × 4	TP 10 plants per pot × 4 FA 10 plants per pot × 4

This procedure was repeated after three weeks maintaining this same manner.

2.5. Preparation of Extracts from Leaf Tissues

The leaves of the *T. patula* and *F. arundinacea* plants were ground (1:10, *w/v*) in an ice-cold mortar using 50 mM sodium phosphate buffer (pH 7.0) containing 0.5 M NaCl, 1 mM EDTA and 1 mM sodium ascorbate. Crude homogenate obtained after filtration was assayed for chlorophyll and α-tocopherol content. The slurry was filtered through two layers of Miracloth. The filtrates of homogenized *T. patula* and *F. arundinacea* leaves were then centrifuged (15,000× *g*, 15 min). After centrifugation, the supernatant was collected and protein content and the degree of lipid peroxidation were measured.

2.5.1. Chlorophyll Content

Chlorophyll content was assayed as described previously [37]. A crude homogenate suspended in 80% acetone was centrifuged (5000× *g*, 5 min), and the absorbance of the clear supernatant was measured at 663 nm and 645 nm using a Helios Gamma spectrophotometer (Thermo Spectronic, Cambridge, UK). The content of chlorophyll was expressed as mg g^{-1} fresh weight.

2.5.2. α-Tocopherol Content

Crude homogenate obtained in the first step of preparation was assayed for α-tocopherol content according to Taylor and Tappel [38]. After saponification of the sample with KOH in the presence of ascorbic acid, α-tocopherol was extracted with n-hexane. Fluorescence of the organic layer was measured at 280 nm (excitation) and 310 nm (emission) using a F-2500 Fluorescence Spectrophotometer (Hitachi, Limited, Tokyo, Japan). The content of α-tocopherol was expressed as μg g^{-1} fresh weight of the original plant tissue.

2.5.3. Protein Content

The protein content was determined according to Bradford [39] with standard curves prepared using bovine serum albumin. Samples were assayed using a spectrophotometer (Helios Gamma, Thermo Spectronic, Cambridge, UK). The protein content was given as mg g^{-1} fresh weight of the original plant tissue.

2.5.4. Degree of Lipid Peroxidation (TBARS)

Concentration of lipid peroxides was estimated spectrofluorometrically (F-2500 Fluorescence Spectrophotometer; Hitachi, Limited, Tokyo, Japan) according to Yagi [40], by measuring the content of 2-thiobarbituric acid reactive substances (TBARS). The concentration of lipid peroxides was calculated in terms of 1,1,3,3-tetraethoxypropane, which was used as a standard and expressed in nmol g^{-1} fresh weight.

2.6. Total Phenolic Compound (TPC) Content

Leaf samples were homogenized (1:5 *w/v*) in a mortar with 80% methanol. The homogenate was centrifuged (20,000× *g*, 20 min) and the diluted (1:19 *v/v*) supernatant was used in the reaction with

Folin-Ciocalteu reagent according to Singleton and Rossi [41]. Absorbance was measured at 725 nm using a Helios Gamma spectrophotometer (Thermo Spectronic, Cambridge, UK). The total phenolic content (TPC) was calculated using a standard curve prepared for gallic acid (GA) and expressed in mg per g fresh weight.

2.7. Anthocyanins Content

To determine anthocyanins content, the pH-differential method described by Giusti and Wrolstad [42] was used. This method relies on the structural transformation of the anthocyanins chromophore as a function of pH and based on measurements using optical spectroscopy. Each analyzed leaf tissue sample was homogenized in two buffers: 0.025 M potassium chloride buffer pH = 1.0 and in 0.4 M sodium acetate buffer pH = 4.5. For each sample the absorbance was measured at the 520 nm (λ_{max}) and at 700 nm using a Helios Gamma spectrophotometer (Thermo Spectronic, Cambridge, UK). Monomeric anthocyanin pigment was calculated as cyanidin-3-glucoside, where MW = 449.2 and $\varepsilon = 25740$ L $\times$ mol^{-1} $\times$ cm^{-1} at 520 nm in 0.1 M HCl [43]. The anthocyanins content was expressed as mg g^{-1} fresh weight.

2.8. Statistical Analysis

All measurements concerning biomass yield and plant biochemical parameters were performed as four independent replicates ($n = 4$). Each replicate consisted of ten plants whose leaves were mixed and a representative sample was taken. Two-way analysis of variance (ANOVA II) was used to test the effect of sediments in soil and inoculation with the p87 and RP92 bacterial strains on plant biochemical parameters (i.e., total chlorophyll content, as well as chlorophyll a/b ratio and the levels of protein, α-tocopherol, TBARS, TPC and anthocyanins). If the analysis of variance found any of the factors to have a significant effect, a subsequent Duncan's test was used. All analyses were performed using STATISTICA 13 software [44].

3. Results

The soil and sediment samples were analyzed for total concentration of trace elements, sixteen PAH compounds and seventeen toxic congeners of PCDD/PCDF. As presented in our previous work, the sediments were moderately contaminated with zinc and heavily contaminated with PAH, PCDD and PCDFs [35].

The addition of bottom sediments and the inoculation of bacterial strains were found to have different influences on the plants depending on the species.

3.1. Plant Morphology and Dry Biomass

T. patula plants growing on sediment-amended soil were characterized by very poor biomass growth and inhibited development (flower buds did not develop) compared to the other variants i.e., those inoculated with bacterial strains p87 and RP92, as well as those grown on non-amended soil (Figure 1). The non-inoculated plants grown on the amended soil had characteristic reddish-purple discolorations and slight necrosis at the edge of the leaves. However, the plants growing on amended soil with bacterial inoculation were well developed; those with p87 displayed a large number of young shoots in the leaves at a later stage of development, and those inoculated with RP92 were characterized by an intense green color. In addition, the leaves of plants grown on sediment-amended soil and inoculated with p87 or RP92 were in good general condition, with very few being wrinkled, or folded with incorrect structure. *T. patula* grown on non-amended soil demonstrated proper development, and bacterial inoculation had no effect on plant morphology. In *F. arundinacea*, neither inoculation with bacterial strains nor the presence of sediment appeared to have any significant influence on morphology.

Figure 1. Morphology of 10-week-old *T. patula* and *F. arundinacea* plants (just before harvest) grown on control soil and soil amended with bottom sediments, with or without (n) inoculation with bacterial strain p87 or RP92.

In *T. patula* plants, both the presence of sediments in the soil (F = 122.756; df = 1, 18; p = 0.0000) and the type of inoculation had an effect on the biomass yield (F = 27.145; df = 2, 18; p = 0.0000; ANOVA II; Figure 2). The addition of urban sediments significantly reduced the biomass yield of *T. patula* plants. If bacterial inoculates were also added to the soil with urban sediments, the yield increased to 213% (p87) and to 238% (RP92) compared to plants growing on non-inoculated soil. The yield of *T. patula* plants growing on soil without sediment was 274% of that of plants growing on soil with sediment. The use of bacterial strains on soil without sediment also increased the yield to 131% (p87) and to 117% (RP92) compared to plants growing on soil with sediment and inoculated with strains p87 and RP92, respectively. Inoculation was found to have no effect on biomass in plants growing on soil without sediments.

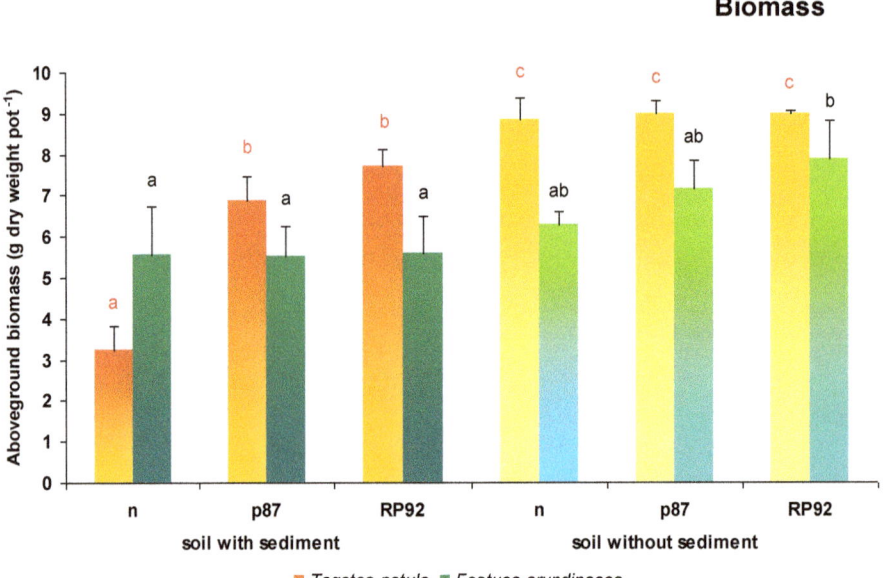

Figure 2. Yield of *T. patula* and *F. arundinacea* grown on control soil and soil amended with bottom sediments, with or without (n) inoculation with bacterial strain p87 or RP92. The same letters denote groups that did not significantly differ (at $p < 0.05$; the Duncan post hoc test). Statistical significance was measured for each plant separately and the results were marked by different colors.

The biomass yield in *F. arundinacea* plants depended only on the presence of sediment in soil (F = 7.4795; df = 1, 18; p = 0.0136). *F. arundinacea* plants did not show differences in biomass yield, either after using urban sediments or inoculation with bacterial strains. The only exception was observed for plants growing on soil without sediment and inoculated with RP92 strain, which demonstrated 41%–42% greater biomass than plants growing on soil with sediments.

3.2. Total Soluble Protein Content

Changes in soluble protein content in *T. patula* leaf tissues were influenced by the presence of sediment in soil and application of inoculation (Figure 3). ANOVA II showed that the presence of sediment (F = 6.98; df = 1, 18; p = 0.0166) and bacterial inoculation (F = 15.102; df = 2, 18; p = 0.000141) both influenced soluble protein content. The Duncan test revealed that the use of bottom sediments caused a decrease in soluble protein content compared to plants growing on soil without sediments: It was 15% lower than for plants growing on non-inoculated soil and 14% lower than in soil inoculated with the p87 strain.

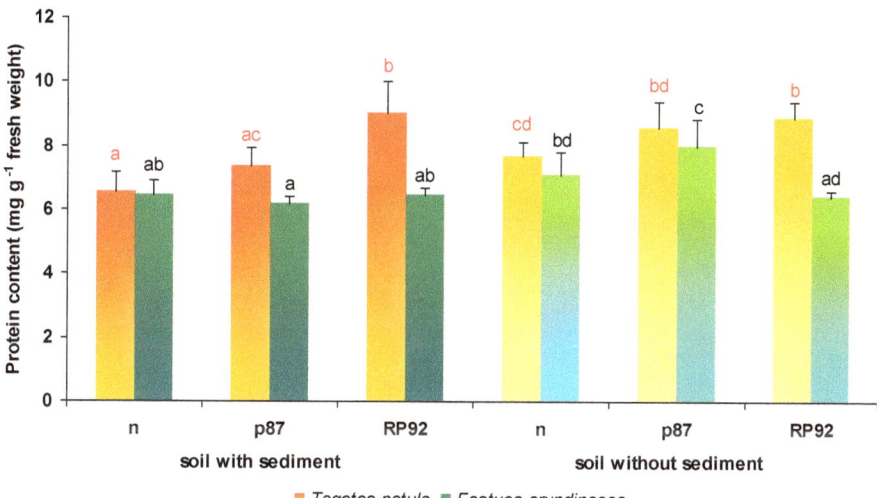

Figure 3. Soluble protein content in leaves of *T. patula* and *F. arundinacea* grown on control soil and soil amended with bottom sediments, with or without (n) inoculation with bacterial strain p87 or RP92. The same letters denote groups that did not significantly differ (at $p < 0.05$; the Duncan post hoc test). Statistical significance was measured for each plant separately and the results were marked by different colors.

The use of the RP92 strain significantly increased the soluble protein content in *T. patula* leaves compared to plants growing on non-inoculated soil, both after the application of sediment (139%) and on the soil without sediment (116%). In contrast, no such dependence was found after the inoculation with the p87 strain. Moreover, the soluble protein content of the plants grown on the soil containing sediments was found to be significantly higher after inoculation with the RP92 strain than the p87 strain (123%). The application of p87 caused a significant increase in protein content, but only among plants growing on the non-enriched substrate, compared to those growing on non-inoculated soil enriched with sediments (131%).

In the case of *F. arundinacea*, ANOVA II showed that only the use of sediments had a significant effect on the soluble protein content (F = 14.737; df = 1, 18; p = 0.0012). The Duncan test showed that the use of the p87 strain on soil to which no sediments were added significantly increased the soluble protein content in leaf tissues compared to all other variants (from 112% to 129%). Interestingly, while the use of the p87 strain with the sediments reduced the total soluble protein content in plant tissues to 87%, inoculation without the simultaneous use of sediments increased the protein content to 112%: both cases compared to the non-inoculated variants.

3.3. Total Chlorophyll Content

The presence of sediment was not found to influence total chlorophyll content in *T. patula* leaves (ANOVA II). In contrast, the type of inoculation dependent changes in the value of this parameter was observed (F = 7.901; df = 2, 18; p = 0.0034). The Duncan post hoc test indicated that the total chlorophyll content of *T. patula* plants growing in sediment-enriched soil without inoculum was significantly lower than in all other experimental variants (Figure 4).

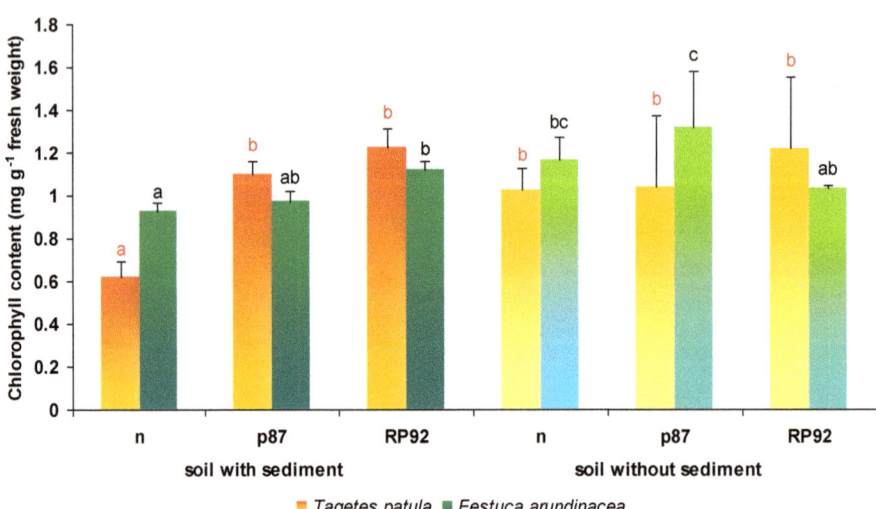

Figure 4. Total chlorophyll content in the leaves of *T. patula* and *F. arundinacea* grown on control and bottom sediments amended soil with or without (n) inoculation with bacterial strain p87 or RP92. The same letters denote groups that did not significantly differ (at $p < 0.05$; the Duncan post hoc test). Statistical significance was measured for each plant separately and the results were marked by different colors.

The total chlorophyll content in *F. arundinacea* was dependent on the presence of sediments in the soil (F = 11.300; df = 1, 18; p = 0.0035); however, it was not dependent on the type of inoculation used. The Duncan's post hoc test revealed that plants growing on non-inoculated soil presented higher total chlorophyll content when the soil was not enriched with sediments (126%). Moreover, for plants growing on soil with sediments, RP92 inoculation increased the total chlorophyll content (121%) compared to plants growing on non-inoculated soil with sediments. The highest total chlorophyll content was found in plants growing on soil without sediments inoculated with the p87 strain, with the content ranging from 117% to 142% compared to the other variants; this value was also higher than in plants growing on non-amended soil inoculated with strain RP92.

3.4. Chlorophyll a/b Ratio

In *T. patula* plants (Figure 5), both the presence of sediments in soil (F = 31.00; df = 1, 18; p = 0.000028) and the type of inoculation (F = 4.7; df = 2, 18; p = 0.0224) influenced the chlorophyll a/b ratio (ANOVA II). For the soil treated with sediment, the non-inoculated plants and plants inoculated with p87 demonstrated significantly lower chlorophyll a/b ratio (from 7% to 3.5%) compared to those inoculated with RP92. The RP92 inoculated plants also demonstrated a lower chlorophyll a/b ratio than the non-inoculated plants on soil without sediments (2.5%) and those treated with PR92 on sediment-amended soil (2%).

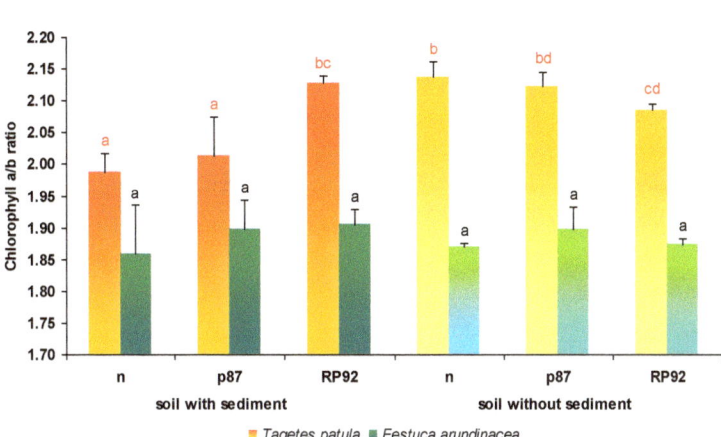

Figure 5. Chlorophyll a/b ratio in leaves of *T. patula* and *F. arundinacea* grown on control and bottom sediment-amended soil with or without (n) inoculation with bacterial strain p87 or RP92. The same letters denote groups that did not significantly differ (at $p < 0.05$; the Duncan post hoc test). Statistical significance was measured for each plant separately and the results were marked by different colors.

In the case of *F. arundinacea* plants, no statistically significant differences were found between variants regarding the chlorophyll a/b ratio (ANOVA II).

3.5. TBARS Content

Neither the presence of sediment nor the type of inoculation influenced the TBARS value in *T. patula* leaves, an indicator of oxidative lipid damage (ANOVA II; Figure 6).

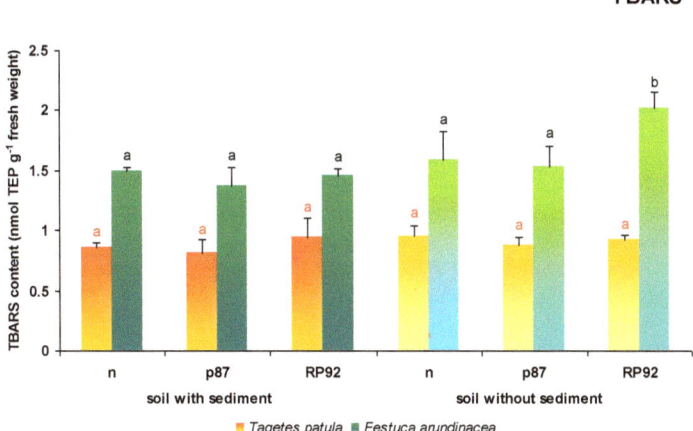

Figure 6. 2-thiobarbituric acid reactive substances (TBARS) content in leaves of *T. patula* and *F. arundinacea* grown on control and bottom sediments amended soil with or without (n) inoculation with bacterial strain p87 or RP92. The same letters denote groups that did not significantly differ (at $p < 0.05$; the Duncan post hoc test). Statistical significance was measured for each plant separately and the results were marked by different colors.

In contrast, in the case of *F. arundinacea*, both the presence of sediment in soil (F = 20.647; df = 1, 18; p = 0.00025) and the type of inoculation influenced TBARS content in leaves (F = 7.863; df = 2, 18; p = 0.0035; ANOVA II). However, leaf tissue TBARS content only increased in the untreated soil inoculated with the RP92 strain, reaching between 127% and 146% of the value found in the other experimental variants (Duncan test).

3.6. α-Tocopherol Content

No statistically significant differences in α-tocopherol content were observed between any pairs of experimental variants in *T. patula* leaves (ANOVA II; Figure 7).

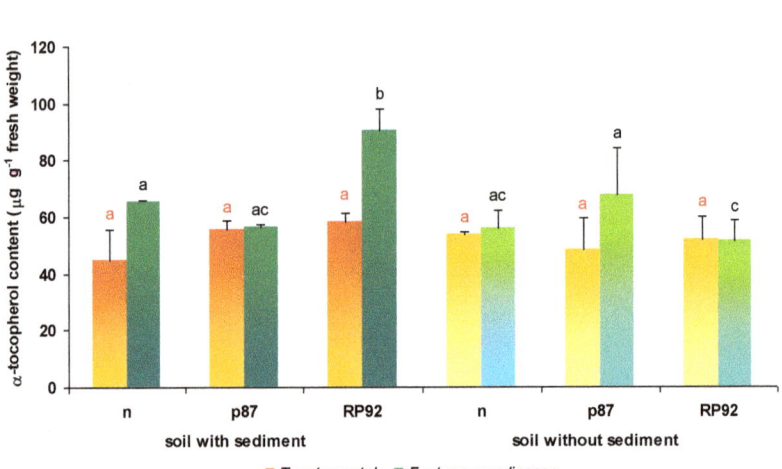

Figure 7. α-tocopherol content of the leaves of *T. patula* and *F. arundinacea* grown on control soil and soil amended with bottom sediments, with or without (n) inoculation with bacterial strain p87 or RP92. The same letters denote groups that did not significantly differ (at $p < 0.05$; the Duncan post hoc test). Statistical significance was measured for each plant separately and the results were marked by different colors.

The content of α-tocopherol, the main antioxidant of lipid fraction in *F. arundinacea* plants depended on both the presence of sediment (F = 13.039; df = 1, 18; p = 0.002) as well as the type of inoculation (F = 3.736; df = 2, 18; p = 0.044). However, the highest α-tocopherol content was demonstrated by the plants growing on the soil enriched with sediment and the inoculated strain RP92 (Duncan test), with its value ranging from 134% to 175% of those found in the other variants. In addition, in soil without sediments, α-tocopherol content was found to be 30% higher in soil treated with the p87 strain than with RP92. The latter also demonstrated 21% lower α-tocopherol content than plants growing on non-inoculated soil with sediments.

3.7. TPC Content

The presence of bottom sediments in the soil had a significant effect on the TPC content in *T. patula* plants (ANOVA II; F = 13.41; df = 1, 18; p = 0.00178; Figure 8); however, the type of inoculation also influenced TPC content: On soil without sediments, plants inoculated with p87 demonstrated 124% higher TPC content and those inoculated with RP92 115% higher, than the p87 and RP92 plants growing in treated soil.

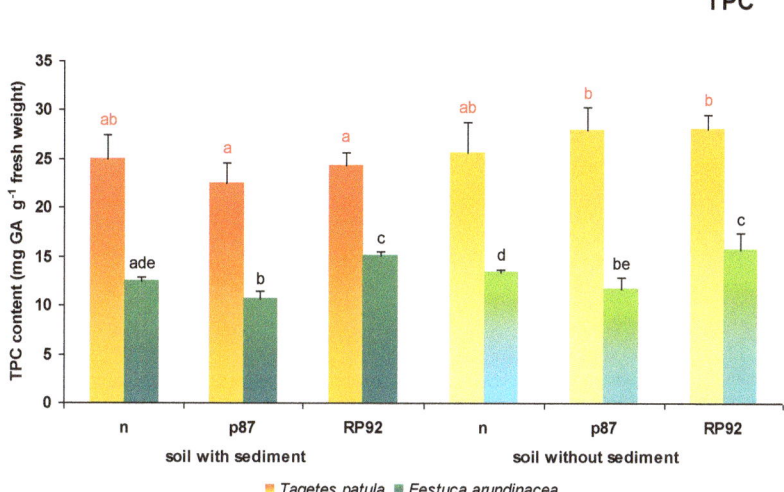

Figure 8. Total phenolic content (TPC) content in leaves of *T. patula* and *F. arundinacea* grown on control and bottom sediment-amended soil with or without (n) inoculation with bacterial strain p87 or RP92. The same letters denote groups that did not significantly differ (at $p < 0.05$; the Duncan post hoc test). Statistical significance was measured for each plant separately and the results were marked by different colors.

In *F. arundinacea* plants, both the presence of sediments in the soil (F = 4.747; df = 1, 18; p = 0.0423) and the type of inoculation had an effect on leaf TPC content (F = 39.556; df = 2, 18; p = 0.00000; ANOVA II). Differences were also found between the inoculated bacterial variants (Duncan post hoc test): Those treated with p87 in sediment-amended soil displayed 85% TPC compared to the plants growing on non-inoculated soil, while those treated with RP92 displayed 120% TPC. Similarly, in *F. arundinacea* plants growing on a substrate without sediments, inoculation with p87 strain decreased (88%), and inoculation with strain RP92 increased (118%) the TPC content in leaves compared to the plants growing on non-inoculated soil.

3.8. Anthocyanins Content

In *T. patula* plants, both the presence of sediments in the soil (F = 66.248; df = 1, 18; p = 0.0000) and the type of inoculation used had an effect on the content of anthocyanins in leaf tissues (F = 51.012; df = 2, 18; p = 0.00000; ANOVA II; Figure 9). On soil treated with sediment, non-inoculated variants demonstrated a 562% increase in anthocyanins content, and those inoculated with p87 displayed a 225% increase, compared to those grown on soil without sediments. In addition, anthocyanins content was lowest in plants growing on soil enriched with sediment and inoculated with RP92 strain: In this case, content was only 3.3% of that found in the variant with sediments, and 6.6% of the variant with sediment and p87 inoculation. Inoculation with bacterial strains did not appear to have any influence on the anthocyanins content of plants growing on soil without sediments.

In the case of *F. arundinacea* plants, neither the presence of sediment nor the type of inoculation used appeared to influence anthocyanins content.

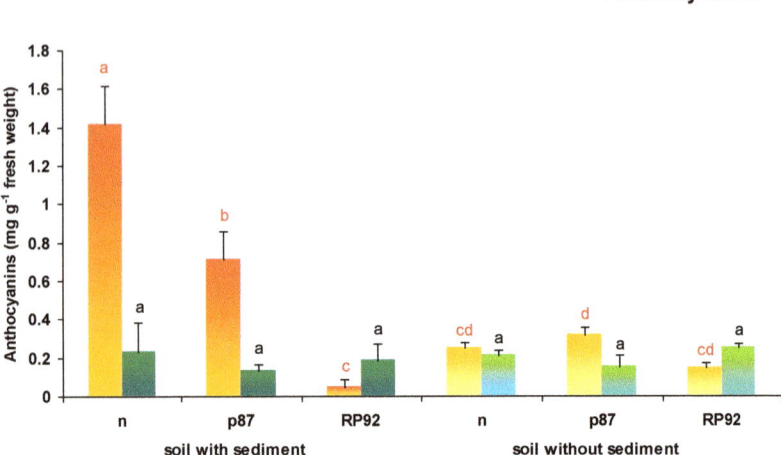

Figure 9. Anthocyanins content in leaves of *T. patula* and *F. arundinacea* grown on control and bottom sediment-amended soil with or without (n) inoculation with bacterial strain p87 or RP92. The same letters denote groups that did not significantly differ (at $p < 0.05$; the Duncan post hoc test). Statistical significance was measured for each plant separately and the results were marked by different colors.

4. Discussion

Based on changes in plant morphology, our findings suggest that the dicotyledonous *T. patula* is more sensitive to the presence of substances included in urban sediments used to amend the soil in the present study than the monocotyledonous *F. arundinacea*. In addition, inoculation of the soil with p87 and RP92 strains of bacteria reduces the toxic effects of this sediment on *T. patula*. Previous studies show that morphological changes resulting from the same stress factor may vary significantly depending on the plant species. For example, one study found two species of Cucurbitaceae, zucchini and cucumber to display different reactions to sewage sludge treatment [45]: Despite their relatedness, the leaves of the zucchini plants displayed chlorotic spots and bleaching, as well as necrosis and non-physiological shapes, while those of the cucumber plants showed no real damage. Nevertheless, despite such different reactions and different resistance to environmental stress factors, both species contributed to a similar degree to the reduction of substrate toxicity [46]. The reduction in the severity of morphological changes observed after the application of bacterial strains p87 and RP92 to sediment-amended soil could be attributed to the fact that some PGPB strains may produce indole acetic acid, gibberellins, cytokinin, antioxidants and some unknown determinants, which may lead to an enhanced uptake of nutrients, thereby improving plant health under stress conditions and eliminating the effects of stress [47].

In addition to morphological changes, inhibition of *T. patula* growth and reduction of biomass yield after urban sediment use indicates their toxic effect on this plant species. Interestingly, the application of p87 and RP92 bacterial strains to soil enriched with sediments led to a 2.1 and 2.4-fold increase in biomass production, respectively, compared to plants growing on non-inoculated soils with sediments. In addition, the use of bacterial strains on untreated soil does not change the yield of biomass obtained: It is possible that if plants are not exposed to stress factors and, as a consequence, they develop properly with efficient biomass growth, the addition of bacterial strains to uncontaminated soil does not stimulate even greater biomass growth. This situation indicates that the presence of bacterial strains in the contaminated soil can trigger mechanisms in soil and/or in plants that allow the stress factor to interact directly with the plant or minimize its effects. The presence of PGPB in the soil facilitates plant growth by operating through two mechanisms: Indirectly by reducing plant pathogens and directly by facilitating nutrient uptake through phytohormone production, by enzymatic reduction of plant ethylene levels and/or by siderophore production [47]. PGPB inoculation with *Pseudomonas mendocina* in *Lactuca sativa* plants affected by salt stress resulted in significantly greater shoot biomass than controls; the authors suggest that inoculation with selected PGPB could be an effective tool for alleviating salinity stress in salt sensitive plants [48]. Moreover, when PGPB was used to seed bacterization, plants show enhanced root and shoot length, biomass and biochemical parameters such as chlorophyll, carotenoid and protein content [49]. Otherwise excellent works on the remediation capabilities of *T. patula* provide no information on the impact of using bacterial strains during remediation, or their significance for these plants [31,32]. Our present findings indicate that as in the case of plant morphology, *F. arundinacea* did not show major changes in biomass yield following sediment treatment; however, the use of RP92 strain on soil without sediment had a promoting effect.

Differences in the biochemical parameters were also observed between *T. patula* and *F. arundinacea*, and these variations were related to the use of sediments and the presence of bacterial inoculums in the soil.

After the application of urban sediments, *T. patula* plants responded with a decrease in total soluble protein, total chlorophyll and chlorophyll a/b ratio. Environmental factors that have an adverse effect on plant metabolism can lead to premature senescence, even before full plant maturity is achieved. One of the first symptoms of the aging process is a reduction in the concentration of total soluble protein in tissues; this has been observed even before the natural decline of photochemical processes in wheat leaves during the aging process [50].

Total soluble protein and total chlorophyll content were also found to be key parameters among physiological and biochemical signs of senescence in a paper examining the influence of nitrogen deficit on aging in sugar beet [51]. Their values were always lower in older leaves of sugar beet than in younger ones, and they decreased almost synchronously during leaf senescence. The authors attribute the observed decrease in total chlorophyll content to the destruction of the quickly renewing protein(s) of the chlorophyll-protein complex, resulting in immediate discoloration of the chloroplast.

The initial steps in the degradation of chlorophylls and chloroplast proteins can occur in intact organelles but a morphological analysis of senescing chloroplasts revealed the later breakdown of the thylakoid membrane system and degradation of thylakoid-bound proteins [52]. Another indicator of the progressive senescence of green plant tissues is the presence of changes in chlorophyll a/b ratio. Nath et al., [53] report a linear decrease in chlorophyll a/b ratio occurring simultaneously with a decrease in the total chlorophyll content in *Arabidopsis thaliana* plants, which they attribute to different rates of degradation of chlorophyll a and chlorophyll b during senescence.

In various plants, the decrease of soluble protein commonly observed in the senescing leaves may also be related to proteolysis in the cell vacuoles [54]. An increase in proteinase activity results in the degradation of proteins to amino acids; these are then transported through the phloem sap to the growing organs. This process of nitrogen remobilization plays an essential role at the whole-plant level during sequential senescence due to the fact that the mechanisms associated with senescence allow the redistribution of nutrients from old organs to sink organs for growth or storage [55]. The correlation between nitrogen remobilization and senescence severity has been observed during the vegetative growth of several plants e.g., *Arabidopsis* [56].

It is important to note that inoculation with the bacterial strain RP92 was found to increase the soluble protein content in the leaves of *T. patula* plants, both on those cultivated on sediment-amended soil, and those on unamended soil. In addition, in *T. patula* cultivated on sediment-amended soil, the use of RP92 almost doubles the chlorophyll content in the leaves and increases the chlorophyll a/b ratio. This may indicate that the strain has a protective effect on *T. patula* against the compounds in the applied sediments, which may act by preventing the plant from entering premature senescence. Of the two applied rhizobacterial strains, it seems that RP92 had a stronger protective effect than p87 on the analyzed parameters.

Among *F. arundinacea* plants, the use of urban sediments did not appear to cause significant changes in the content of soluble protein, but instead reduced the content of chlorophyll in green tissues. The use of bacterial strains generally had no effect on the soluble protein content of the tissues after sediment application, and in the case of plants growing on soil without sediments, only the simultaneous introduction of the p87 strain increased soluble protein content. Of the tested groups, the *F. arundinacea* growing on non-amended soil supplemented with p87 displayed the highest total chlorophyll content. In addition, RP92 supplementation also partially negated the unfavorable impact of sediments by increasing the content of the green assimilation dye. However, changes in total chlorophyll content were minor and had no effect on the chlorophyll a/b ratio.

One of the defenses employed by plants against environmental stress concerns the occurrence of secondary oxidative stress, characterized by an imbalance between the formation of reactive oxygen species (ROS) and their utilization by antioxidative systems. ROS such as superoxide anions ($O_2^-·$), hydroxyl radicals (·OH), peroxyl radicals (ROO·), singlet oxygen ($^1O_2$) and hydrogen peroxide (H_2O_2), are characterized by very high reactivity and cause oxidative damage to proteins, lipids, nucleic acids and other biologically important molecules [57]. Such damage can lead to dysfunctions and irreversible degradation of molecules that are part of cellular structures and, consequently, inhibit the growth and development of plants. One group of plant secondary metabolites with non-enzymatic antioxidant properties is those based on phenolic compounds. Phenols exert their antioxidant properties through their ability to quench free radical reactions due to their potential to chelate metal ions through the properties of their nucleophilic aromatic ring [58].

Phenolic compounds are also believed to inhibit membrane lipid peroxidation by "catching" alkoxyl radicals (RO·). Some phenols, particularly flavonoids, stabilize membranes by decreasing this fluidity due to the ability of phenolics to bind to some of the integral membrane proteins and phospholipids. Decreased fluidity in turn limits the diffusion of free radicals and reduces the peroxidation of membrane lipids. Some phenolic compounds (phenylpropanoids) are able to perform protective functions in plant cells by forming lignins and suberin, which mechanically reinforce the cell wall [58,59].

In the case of the *T. patula* plants studied in this work, the addition of urban sediments to the soil did not affect the TPC content in the leaves. Differences associated with the addition of sediments to the soil were only visible in the variants in which bacterial strains were also applied. Interestingly, regarding plants treated with p87 and RP92, those cultivated on soil without sediments displayed greater TPC than those grown on soil with sediments. It is possible that in *T. patula* plants, these small changes in the antioxidant component of TPC were sufficient to maintain the correct redox balance and prevent oxidative changes following sediment treatment or bacterial supplementation, as suggested by the lack of changes in the degree of lipid peroxidation measured as TBARS content. No significant changes in TBARS content have also been observed in previous studies conducted on cucumber plants growing on soil with the addition of urban sediments [60].

Similarly, no TPC changes were associated with the application of urban sediments for *F. arundinacea* plants. Compared to the non-inoculated plant cultures, TPC decreased following p87 supplementation but increased following RP92 supplementation, both findings being irrespective of sediment addition. In this case, it may be surprising that the *F. arundinacea* plants growing on unamended soil and inoculated with RP92 demonstrated increased TPC content and elevated TBARS, which may indicate the induction of oxidative damage of lipids. It is possible that in this case the increase in the content of TPC did not compensate for the increase in oxidative reactions.

It cannot be ruled out that insufficiencies may exist in other elements of the antioxidant system, for example, changes in the α-tocopherol content of *F. arundinacea* plants. α-tocopherol is the main antioxidant of the lipid fraction of the cell; it is synthesized exclusively in oxygenic photosynthetic organisms and constitutes more than 90% of the foliar tocopherols. This lipophilic antioxidant has two main antioxidant functions; firstly, the quenching of singlet oxygen generated mostly by triplet chlorophyll in photosystem II; and secondly, the scavenging of harmful radicals by donating electrons to various receivers, such as to a lipid peroxyl radical to prevent membrane lipid peroxidation reactions [61]. In *F. arundinacea* plants, α-tocopherol content was found to increase significantly in plants grown on amended soil inoculated with RP92; however, in the leaves of inoculated plants growing on unamended soil, the α-tocopherol content is similar that of plants on non-inoculated soils. This could be a reason for the oxidative damage of lipids formed in this experimental variant and hence, the observed increase in TBARS value.

Anthocyanins are blue, red or purple phenolic compounds that act as water-soluble pigments. They accumulate in the vacuoles of a wide range of cells and give color to flowers and fruits. In addition, when present in the leaves of plants exposed to environmental stressors, they may exert a protective influence by mitigating the effects of stress. At present, their main functions are considered to be antioxidants and sunscreens, mediators of ROS-induced signaling cascades, chelating agents for metals and metalloids, and delayers of leaf senescence, especially in plants growing under conditions of nutrient deficiency [62,63]. Foliar anthocyanins can protect chloroplasts from the adverse effect of excess light and have the potential to reduce both the incidence and severity of photooxidative damage. Moreover, anthocyanic leaves often display characteristics typical of leaves growing under shaded conditions, such as a lower chlorophyll a/b ratio compared to green leaves of the same plant species that do not contain high levels of anthocyanins [64]. The protective function of anthocyanins can be of great importance when leaves grown under conditions of high light intensity simultaneously face other environmental stressors such as cold, drought, salinity or wounding, when the capacity of plants to usefully process radiant energy is severely constrained. The biosynthesis of anthocyanins in such conditions can represent one way to reduce the risk of photoinhibition [65].

Among all studied biochemical parameters in *T. patula*, anthocyanins production demonstrated the most pronounced changes after exposure to urban sediments and inoculation with bacterial strains. After the application of urban sediments, the content of anthocyanins was 5.6 times higher than in plants grown on soil without sediments. Considering the plant functions performed by anthocyanins, and considering that anthocyanins are also markers of environmental stress, such a large increase in their content may indicate the presence of a strong environmental stressor associated with the presence of substances from urban sediments in the soil. In this case, the increase in the content of anthocyanins could be associated with their antioxidative function. However, it cannot be ruled out that if urban sediments were used, the anthocyanins may have been more employed for their chelating properties, although this depends on the composition of the sediments and the potential risk of metal ions in the substrate, as well as the availability of these metals. Previous studies have found urban sediments to be moderately contaminated with zinc, and that other metals tested are present with acceptable limits [35]. This may indicate that a strong reaction associated with a high accumulation of anthocyanins in *T. patula* tissues may be associated with the presence of other substances acting as stress factors such as PAH, PCDD or PCDFs. It cannot be ruled out that the increased content of anthocyanins in *T. patula* protects green plant tissues from photoinhibition, especially as, as already mentioned, the chlorophyll a/b ratio was found to decrease in plants exposed to urban sediments. Our findings also show that the bacterial strains p87 and RP92 demonstrated a positive effect in mitigating the effect of environmental stress, as indicated by an associated decrease in the content of anthocyanins, which serves as an indicator of the occurrence of stress factors. In addition, the RP92 strain seems to have a stronger effect in this respect.

5. Conclusions

The dicotyledonous *T. patula* plant was more sensitive than the monocotyledonous *F. arundinacea* to the toxic substances contained in sediments from urban reservoirs added to the soil; this was manifested by the inhibition of plant development, poor biomass production and visible morphological changes. Soil inoculation with rhizobacterial strains reduced the toxic effect of the added sediment. Of the strains used, it seemed that RP92 had a more favorable effect on the physiological condition of *T. patula* plants than p87. In the case of *F. arundinacea* plants, the addition of sediments to the soil did not adversely affect biomass production and development of this plant species, and inoculation with the p87 and RP92 strains appeared to have similar effects. Our findings suggest that among the examined biochemical parameters, leaf anthocyanins content was an excellent marker for the emergence of environmental stress, especially after applying urban sediment. Both tested plant species were suitable for planting on soils enriched with urban sediments, and the addition of bacterial inoculums promoted plant growth and reduced the damage caused by the presence of xenobiotics contained in sediments.

Author Contributions: A.W. (Anna Wyrwicka), M.U. and G.S. conceived and designed the experiments; A.W. (Anna Wyrwicka), M.U., G.S. and S.S. organized the bottom sediments sampling procedures; G.S. and S.S. organized the soil sampling procedures and performed the experimental work on plant growth; A.W. (Anna Wyrwicka) and M.U. performed the plant biometric parameters analyses; A.W. (Anna Wyrwicka) and A.W. (Aleksandra Witusińska) performed all plant biochemical parameters analyses; A.W. (Anna Wyrwicka), J.C.-K., M.P. performed the data elaboration and statistical analysis of data; A.W. (Anna Wyrwicka) wrote the paper with the contribution of M.U. and G.S.; A.W. (Anna Wyrwicka) writing—review and editing; A.W. (Anna Wyrwicka) graphic; P.S.K. provided bacterial strains used in the experiment.

Funding: This research was supported by the European Commission under the Seventh Framework Program for Research (FP7-KBBE-266124, Greenland) and University of Lodz Grant No. B17 11 000 000 052.01. There was no additional external funding received for this study. The funders had no role in study design, data collection and analysis, decision to publish, or preparation of the manuscript.

Conflicts of Interest: The authors declare no conflict of interest. The funders had no role in the design of the study; in the collection, analyses, or interpretation of data; in the writing of the manuscript, or in the decision to publish the results.

References

1. Yazdani, M.; Monavari, S.M.; Omrani, G.A.; Shariat, M.; Hosseini, S.M. Landfill site suitability assessment by means of geographic information system analysis. *Solid Earth* **2015**, *6*, 945–956. [CrossRef]
2. Förstner, U.; Salomons, W. Sediment research, management and policy. A decade of JSS. *J. Soils Sed.* **2010**, *10*, 1440–1452. [CrossRef]
3. Maj, K.; Koszelnik, P. Possibility of use of bottom sediments derived from the San River. *Ecol. Eng.* **2016**, *48*, 147–152. [CrossRef]
4. Zalewski, M.; Wagner, I.; Frątczak, W.; Mankiewicz-Boczek, J.; Paniewski, P. Blue-green city for compensating global climate change. *Parliam. Mag.* **2012**, *350*, 2–3.
5. Apitz, S.E.; Power, E.A. From risk assessment to sediment management. An international perspective. *J. Soils Sed.* **2002**, *2*, 61–66. [CrossRef]
6. McCauley, D.J.; DeGraeve, G.M.; Linton, T.K. Sediment quality guidelines and assessment: Overview and research needs. *Environ. Sci. Policy* **2000**, *3*, S133–S144. [CrossRef]
7. Michalec, B. Qualitative and quantitative assessment of sediments pollution with heavy metals of small water reservoirs. In *Soil Health and Land Use Management*; Hernandez Soriano, M.C., Ed.; InTech: Rijeka, Croatia, 2012; pp. 255–278.
8. Wyrwicka, A.; Steffani, S.; Urbaniak, M. The effect of PCB-contaminated sewage sludge and sediment on metabolism of cucumber plants (*Cucumis sativus* L.). *Ecohydrol. Hydrobiol.* **2014**, *14*, 75–82. [CrossRef]
9. Mamedov, A.I.; Bar-Yosefl, B.; Levkovich, I.; Rosenberg, R.; Silber, A.; Fine, P.; Levy, G.J. Amending soil with sludge, manure, humic acid, orthophosphate and phytic acid: Effects on infiltration, runoff and sediment loss. *Land Degrad. Dev.* **2016**, *27*, 1629–1639. [CrossRef]
10. Urbaniak, M.; Wyrwicka, A.; Zieliński, M.; Mankiewicz-Boczek, J. Potential for phytoremediation of PCDD/PCDF-contaminated sludge and sediments using Cucurbitaceae plants: A pilot study. *Bull. Environ. Contam. Toxicol.* **2016**, *97*, 401–406. [CrossRef] [PubMed]
11. Herzel, H.; Krüger, O.; Hermann, L.; Adama, C. Sewage sludge ash—A promising secondary phosphorus source for fertilizer production. *Sci. Total Environ.* **2016**, *542*, 1136–1143. [CrossRef]
12. Tontti, T.; Poutiainen, H.; Heinonen-Tanski, H. Efficiently treated sewage sludge supplemented with nitrogen and potassium in a good fertilizer for cereals. *Land Degrad. Dev.* **2017**, *28*, 742–751. [CrossRef]
13. Yazdani, M.; Monavari, S.M.; Omrani, G.A.; Shariat, M.; Hosseini, S.M.; Garcia-Pausas, J.; Rabissi, A.; Rovira, P.; Romanyà, J. Organic fertilization increases C and N stocks and reduces soil organic matter stability in Mediterranean vegetable gardens. *Land Degrad. Dev.* **2017**, *28*, 691–698.

14. Wyrwicka, A.; Urbaniak, M. The biochemical response of willow plants (*Salix viminalis* L.) to the use of sewage sludge from various sizes of wastewater treatment plant. *Sci. Total Environ.* **2018**, *615*, 882–894. [CrossRef] [PubMed]
15. Tejada, M.; Gómez, I.; Fernández-Boy, E.; Díaz, M.-J. Effects of sewage sludge and *Acacia dealbata* composts on soil biochemical and chemical properties. *Commun. Soil Sci. Plant Anal.* **2014**, *45*, 570–580. [CrossRef]
16. Lloreta, E.; Pascuala, J.A.; Brodieb, E.L.; Bouskillb, N.J.; Insamd, H.; Fernández-Delgado Juárezd, M.; Goberna, M. Sewage sludge addition modifies soil microbial communities and plant performance depending on the sludge stabilization process. *Appl. Soil Ecol.* **2016**, *101*, 37–46. [CrossRef]
17. Cunningham, S.D.; Berti, W.R. Remediation of contaminated soils with green plants: An overview. *In Vitro Cell. Dev. Biol. Plant* **1993**, *29P*, 207–212. [CrossRef]
18. Cunningham, S.D.; Berti, W.R.; Huang, J.W. Phytoremediation of contaminated soils. *Trends Biotechnol.* **1995**, *13*, 393–397. [CrossRef]
19. Anderson, T.A.; Guthire, E.A.; Walton, B.T. Bioremediation in the rhizosphere: Plant roots and associated microbes clean contaminates soil. *Environ. Sci. Technol.* **1993**, *27*, 2630–2636. [CrossRef]
20. Glenn, E.P.; Jordan, F.; Waugh, W.J. Phytoremediation of a nitrogen contaminated desert soil by native shrubs and microbial processes. *Land Degrad. Dev.* **2017**, *28*, 361–369. [CrossRef]
21. Wiesmeier, M.; Lungu, M.; Hübner, R.; Cerbari, V. Remediation of degraded arable steppe soils in Moldova using vetch as green manure. *Solid Earth* **2015**, *6*, 609–620. [CrossRef]
22. Chen, Y.-C.; Banks, M.K. Bacterial community evaluation during establishment of tall fescue (*Festuca arundinacea*) in soil contaminated with pyrene. *Int. J. Phytoremediat.* **2004**, *6*, 227–238. [CrossRef] [PubMed]
23. Bisht, S.; Pandey, P.; Bhargava, B.; Sharma, S.; Kumar, V.; Sharma, K.D. Bioremediation of polyaromatic hydrocarbons (PAHs) using rhizosphere technology. *Braz. J. Microbiol.* **2015**, *46*, 7–21. [CrossRef] [PubMed]
24. Gałązka, A.; Grządziel, J.; Gałązka, R.; Ukalska-Jaruga, A.; Strzelecka, J.; Smreczak, B. Genetic and functional diversity of bacterial microbiome in soils with long term impacts of petroleum hydrocarbons. *Front. Microbiol.* **2018**, *9*, 1923. [CrossRef] [PubMed]
25. Babalola, O.O. Beneficial bacteria of agricultural importance. *Biotechnol. Lett.* **2010**, *32*, 1559–1570. [CrossRef] [PubMed]
26. Becerra-Castro, C.; Monterroso, C.; Prieto-Fernández, Á.; Rodríguez-Lamas, L.; Loureiro-Viñas, M.; Acea, M.J.; Kidd, P.S. Pseudometallophytes colonising Pb/Zn mine tailings: A description of the plant-microorganism-rhizosphere soil system and isolation of metal-tolerant bacteria. *J. Hazard. Mater.* **2012**, *217–218*, 350–359. [CrossRef] [PubMed]
27. Becerra-Castro, C.; Kidd, P.S.; Prieto-Fernández, Á.; Weyens, N.; Acea, M.; Vangronsveld, J. Endophytic and rhizoplane bacteria associated with *Cytisus striatus* growing on hexachlorocyclohexane-contaminated soil: Isolation and characterization. *Plant Soil* **2011**, *340*, 413–433. [CrossRef]
28. Balseiro-Romero, M.; Gkorezis, P.; Kidd, P.S.; Vangronsveld, J.; Monterroso, C. Enhanced degradation of diesel in the rhizosphere of *Lupinus luteus* after inoculation with diesel-degrading and PGP bacterial strains. *J. Environ. Qual.* **2016**, *45*, 924–932. [CrossRef]
29. Padden, A.N.; Rainey, F.A.; Kelly, D.P.; Wood, A.P. *Xanthobacter tagetidis* sp. nov., an organism associated with *Tagetes* species and able to grow on substituted. *Int. J. Syst. Bacteriol.* **1997**, *47*, 394–401. [CrossRef]
30. Cicevan, R.; Al Hassan, M.; Sestras, A.F.; Prohens, J.; Vicente, O.; Sestras, R.E.; Boscaiu, M. Screening for drought tolerance in cultivars of the ornamental genus *Tagetes* (Asteraceae). *PeerJ* **2016**, *4*, e2133. [CrossRef]
31. Sun, Y.; Xu, Y.; Zhou, Q.; Wang, L.; Lin, D.; Liang, X. The potential of gibberellic acid 3 (GA3) and Tween-80 induced phytoremediation of co-contamination of Cd and Benzo[a]pyrene (B[a]P) using *Tagetes patula*. *J. Environ. Manag.* **2013**, *114*, 202–208. [CrossRef]
32. Bareen, F.; Nazir, A. Metal decontamination of tannery solid waste using *Tagetes patula* in association with saprobic and mycorrhizal fungi. *Environmentalist* **2010**, *30*, 45–53. [CrossRef]
33. Khashij, S.; Karimi, B.; Makhdoumi, P. Phytoremediation with *Festuca arundinacea*: A Mini Review. *Int. J. Health Life Sci.* **2018**, *4*, e86625. [CrossRef]
34. Sun, M.; Fu, D.; Teng, T.; Shen, Y.; Luo, Y.; Li, Z.; Christie, P. In situ phytoremediation of PAH-contaminated soil by intercropping alfalfa (*Medicago sativa* L.) with tall fescue (*Festuca arundinacea* Schreb.) and associated soil microbial activity. *J. Soils Sediments* **2011**, *11*, 980–989. [CrossRef]

35. Siebielec, S.; Siebielec, G.; Urbaniak, M.; Smreczak, B.; Grzęda, E.; Wyrwicka, A.; Kidd, P.S. Impact of rhizobacterial inoculants on plant growth and enzyme activities in soil treated with contaminated bottom sediments. *Int. J. Phytoremediat.* **2019**, *21*, 325–333. [CrossRef] [PubMed]
36. Mergeay, M.; Nies, D.; Schlegel, H.G.; Gerits, J.; Charles, P.; Van Gijsegem, F. Alcaligenes eutrophus CH34 is a facultative chemolithotroph with plasmid-bound resistance to heavy metals. *J. Bacteriol.* **1985**, *162*, 328–334. [PubMed]
37. Porra, R.J.; Thompson, W.A.; Kriedmann, P.E. Determinate of accurate extinction coefficients and simultaneous equations for assaying chlorophylls a and b extracted with four different solvents: Verification of the concentration of chlorophyll standards by atomic absorption spectroscopy. *Biochim. Biophys. Acta/Gen. Subj.* **1989**, *975*, 384–390. [CrossRef]
38. Taylor, S.L.; Tappel, A.L. Sensitive fluorometric methods for tissue tocopherol analysis. *Lipids* **1976**, *11*, 530–538. [CrossRef] [PubMed]
39. Bradford, M.M. A rapid and sensitive method for the quantification of microgram quantities of protein utilizing the principle of protein-dye binding. *Anal. Biochem.* **1976**, *72*, 248–254. [CrossRef]
40. Yagi, K. Assay for serum lipid peroxide level its clinical significance. In *Lipid Peroxides in Biology and Medicine*; Yagi, K., Ed.; Academic Press Inc.: London, UK; New York, NY, USA, 1982; pp. 223–241.
41. Singleton, V.L.; Rossi, J.A. Colorimetry of total phenolics with phosphomolybdic-phosphotungstic acid reagents. *Am. J. Enol. Vitic.* **1965**, *16*, 144–158.
42. Giusti, M.; Wrolstad, R.E. Characterization and measurement of anthocyanins by UV-visible spectroscopy. *Curr. Protoc. Food Anal. Chem.* **2001**, F1.2.1–F1.2.13. [CrossRef]
43. McClure, J.W. Photocontrol of *Spirodela intermedia* flavonoids. *Plant Phys.* **1967**, *43*, 193–200. [CrossRef] [PubMed]
44. Dell Inc. Dell Statistica (Data Analysis Software System). version 13. 2016. Available online: http://software.dell.com (accessed on 18 May 2017).
45. Wyrwicka, A.; Urbaniak, M. The different physiological and antioxidative responses of zucchini and cucumber to sewage sludge application. *PLoS ONE* **2016**, *11*, e0157782. [CrossRef] [PubMed]
46. Urbaniak, M.; Zieliński, M.; Wyrwicka, A. The influence of the Cucurbitaceae on mitigating the phytotoxicity and PCDD/PCDF content of soil amended with sewage sludge. *Int. J. Phytoremediat.* **2017**, *19*, 207–213. [CrossRef] [PubMed]
47. Shrivastava, P.; Kumar, R. Soil salinity: A serious environmental issue and plant growth promoting bacteria as one of the tools for its alleviation. *Saudi J. Biol. Sci.* **2015**, *22*, 123–131. [CrossRef] [PubMed]
48. Kohler, J.; Hernandez, J.A.; Caravaca, F.; Roldan, A. Induction of antioxidant enzymes is involved in the greater effectiveness of a PGPR versus AM fungi with respect to increasing the tolerance of lettuce to severe salt stress. *Environ. Exp. Bot.* **2009**, *65*, 245–252. [CrossRef]
49. Tiwari, S.; Singh, P.; Tiwari, R.; Meena, K.K.; Yandigeri, M.; Singh, D.P.; Arora, D.K. Salt-tolerant rhizobacteria-mediated induced tolerance in wheat (*Triticum aestivum*) and chemical diversity in rhizosphere enhance plant growth. *Biol. Fertil. Soils* **2011**, *47*, 907–916. [CrossRef]
50. Camp, P.J.; Huber, S.C.; Burke, J.J.; Moreland, D.E. Biochemical changes that occur during senescence of wheat leaves. *Plant Physiol.* **1982**, *70*, 1641–1646. [CrossRef] [PubMed]
51. Romanova, A.K.; Semenova, G.A.; Ignat'ev, A.R.; Novichkova, N.S.; Fomina, I.R. Biochemistry and cell ultrastructure changes during senescence of *Beta vulgaris* L. leaf. *Protoplasma* **2016**, *253*, 719–727. [CrossRef]
52. Hörtensteiner, S.; Feller, U. Nitrogen metabolism and remobilization during senescence. *J. Exp. Bot.* **2002**, *53*, 927–937. [CrossRef]
53. Nath, K.; Phee, B.-K.; Jeong, S.; Lee, S.Y.; Tateno, Y.; Allakhverdiev, S.I.; Lee, C.-H.; Nam, H.G. Age-dependent changes in the functions and compositions of photosynthetic complexes in the thylakoid membranes of *Arabidopsis thaliana*. *Photosynth. Res.* **2013**, *117*, 547–556. [CrossRef]
54. Dubinina, I.M.; Burakhanova, E.A.; Kudryavtseva, L.F. Vacuoles of mesophyll cells as a transient reservoir for assimilates. *Russ. J. Plant Physiol.* **2001**, *48*, 32–37. [CrossRef]
55. Girondé, A.; Poret, M.; Etienne, P.; Trouverie, J.; Bouchereau, A.; Le Cahérec, F.; Leport, L.; Orsel, M.; Niogret, M.-F.; Deleu, C.; et al. A profiling approach of the natural variability of foliar N remobilization at the rosette stage gives clues to understand the limiting processes involved in the low N use efficiency of winter oilseed rape. *J. Exp. Bot.* **2015**, *66*, 2461–2473. [CrossRef] [PubMed]

56. Diaz, C.; Lemaître, T.; Christ, A.; Azzopardi, M.; Kato, Y.; Sato, F.; Morot-Gaudry, J.F.; Le Dily, F.; Masclaux-Daubresse, C. Nitrogen recycling and remobilization are differentially controlled by leaf senescence and development stage in *Arabidopsis* under low nitrogen nutrition. *Plant Physiol.* **2008**, *147*, 1437–1449. [CrossRef] [PubMed]
57. Halliwell, B.; Gutteridge, J.M.C. *Free Radicals in Biology and Medicine*, 5th ed.; Oxford University Press: Oxford, UK, 2015.
58. Chalker-Scott, L.; Fuchigami, L.H. The role of phenolic compounds in plant stress responses. In *Low Temperature Stress Physiology in Crops*; Li, P.H., Ed.; CRC Press, Inc.: Boca Raton, FL, USA, 1989; pp. 67–79.
59. Cavaiuolo, M.; Cocetta, G.; Ferrante, A. The antioxidants changes in ornamental flowers during development and senescence. *Antioxidants* **2013**, *2*, 132–155. [CrossRef] [PubMed]
60. Wyrwicka, A.; Urbaniak, M.; Przybylski, M. The response of cucumber plants (*Cucumis sativus* L.) to the application of PCB-contaminated sewage sludge and urban sediment. *PeerJ* **2019**, *7*, e6743. [CrossRef] [PubMed]
61. Espinoza, A.; San Martín, A.; López-Climent, M.; Ruiz-Lara, S.; Gómez-Cadenas, A.; Casaretto, J.A. Engineered drought-induced biosynthesis of α-tocopherol alleviates stress-induced leaf damage in tobacco. *J. Plant Physiol.* **2013**, *170*, 1285–1294. [CrossRef]
62. Landi, M.; Tattini, M.; Gould, K.S. Multiple functional roles of anthocyanins in plant-environment interactions. *Environ. Exp. Bot.* **2015**, *119*, 4–17. [CrossRef]
63. Khoo, H.E.; Azlan, A.; Tang, S.T.; Lim, S.M. Anthocyanidins and anthocyanins: Colored pigments as food, pharmaceutical ingredients, and the potential health benefits. *Food Nutr. Res.* **2017**, *61*, 1361779. [CrossRef]
64. Manetas, Y.; Petropoulou, Y.; Psaras, G.K.; Drinia, A. Exposed red (anthocyanic) leaves of *Quercus coccifera* display shade characteristics. *Funct. Plant Biol.* **2003**, *30*, 265–270. [CrossRef]
65. Pietrini, F.; Iannelli, M.A.; Massacci, A. Anthocyanin accumulation in the illuminated surface of maize leaves enhances protection from photo-inhibitory risks at low temperature, without further limitation to photosynthesis. *Plant Cell Environ.* **2002**, *25*, 1251–1259. [CrossRef]

© 2019 by the authors. Licensee MDPI, Basel, Switzerland. This article is an open access article distributed under the terms and conditions of the Creative Commons Attribution (CC BY) license (http://creativecommons.org/licenses/by/4.0/).

Article

Effect of Different Copper Levels on Growth and Morpho-Physiological Parameters in Giant Reed (*Arundo donax* L.) in Semi-Hydroponic Mesocosm Experiment

Fabrizio Pietrini [1], Monica Carnevale [2], Claudio Beni [2], Massimo Zacchini [1], Francesco Gallucci [2] and Enrico Santangelo [2,*]

[1] Istituto di Ricerca sugli Ecosistemi Terrestri (IRET)—Consiglio Nazionale delle Ricerche (CNR)-via Salaria km 29.300, 00015 Monterotondo, Rome, Italy

[2] Consiglio per la Ricerca in Agricoltura e l'Analisi dell'Economia Agraria (CREA)—Centro di Ricerca Ingegneria e Trasformazioni Agroalimentari (CREA-IT)-Via della Pascolare 16, 00015 Monterotondo, Rome, Italy

* Correspondence: enrico.santangelo@crea.gov.it; Tel.: +39-06-9067-5252; Fax: +39-06-9062-5591

Received: 4 July 2019; Accepted: 2 September 2019; Published: 4 September 2019

Abstract: In Mediterranean countries, the use of copper-based fungicides in agriculture is causing a concerning accumulation of copper in the upper layer (0–20 cm) of soils and water bodies. Phytoremediation by energy crops offers the chance to associate the recovering of polluted environments with the production of biomass for bioenergy purposes. The purpose of this work was to evaluate the morpho-physiological response of giant reed (*Arundo donax* L.), a well-known energy crop, when treated with increasing concentrations of Cu (0, 150, and 300 ppm) in a semi-hydroponic growing system (mesocosm) for one month. The plant morphology (height and base diameter of the stem, number of stems) was not affected by the treatments. The presence of Cu led to the disequilibrium of Fe and Zn foliar concentration and caused an impairment of photosynthetic parameters: at 150 and 300 ppm the chlorophyll content and the ETR were significantly lower than the control. The study demonstrated that, although the presence of Cu may initially affect the plant physiology, the *Arundo* plants can tolerate up to 300 ppm of Cu without any adverse effect on biomass production, even when grown in semi-hydroponic conditions.

Keywords: phytoremediation; heavy metals; energy crops; pollution; water contamination; chlorophyll fluorescence

1. Introduction

Soil contamination by heavy metals and organic pollutants is a major threat at both European and national level [1–3]. In Italy, more than one million hectares (corresponding to about 3% of the national territory), distributed in 57 different sites (sites of national interest, SIN) belong to the national list of polluted sites. The SINs include all the main Italian industrial areas and, according to recent estimates, their remediation should require 30 million € [4]. A recent European survey [2] reported that copper (Cu), mercury (Hg), and lead (Pb) are the main metals diffused at critical levels in the first 20 cm of the Italian farming soils.

The accumulation of copper in soils has mainly an anthropogenic origin as a result of mining or agro-industrial activities. Inorganic copper is used as a broad-spectrum fungicide and bactericide in horticulture and viticulture because it combines efficacy and low cost [5]. The use of products containing copper salts (e.g., pesticides applied in vineyards and orchards) have caused high levels of accumulations in the upper layer (0–20 cm) of agricultural soils in Mediterranean countries (France, Italy, Portugal, and Spain). Due to the long persistence in soil and the toxicity to aquatic and terrestrial organisms, a regulatory process included the copper compounds among the candidates for substitution [6].

Copper is an essential micronutrient for plant growth and development. It acts as a catalyst in photosynthesis and respiration and plays an important role in the formation of lignin in the cell wall. Nevertheless, at high concentrations copper can become extremely toxic for plants causing symptoms such as chlorosis and necrosis, stunting, leaf discoloration, and inhibition of root growth [7,8]. In the presence of copper excess, plants undergo oxidative stress due to the overproduction of reactive oxygen species [5,9] resulting in the impairment of the main processes associated with photosynthesis [10,11]. Depending on the plant genotype and copper concentration, many plant species have been recognized as a valuable biological tool for the phytoremoval of copper from contaminated soil and water [11,12].

Studies in controlled or semi-controlled conditions are the fundamental preliminary step to assess the potential of plant for phytoremediation. The plant responses to heavy metal exposure are commonly evaluated under hydroponic growth conditions or pot in growth chambers or greenhouse [13] and mesocosms in an on-field environment [14]. Since the nutritional factors are maintained at optimal levels for plant growth, such systems represent suitable tools to assess the physiological impact of contaminants to define the maximum potential phytoremediation. The use of mesocosms offers some advantages related to the absence of the buffering effect of the soil, the high volume available for the growth of the rhizosphere, the complete availability, and readiness of the contaminant.

The giant reed (*Arundo donax* L.) has been proposed as a promising candidate for phytoremediation due to its favorable characteristics as a biomass crop [15–19]: rapid growth and high production of biomass; simple agronomic management and easy harvesting of biomass; good tolerance and ability to assimilate the metals, preferably in the aboveground biomass [20,21]. The present work aimed to evaluate the physiological response of *Arundo* in a semi-hydroponic growing system in on-field environmental conditions (mesocosm) contaminated with different concentrations of Cu (0, 150, and 300 ppm). In Italian areas where vineyards and orchards are largely diffused, the Cu concentration very often overcomes 200 mg kg^{-1} in the topsoil, exceeding the Italian threshold (120 mg kg^{-1}) for residential/recreational use [1,22]. The concentrations tested in the present study are higher than these limits. Hence, the growth of the *Arundo* in a system where high amounts of Cu are freely available, allowed us to test its tolerance capacity and to understand how to use the species in such contaminated soils.

2. Materials and Methods

2.1. Plant Growth and Contamination

Arundo plants were grown under on-field environmental conditions in mesocosms (PVC) of 1 m^3 (0.785 m^2 × 1.3 m), filled with an upper layer (75 cm high) of perlite and a bottom layer (25 cm high) of gravel (15–30 mm of diameter). The mesocosms were positioned on an external research platform on a concrete basement and filled with 400 L of water (Figure 1). A recovery tank (50 L) collected the drainage and transferred it, with a pump, on the surface of the substrate, through a tube in PVC arranged in a ring and equipped with nebulizers. The solution drained again by gravity to the recovery tank in 150 min. A 10 L tank, equipped with a floating and a water flow meter, adjustable in height, allowed to fix and maintain the groundwater level.

Four homogeneous *Arundo* cuttings (considered as replicates) per mesocosm were transplanted on October 2017 (Figure 2). The leaves of the plants were sprayed monthly from October to December with a foliar fertilizer (P-K 30–20, plus negligible amounts of microelements from 0.1 to 0.05%). The same fertilizer was dissolved also in the tank. In June, the developed plants underwent the treatment with copper sulfate penta-hydrated (Carlo Erba Reagent, CAS 7758-99-8S). We compared three concentrations: 0 (control), 150, and 300 ppm. Two mesocosms were chosen as controls, while the remaining two mesocosms were treated, respectively, with 150 and 300 ppm. The mesocosms were randomly assigned to different treatments. The morphological, nutritional, and physiological characterization of the plants took place at 7, 14, 21, and 28 days after the contamination. In Figure 3 the main meteorological variables of the period are shown.

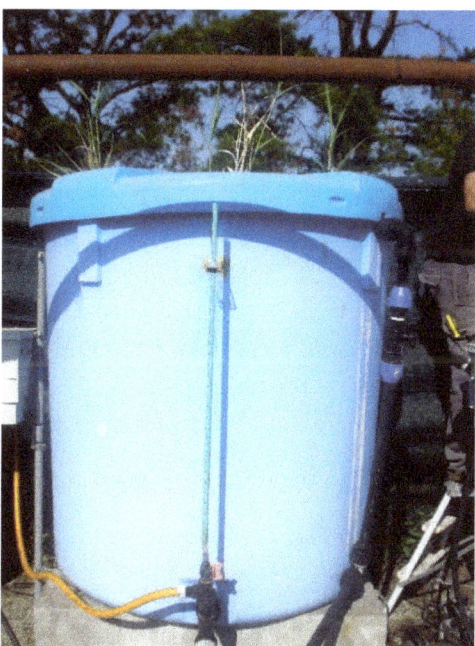

Figure 1. Mesocosm used for growing the *Arundo* plants.

Figure 2. Transplant of the rooted cuttings inside the mesocosm.

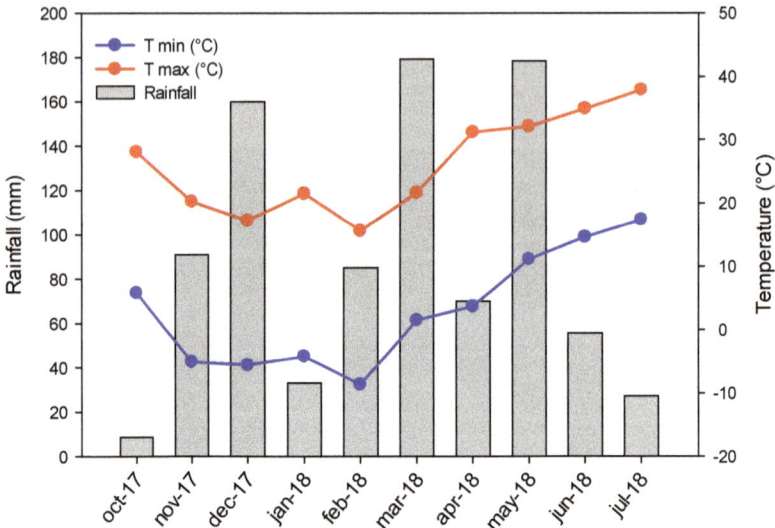

Figure 3. Rainfall, and the minimum and maximum temperature during the grown of *Arundo* plants.

2.2. Morphological Characterization of the Plants

The morphological traits measured at the scheduled time (7, 14, 21, and 28 days after the copper contamination) were the height and the base diameter of the main stem, and the number of the stems of each plant. The four plants of each mesocosm were treated as replicates.

2.3. Nutritional Characterization of the Plants

Leaf samples were collected at each time-interval to examine the content of Cu, Fe, and Zn at the laboratory LAS-ER-B of CREA-IT (Monterotondo). For each treatment, one leaf per plant heavier than one gram was analyzed. The samples were dried at 105 °C for 24 h and homogenized to have a uniform distribution of the elements. At the end, about 0.5 g of each sample was weighed and placed into a microwave Milestone START D with the addition of 6 ± 0.1 mL of HNO_3 65% and 3 ± 0.1 mL of H_2O_2 30%. The digestion was accomplished at 180 °C, 650 W for 42 min. In the end, the samples were filtered and diluted with deionized water. Each sample was analyzed in triplicate and with a blank. The content of microelements (mg kg^{-1} dry weight) was measured with ICP-MS (Agilent 7700).

2.4. Determination of Physiological Parameters

The leaf chlorophyll content was estimated at 7, 14, 21, and 28 days after the copper contamination by a SPAD-502 Chlorophyll meter (Minolta Inc., Osaka, Japan), as reported by [14]. The measurements were taken from at least two fully developed leaves per plant. Four SPAD readings were taken from the widest portion of the leaf lamina while avoiding major veins. The four SPAD readings were averaged to represent the SPAD value of each leaf. SPAD values were converted to chlorophyll content ($\mu g\ cm^{-2}$) using the equation [23]:

$$\text{Chlorophyll content} = (99 \times \text{SPAD value})/(144 - \text{SPAD value}) \qquad (1)$$

The chlorophyll fluorescence parameters were measured on the same leaves used for the chlorophyll content. Chlorophyll a fluorescence transient measurement (OJIP transients) was carried out using the PEA fluorimeter (Plant Efficiency Analyzer, Hansatech Instruments Ltd., King's Lynn, UK). Plant materials were dark-adapted (with leaf clips) for about 1 h before measurements. Chlorophyll fluorescence transient was induced by applying a pulse of saturating red light (peak at 650 nm, 3000 µmol m^{-2} s^{-1}). Changes in fluorescence were measured for 1 s, starting from 50 µs after the onset of illumination. During the first 2 ms changes were recorded every 10 µs and every 1 ms afterward. The obtained data were used in the JIP test [24] to calculate several (the following) bioenergetic parameters of PSII photochemistry (Table 1).

Table 1. Selected JIP-test parameters calculated on the basis of fast fluorescence kinetics.

Fluorescence Parameters	Description
F_0	fluorescence intensity at 50 µs (O step)
F_{300}	fluorescence intensity at 300 µs
F_J	fluorescence intensity at 2 ms (J step)
F_I	fluorescence intensity at 30 ms (I step)
F_m	maximal fluorescence intensity (P step)
$F_v = F_m - F_0$	maximal variable fluorescence
$V_J = (F_J - F_0)/(F_m - F_0)$	variable fluorescence at J step;
$M_0 = 4 (F_{300} - F_0)/(F_m - F_0)$	approximated initial slope of the fluorescence transient, expressing the rate of RCs' closure
$ABS/RC = M_0 \times (1/V_J) \times [1/(F_v/F_m)]$	absorption per active reaction center
$TR_0/RC = M_0 \times (1/V_J)$	trapping per active reaction center
$ET_0/RC = M_0 \times (1/V_J) \times (1 - V_J)$	electron transport per active reaction center
$DI_0/RC = (ABS/RC) - (TR_0/RC)$	dissipation per active reaction center
$TR_0/ABS = F_v/F_m = \varphi P_0 = (F_m - F_0)/F_m$	maximum quantum yield of PSII photochemistry
$ET_0/TR_0 = \psi_0 = (F_m - F_J)/(F_m - F_0)$	probability that a trapped exciton moves an electron into the electron transport chain beyond Q_A
$ET_0/ABS = \varphi E_0 = \varphi P_0 \times \psi E_0$	quantum yield of electron transport
$F_v/F_0 = TR_0/DI_0 = (F_m - F_0)/F_0$	maximum ratio of quantum yields of photochemical and concurrent non-photochemical processes in PSII
$F_0/F_m = DI_0/ABS = \varphi D_0$	maximum quantum yield for energy dissipation at the antenna level
$PI_{ABS} = [\varphi P_0 (V_J/M_0)] \times [\varphi P_0/(1 - \varphi P_0)] \times [\psi E_0/(1 - \psi E_0)]$	performance index (potential) for energy conservation from photons absorbed by PSII to the reduction of intersystem electron acceptors

Moreover, always on the same leaves, the electron transport rate (ETR) was measured using the MINI-PAM fluorimeter (Walz, Effeltrich, Germany) equipped with a leaf clip holder (Model 2030-B, Walz). The electron transport rate (ETR) was determined by adapting the leaves for at least 10 min to a photosynthetic photon flux density (PPFD) of 1000 µmol m^{-2} s^{-1}. The value of ETR was calculated as follows:

$$ETR = \Phi PSII \times PPFD \times f \times Abs, \qquad (2)$$

where $\Phi PSII$ is the quantum efficiency of PSII photochemistry in light-adapted leaves [25], f is a factor that accounts for the partitioning of energy between PSII and PSI and is assumed to be 0.5, indicating that excitation energy is distributed equally between the two photosystems [26] and Abs is the fraction of PPFD absorbed by the leaf. The Abs value depended on the chlorophyll content (µg cm^{-2}) and it was calculated by applying the modified equation of [27] as follows:

$$Abs = Chl/(Chl + 6.66), \qquad (3)$$

where 6.66 is an empirical constant with the dimension of µg cm^{-2}. The ETR value represents the overall photosynthetic capacity in vivo and is used as a proxy for photosynthesis in field investigations.

2.5. Statistical Analysis

The PAST software (version 3.22, 2018, Øyvind Hammer, University of Oslo, Norway) was used for the analysis of morphological characters. The data were verified for normality with the Shapiro–Wilcoxon test and in case of deviation, they were analyzed with the non-parametric Kruskal–Wallis test. With normally distributed and homoscedastic data we proceeded to Analysis of variance ANOVA (one or two way). The separation of the means was performed using Tukey's HSD test unless otherwise stated. A principal component analysis (PCA) using the same software was run on both the morphological data and the microelements content.

3. Results and Discussion

3.1. Morphological Characterization

Copper plays a role in plant physiology and at low levels, (3–20 mg kg^{-1} DW), is required for plant development [28,29]. It is a constituent of proteins and enzymes. These act in cell compartments like endoplasmic reticulum, mitochondria, and chloroplasts [30]. The copper supplied to the substrate favored the plant growth soon after treatment. In the present study, the height of the main stem was always higher in the treated plants than in the control. But, at each time interval, the difference was never significant (Table 2). The number of stems in plants grown on 300 ppm Cu was two-fold higher than in control plants one week after treatment (Table 2). This was the only case where the difference among treatments was significant. After 28 days, the gap between 300 ppm and the other two treatments diminished.

The basal diameter of the main stem showed a similar trend (Table 2). The highest growth was observed at 300 ppm during the first two weeks. Then, the diameter remained around 12 mm while in the plants of the control and 150 ppm increased up to 11 mm. However, it should be stressed that the percentage variation between the value of the variables at day 0 (starting of the contamination) and at the end of the test was always higher in untreated than treated plants (Table 2). Therefore, this aspect suggests that an inhibitory effect of copper on the growth rate of the plants occurred and that such effect could be more evident in the long term. It is also possible that a part of Cu applied was no more bioavailable due to precipitation of CuS or other compounds, but these have not been considered.

Table 2. Time course of the plant height (main stem length), the number of stems and the basal diameter of the main stem in *Arundo* plants treated with different copper concentrations (0, 150, and 300 ppm) for 28 days. The values represent means (n = 4) ± S.E., respectively. One-way ANOVA was applied, and different letters indicate significant difference according to Tukey test ($p \leq 0.05$).

Variable	Cu Level (ppm)	Day 0 (Contamination Start)	Days after Contamination				day$_{28}$–day$_0$ (%)
			7	14	21	28	
Stem height (cm)	0	44.8 ± 9.5	56.2 ± 7.2	67.1 ± 8.3	78.4 ± 12.4	110.7 ± 12.9	147.0
	150	67.5 ± 3.5	75.0 ± 2.7	95.0 ± 7.3	99.2 ± 9.0	123.2 ± 9.6	82.5
	300	71.5 ± 14.1	82.0 ± 11.5	100.7 ± 19.0	122.0 ± 18.4	143.5 ± 15.9	100.7
Basal diameter (mm)	0	5.9 ± 2.4	8.9 ± 1.5	8.7 ± 0.6	11.4 ± 1.3	11.5 ± 1.6	94.9
	150	9.5 ± 1.0	10.7 ± 1.1	11.0 ± 1.3	10.0 ± 0.7	11.0 ± 1.7	15.8
	300	7.7 ± 1.9	10.7 ± 0.9	13.7 ± 1.7	13.0 ± 1.1	12.2 ± 1.0	58.4
Number of stems	0	2.9 ± 1.1	3.1 ± 0.6 b	4.3 ± 0.9	4.6 ± 1.3	5.5 ± 1.4	89.7
	150	3.5 ± 0.6	4.2 ± 0.2 b	4.0 ± 0.4	4.5 ± 0.3	4.5 ± 0.3	28.6
	300	4.2 ± 2.4	6.0 ± 1.1 a	5.5 ± 0.6	4.8 ± 0.7	5.3 ± 0.9	26.2

Elhawat et al. [31] observed a behavior resembling what observed in the present study. *Arundo* plantlets of two ecotypes grown in vitro and exposed to increasing Cu levels (from 0 to 26.8 mg L^{-1}) did not show evident symptoms of Cu toxicity. At the highest concentration, the fresh mass of the plant stems increased, but the root length and the number of new buds per plant were higher in one genotype. This implied the activation of tolerance mechanisms slightly different among genotypes [19,31]. In a study of rhizofiltration [29], the *Arundo* plants revealed a high efficiency in removing the Cu from Cu-rich Bordeaux mixture effluents in pilot-scale constructed wetlands (CW). After one month of exposure, the shoot and root dry weight increased on average by 47% and 23%. The authors hypothesized that the Cu excess prompted some detoxification mechanisms, but the metabolic cost lowered the efficiency of other processes like photosynthesis [29]. Our data are in accord with such a hypothesis.

Regardless of the concentrations, a direct relation linked the stem height and the corresponding basal diameter (Figure 4). The result was in agreement with other observations [32] in the Italian environment.

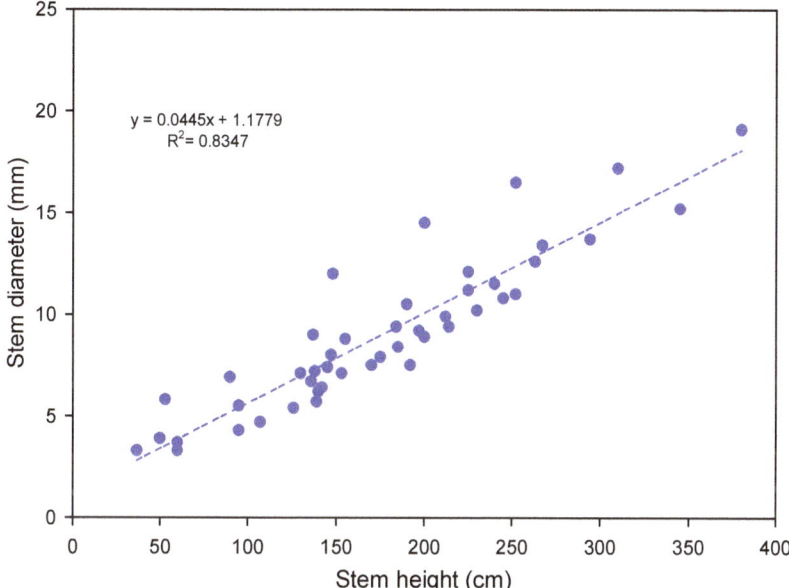

Figure 4. Correlation between height and basal diameter of the main stem measured on semi-hydroponic grown *Arundo donax* plants.

Copper sulfate is a compound used to restrain several pathogens of vegetables, vineyard, and orchard trees. In this study, the goal was to reproduce a condition of a farming system or an ecosystem close to farming activities. In soil, the sulfate is not a limiting factor, as the plants absorb it from the circulating solution or fertilizers. A direct involvement of the ion cannot be excluded but does not appear so influential. On the other side, unless of lithological origin, the content of Cu must derive just from farming activities. Within the range of the concentrations used in this work, the Cu may have played a role in plant growth. Yet, a more in-depth analysis of the effect on the photosynthetic process has shown a negative influence in the short period.

3.2. Nutritional Characterization

As described before, the contaminating solution in the tanks was pushed by a pump on the top layer of perlite and entered again in the system. Within one week, the solution of the tanks re-circulated (several times) in the complex root system-perlite. This explains the low concentration of Cu within the treated tanks (Figure 5). The values reflected quite well the levels of the treatments, and hence the correct application of the element to the substrates. The concentration of copper was significantly higher in the tanks of the treated mesocosms than in the tank of control (Figure 5). Such a difference was present both at the start and the end of the test. One week after the contamination the concentration of copper inside the tanks of 300 ppm treatment was the highest. After 28 days, the difference between the treated and the untreated mesocosms remained. But the copper concentration was comparable in the tanks of 150 and 300 ppm. Thus, it is plausible to assume a greater absorption by the plants exposed to 300 ppm Cu than those treated with 150 ppm Cu.

Net of copper found in the leaves, most of the element was probably accumulated at the root level and hypothetically onto perlite. Many authors observed that the accumulation of metals in *Arundo* occurs mainly in the roots and rhizomes [15,16,29]. Here, mechanisms to protect the plant from heavy metal toxicity are present [33]. Other elements, like Zn, can move from the belowground to aboveground organs [15,16]. Expanded and unexpanded perlite have some properties favoring the adsorption of metal ions [34–36]. The adsorption mechanism is a function of factors as pH, adsorbent dosage, temperature, and contact time [36,37]. For Cu ions, the adsorption percentage on perlite or expanded perlite may reach 80%–90%. This level required an adsorbent dosage greater than 15–20 g L^{-1}, a pH higher than 4, and a temperature between 20 and 30 °C [35–37]. Some works on constructed wetlands (CW) supported the involvement of the substrate in the element removal. The unplanted CW showed higher efficiency in removing the heavy metals compared to the CW units planted with macrophyte [29,38–40].

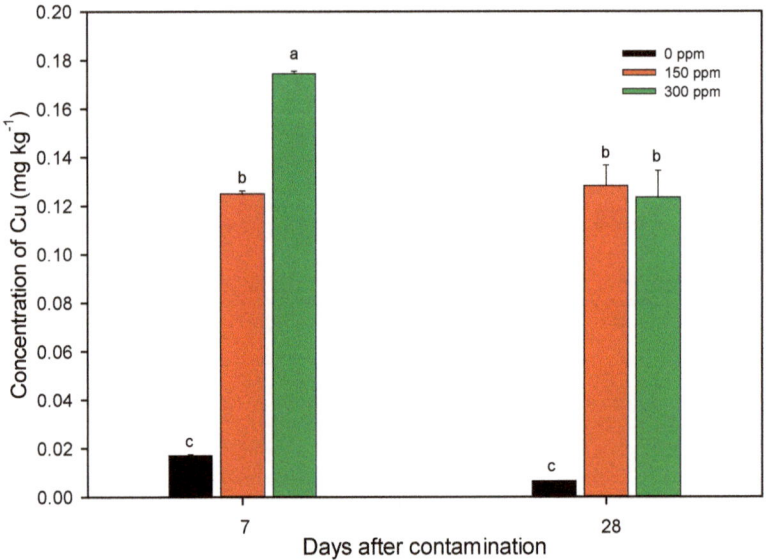

Figure 5. Concentration (average ± S.E.) of copper (mg kg^{-1}) in the collection tank. Different letters indicate a statistically significant difference for $p < 0.01$ after Tukey test.

The copper concentrations in the tank were greatly lower than those at the leaf level (Figure 6). This suggests that the plants adsorbed the element from the complex root system-perlite and transferred it to the leaf (Figure 6). After 7, 21, and 28 days from the treatment, the amount of copper in the treated plants was significantly higher than the control. Yet, the concentration of the element was not dependent on the Cu dosages. After one and four weeks from the contamination, the Cu level at 150 ppm Cu was higher than at 300 ppm Cu. During the study, the concentration of Cu in the leaves of the control plants remained within the normal range (3–20 mg kg^{-1} DW) [28,29]. On the other side, in the treated plants the Cu concentration overcome the upper threshold after one and three weeks and dropped close to the control value at 28 days. Plants exposed to Cu activate a mechanism of detoxification to counteract the adverse effect of the oxidative stress [28,41]. The reduction of the oxidative damage allows recovering a steady growth as observed at the end of the study.

High Cu concentrations cause a competition at the rhizosphere level between Cu and other elements, in particular, Fe and Zn [42]. Accordingly, a physiological imbalance induced by copper was observed (Figure 7). One week after the treatment, the Fe content in the leaves at 300 ppm Cu was the highest. In the following period, the Fe values in the control plants were always higher than the treated plants. Similar behavior occurred for zinc (Figure 8). One week after the contamination, the content of Zn in the leaves was significantly higher in the treated plants. At 150 ppm of Cu, the Zn content in the leaves was twofold higher than the treatment with 300 ppm and about 3–4 times greater than the control. In the following three weeks, the content of Zn in the leaves was always significantly higher in the control plants.

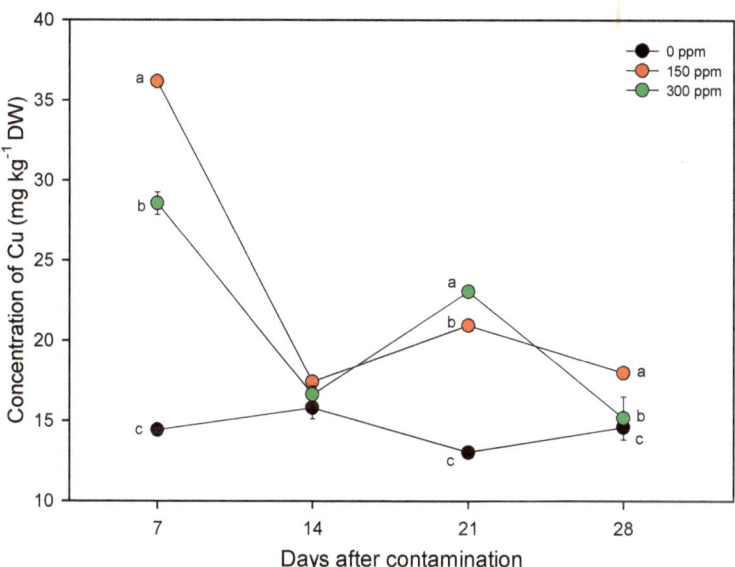

Figure 6. Time course of the copper concentration in the leaves of *Arundo* plants treated with different copper concentrations (0, 150, and 300 ppm) for 28 days. Data points and vertical bars represent means (n = 3) ± S.E., respectively (when not reported S.E. is smaller than symbol size). One-way ANOVA was applied and for a given duration different letters indicate significant difference according to Tukey test ($p \leq 0.01$).

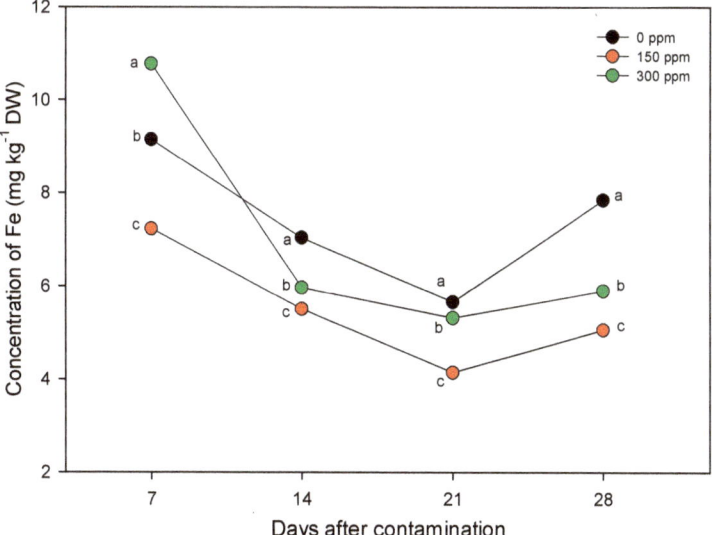

Figure 7. Time course of the iron concentration in the leaves of *Arundo* plants treated with different copper concentrations (0, 150, and 300 ppm) for 28 days. Data points and vertical bars represent means (n = 3) ± S.E., respectively (when not reported S.E. is smaller than symbol size). One-way ANOVA was applied and for a given duration different letters indicate significant difference according to Tukey test ($p \leq 0.01$).

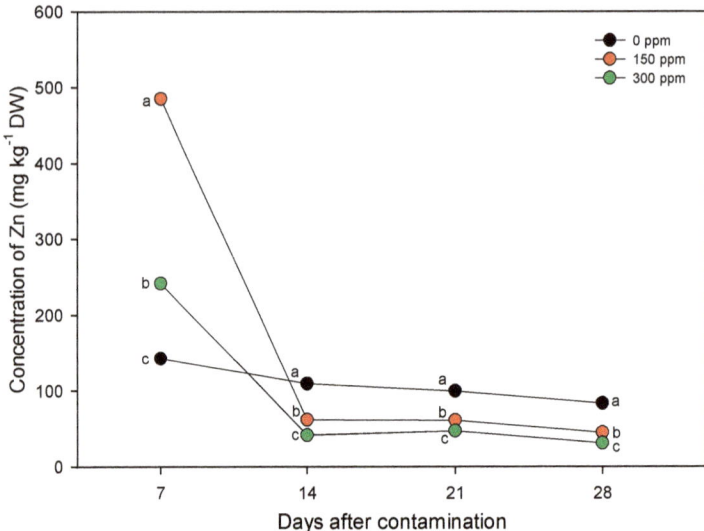

Figure 8. Time course of the zinc concentration in the leaves of *Arundo* plants treated with different copper concentrations (0, 150, and 300 ppm) for 28 days. Data points and vertical bars represent means (n = 3) ± S.E., respectively (when not reported S.E. is smaller than symbol size). One-way ANOVA was applied and for a given duration different letters indicate significant difference according to Tukey test ($p \leq 0.01$).

Excessive Cu uptake modifies the mineral homeostasis. The effects may vary in response to factors as plant species, exposure time, and growth conditions [28]. According to Lequeux et al. [43], it can be difficult to individuate a clear trend for some elements like Mg, S, Fe, and Zn. As observed by Ambrosini et al. and Azez et al. [10,44], the presence of an increasing amount of Cu in the soil may reduce the Fe and Zn availability. An excessive Cu uptake caused a decrease of Fe and Zn content in plant tissues of *Arundo* [29], rapeseed and Indian mustard [45] in a short–medium period. In poplar [46] the supply of Cu decreased the Zn content and increased the Fe concentration. In this context, one of the most common responses in plants exposed to Cu excess is the hindrance of Fe and Zn uptake. Iron is known as an antagonist of Cu during the uptake [45]. Instead, Zn has a similar ion strength of Cu and it competes for the metal transporter molecules [44,45]. The outcome of the present work confirmed such behavior from 14th to 28th day, while in the first week Fe and Zn accumulated in the leaves. A contradictory behavior was observed also by Lequeux et al. [43] which reported an increase in Fe and Zn concentrations in roots of Cu^{2+}-treated plants of *Arabidopsis* grown hydroponically. In this case, the author hypothesized the effect of Fe–EDTA in the nutrient solution. The displacement of Fe by Cu ions from Fe–EDTA complexes could allow a higher availability of Fe ions [43]. Thus, in our case, the substrate (which, as discussed previously, could not be completely inert) might have played a role in the first phases after the contamination. An alteration of the adsorption of metal ions may have led to the increase of Fe and Zn similarly to what observed by Lequeux et al. [43].

To better highlight the response at growth and nutritional level, a multivariate analysis was conducted (Figure 9). The PCA showed a clear differentiation between the Cu-treated and the control plants. The areas of the Cu treatments partially overlapped in the upper quadrants. Instead, the area of the control plants was in the opposite quadrants. The first component accounted for a large part of the morphological traits. These, in turn, were influenced by the highest Cu concentration (300 ppm). The second component was associated with microelement content. The Cu contamination caused differentiation of Cu, Fe, and Zn content in the treated plants, with a specific effect on the plants grown at 150 ppm Cu.

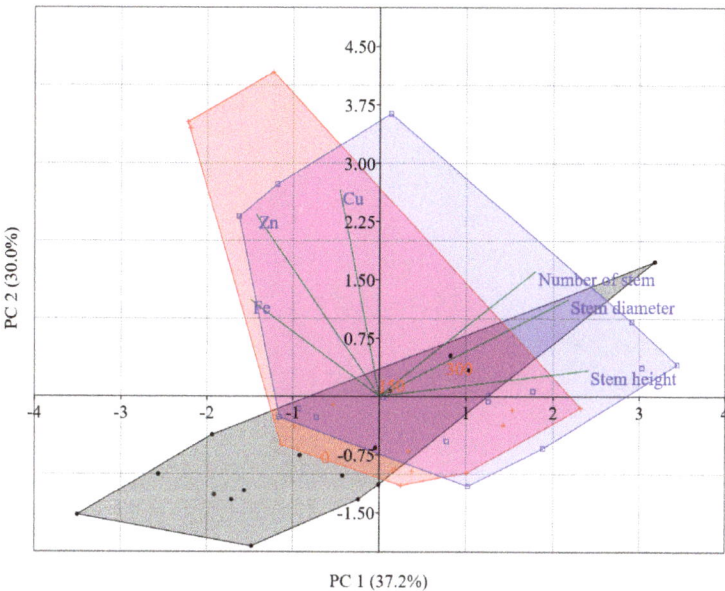

Figure 9. Scores and loadings of principal component analysis (PCA) carried out on the set of plant morphological data and microelements present in the leaves (Cu, Fe, and Zn) collected over a period of 28 days from Cu contamination.

3.3. Determination of Physiological Parameters

The leaf chlorophyll content is one of the most important factors determining the photosynthetic potential and primary production [47]. It can be also used as an indicator of phytotoxicity, allowing to analyze the effect of pollutants on photosynthetic and respiratory rates [48,49]. Figure 10 shows the chlorophyll content along the experimental time-course. Increasing the concentration of copper caused a reduction of the chlorophyll content. The greatest decrease was registered after the first week (around 57%–65%), while in the following weeks the treated plants recovered at least in part. At 28 days, the difference with the control was around 30% (150 ppm) and 40% (300 ppm).

The negative effect on chlorophyll content of Cu excess has been reported for different species, growth system, and Cu concentrations [10,29,41,50,51]. As observed by Oustriere et al. [29], the metabolic cost for detoxifying and limiting the adverse effects of Cu can reduce the resources for other physiological processes. Copper interferes with chlorophyll organization and functionality. Structural damages of the photosynthetic apparatus involved the thylakoid component [28]. Our data confirmed the influence that an unbalanced Cu uptake has on chlorophyll content. Plants exposed to copper show leaf chlorosis and, with increased exposure, necrosis can appear in the leaf tips and margins [31,46,51]. Even so, no necrotic spots appeared during the experiment on leaves of Cu-treated plants. Thus, from one side the growth was not affected or slightly affected by the presence of Cu at 150 and 300 ppm (Table 2). On the other side, the Cu treatments reduced the efficiency of the photosynthetic process (Figure 10 and following). The existence of a threshold of toxic concentration (variable for species and growth system) appears plausible. Below a certain value of Cu, the synthesis of low molecular weight stress proteins reinforced the action of the antioxidant enzymes [51]. The homeostatic control of copper excess limited the damages and maintained a normal growth rate.

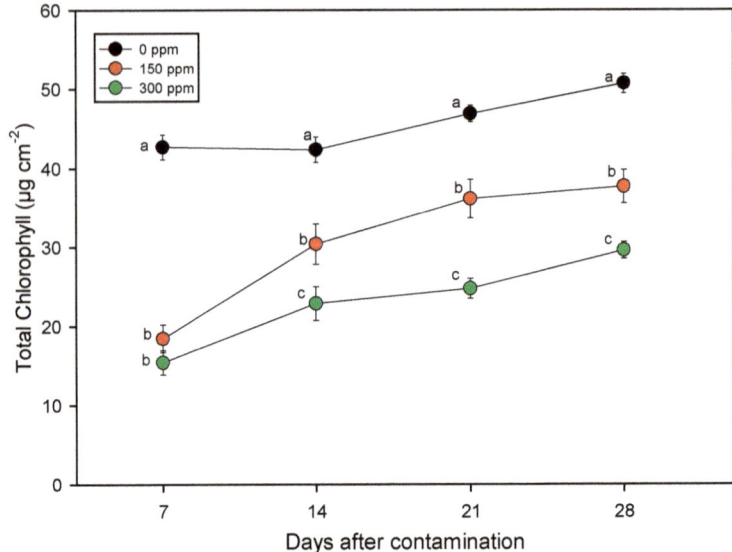

Figure 10. Time course of the chlorophyll content in *Arundo* plants treated with different copper concentrations (0, 150, and 300 ppm) for 28 days. Data points and vertical bars represent means (n = 8) ± S.E., respectively. One-way ANOVA was applied and for a given duration different letters indicate significant difference according to LSD test ($p \leq 0.05$).

In Figure 11, the trend values of the electron transport rate (ETR) along the experimental time-course are reported. ETR is an important parameter that refers to the apparent photosynthetic electron transport rate. It reflects the efficiency of electron capture by the PSII reaction center giving a clue of overall photosynthesis [26]. Deficiency or excess of copper alters the photosynthetic ETR [30,52]. Our data showed that, as in the chlorophyll content, ETR was transiently modified by copper treatments (Figure 11). Even in this case, the highest ETR decrease occurred after the first week (around 18%–21%). In the following weeks, the plants treated with Cu improved their ETR, showing at the 28th-day values around 5% (150 ppm) and 11% (300 ppm) lower than the control.

Finally, we analyze the bioenergetic parameters obtained from the JIP-test. This provides information about the effect of the treatment on the processes involved in the light absorption and its conversion to biochemical energy. The measurements of structural and functional parameters were normalized against the values of the control plants and reported in a radar plot (Figure 12). Chlorophyll a fluorescence-transient analysis is an efficient tool for studying physiological aspects of structure and activity, mainly in the PSII [24]. It has been widely used to assess the damages to the photosynthetic system by various types of stress [26].

The exposure of plants to copper concentrations of 150 and 300 ppm caused an alteration of most of the parameters analyzed except for F_0 (Figure 12A–D). The absence of F_0 variation indicates a good ability of the treated plants to maintain the efficiency of energy transfer between the pigments of the antenna and the PSII reaction center without structural damages at the photosystems level. In fact, an increase in F_0 can be interpreted as indicating irreversible damage to PSII caused by uncontrolled dissipation of heat that produces an excess of excitation energy [53,54]. On the contrary, a decrease of F_0 is a symptom of a high-energy dissipation in the minor antenna [55]. The values of the chlorophyll fluorescence followed those of the chlorophyll content and the ETR. After the first week, the treated plants showed significant differences with the control plants. Thereafter, there was an improvement, but in some cases, the difference remained significant until the end.

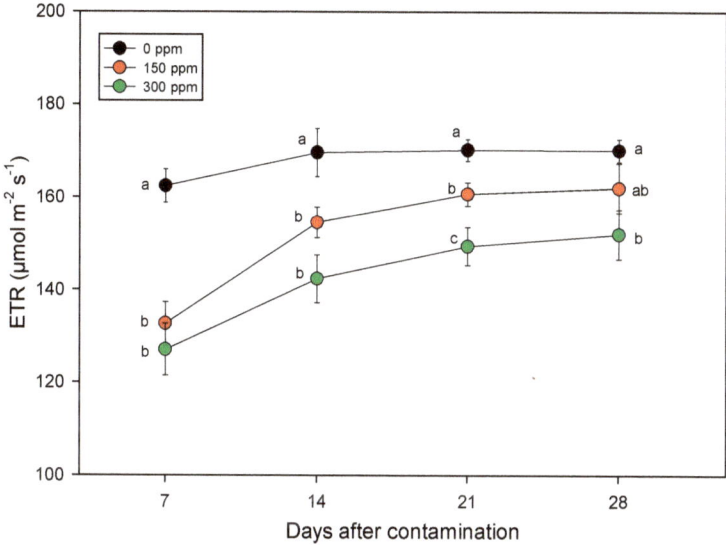

Figure 11. Time course of the electron transport rate (ETR) in *Arundo* plants treated with different copper concentrations (0, 150, and 300 ppm) for 28 days. Data of electron transport rate (ETR) measured at steady-state with a photosynthetic photon flux density (PPFD) of 1000 μmol m^{-2} s^{-1} is shown after 7, 14, 21, and 28 days after contamination. Data points and vertical bars represent means (n = 8) ± S.E., respectively. One-way ANOVA was applied and for a given duration different letters indicate significant difference according to LSD test ($p \leq 0.05$).

After a week, the treatment at 150 and 300 ppm determined an increase of the specific energy fluxes, absorbed (ABS/RC), captured (TR$_0$/RC), and dissipated (DI$_0$/RC) from the active reaction centers of PSII, and a reduction in the electron transport (ET$_0$/RC) (Figure 12A). The increase in ABS/RC could be attributed to the inactivation of reaction centers and a decrease in active Q$_A$ reducing centers [56], while the enhancement in TR$_0$/RC resulted in higher inhibition of reoxidation of Q$_{A-}$ to Q$_A$ [57]. Consequently, the increased value of TR$_0$/RC would result in lower electron transport per reaction center (ET$_0$/RC). Moreover, a reduction of the maximum quantum yield of primary photochemical reactions (TR$_0$/ABS) and the maximum quantum yield for electron transport (ET$_0$/ABS) observed in treated plants was associated with an increase of the maximum quantum yield for energy dissipation at antenna level (F$_0$/F$_m$) (Figure 12A). Similarly, the corresponding reduction of ET$_0$/ABS and ET$_0$/TR$_0$ in treated plants could probably be due to an inhibition of electronic transport beyond Q$_A$. Therefore, the decrease in F$_V$/F$_0$ found after 7 days at 150 and 300 ppm of Cu, which indicates the efficiency of water splitting (and consequently oxygen production) by the PSII, agrees with the data showing a reduced photosynthetic activity. Finally, the sharp decrease in the PI$_{ABS}$ viability index, confirms the inhibitory effect on photochemical processes in Cu-treated plants [58] (Figure 12A). Our results are in line with the literature reporting the effects of copper on different plant species [59–61]. Nevertheless, it should be emphasized that the copper concentrations used in this study are notably higher than those usually utilized in similar experiments.

During the following weeks (Figure 12B–D) the differences between Cu treated and control plants were partially reduced, highlighting the ability to recover the efficiency of photosynthetic energy conversion, especially in *Arundo* plants exposed to lower copper concentration (150 ppm). In general, the exposure of *Arundo* plants at two concentrations of copper (150 and 300 ppm) led to a reduction of the chlorophyll content as well as the parameters related to the photosynthetic activity, mainly after the first week of treatment. However, starting from the following week (14 days) and up to the end of the experiment (28 days), these parameters showed an increase in their values resulting in a partial recovery of the functionality of the photosynthetic apparatus and chlorophyll content, especially in the plants treated with 150 ppm of copper.

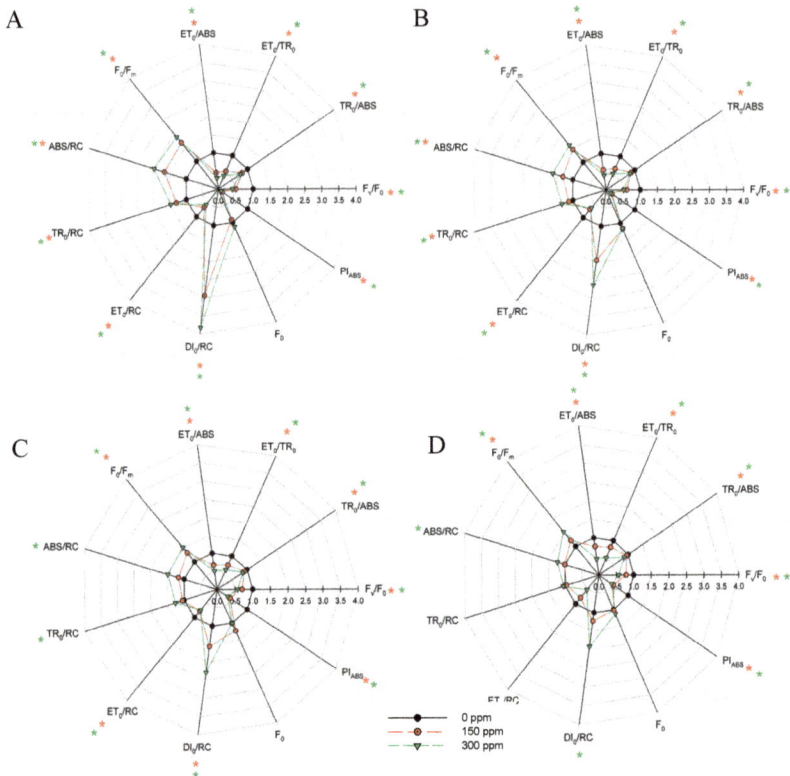

Figure 12. Changes in the shape of the radar plot JIP-test parameters induced by different copper concentrations applied to *Arundo* plants and measured after 7 (**A**), 14 (**B**), 21, (**C**) and 28 (**D**) days of treatment. The data are the average of eight replicates and report the values with respect to plants grown in the absence of contamination (control = 1). Red and green asterisks indicate significant differences (LSD test, $p \leq 0.05$) between control plants (0 ppm) and those exposed to 150 or 300 ppm copper, respectively.

Based on the results, it can be hypothesized that in plants of *Arundo* the high concentrations of copper induced alterations in photochemical processes at the chloroplast level [62]. The initial reduction of the chlorophyll content, observed in the treated plants, could be linked to reduced absorption of iron with which the copper interferes [50]. At the same time, when the quantity of light energy absorbed by the pigments exceeds that used for photosynthesis, the absorbed energy accelerates the photoinhibition process (i.e., the inhibition of photosynthesis caused by excess light) [63]. To cope with excess light energy, plants have developed a protection mechanism that dissipates the energy absorbed in the form of heat, counteracting its negative effects [64]. Therefore, reversible photoinhibition is indicative of a protective mechanism aimed at dissipating excess light energy, while irreversible photoinhibition indicates damage to the photosynthetic systems [65]. Our data highlighted that the parameters directly related to the energy dissipation rate from the PSII (DI_0/RC and F_0/F_m), showed an increase of the values, supporting the hypothesis that the treatment with copper induced photoinhibition of photosynthesis, more pronounced during the first week of treatment. However, during the following weeks, the reduction of the differences in the parameters analyzed between Cu treated and control plants showed that the defense mechanisms used to dissipate the excess energy allowed the plants to recover, at least partially, their photosynthetic performances.

4. Conclusions

The work provided some evidence about the ability of *Arundo* plants to grow in the presence of increasing concentrations of assimilable Cu supplied in semi-hydroponic conditions (mesocosm).

The physiological indexes associated with the photosynthetic machinery resulted altered within the first week from the contamination. The supply of Cu at 150 and 300 ppm caused a sensible decrease (around 57%–65%) of chlorophyll content and ETR (18%–21%). The assimilation of Cu altered the uptake of Fe and Zn which increased their content, at least within the first week. One of the key outcomes was the absence of phenotypic alteration. The plants did not show evident symptoms of stress, and the values of height and basal diameter of the stem or the number of stems were comparable among the control and the plants treated with 150 and 300 ppm. Thus, a sort of counteracting mechanism seems to act at the studied conditions. Altered absorption of Cu affects photosynthesis in a short time, but, below a Cu threshold, the antioxidative defense system may limit the damaging effects and avoid the irreversible inactivation of the photosynthetic system. In this way, even in the presence of lower photosynthetic efficiency, the biomass production and the plant growth were poorly affected by the contamination.

The role played by the growth system used in this trial cannot be overlooked. The advantage of mesocosms should rely on removing the buffering effect of the soil and in reproducing an environment resembling channels, rivers, lakes, ponds, and marshes where the plant can find optimal growth conditions. However, the role of the inert substrate should be carefully evaluated, because possible interaction with the elements dissolved in the solution may alter their dynamics.

Based on such analysis, the data confirmed previous indications about the suitability of using the *Arundo* species for phytoremediation. From a practical point of view, it must be considered that the Italian law (DL 152/06) sets for Cu the limit of contamination for the soils of residential areas at the concentration of 120 mg kg^{-1}. Such a threshold appears compatible with the growth of *Arundo* plants both in soil or in aquatic environments. However, the behavior of the plant in the long term should be verified in further studies.

Author Contributions: Conceptualization, E.S. and C.B.; methodology, E.S., C.B., M.Z. and F.P.; formal analysis, E.S. and F.P.; investigation, E.S., C.B., F.P., M.C. and F.G.; data curation, E.S., C.B., M.Z. and F.P.; writing—original draft preparation, E.S. and F.P.; writing—review and editing, E.S., C.B., M.Z. and F.P.; supervision, E.S.

Funding: This research was funded by the Ministry of Agriculture, Food, Forestry and Tourism (MiPAAFT) as under the AGROENER project (D.D. N 26329 of 01/04/2016).

Conflicts of Interest: The authors declare no conflict of interest.

References

1. Cicchella, D.; Giaccio, L.; Dinelli, E.; Albanese, S.; Lima, A.; Zuzolo, D.; Valera, P.; De Vivo, B. GEMAS: Spatial distribution of chemical elements in agricultural and grazing land soil of Italy. *J. Geochem. Explor.* **2015**, *154*, 129–142. [CrossRef]
2. Toth, G.; Hermann, T.; Da Silva, M.; Montanarella, L. Heavy metals in agricultural soils of the European Union with implications for food safety. *Environ. Int.* **2016**, *88*, 299–309. [CrossRef] [PubMed]
3. Toth, G.; Hermann, T.; Szatmari, G.; Pasztor, L. Maps of heavy metals in the soils of the European Union and proposed priority areas for detailed assessment. *Sci. Total Environ.* **2016**, *565*, 1054–1062. [CrossRef] [PubMed]
4. Marchiol, L.; Fellet, G. Agronomy towards the Green Economy. Optimization of metal phytoextraction. *Ital. J. Agron.* **2011**, *6*, 189–197. [CrossRef]

5. Miotto, A.; Ceretta, C.A.; Brunetto, G.; Nicoloso, F.T.; Girotto, E.; Farias, J.G.; Tiecher, T.L.; De Conti, L.; Trentin, G. Copper uptake, accumulation and physiological changes in adult grapevines in response to excess copper in soil. *Plant Soil* **2014**, *374*, 593–610. [CrossRef]
6. Armentano, G. Rame, si va verso una riduzione delle dosi annue. *Inf. Agrar.* **2017**, *25*, 8.
7. Borghi, M.; Tognetti, R.; Monteforti, G.; Sebastiani, L. Responses of two poplar species (*Populus alba* and *Populus x canadensis*) to high copper concentrations. *Environ. Exp. Bot.* **2008**, *62*, 290–299. [CrossRef]
8. Cuypers, A.; Vangronsveld, J.; Clijsters, H. Biphasic effect of copper on the ascorbate-glutathione pathway in primary leaves of *Phaseolus vulgaris* seedlings during the early stages of metal assimilation. *Physiol. Plant.* **2000**, *110*, 512–517. [CrossRef]
9. La Torre, A.; Iovino, V.; Caradonia, F. Copper in plant protection: Current situation and prospects. *Phytopathol. Mediterr.* **2018**, *57*, 201–236.
10. Ambrosini, V.G.; Rosa, D.J.; Bastos de Melo, G.W.; Zalamena, J.; Cella, C.; Simão, D.G.; Souza da Silva, L.; Pessoa dos Santos, H.; Toselli, M.; Tiecher, T.L.; et al. High copper content in vineyard soils promotes modifications in photosynthetic parameters and morphological changes in the root system of 'Red Niagara' plantlets. *Plant Physiol. Biochem.* **2018**, *128*, 89–98. [CrossRef]
11. Pietrini, F.; Di Baccio, D.; Iori, V.; Veliksar, S.; Lemanova, N.; Juškaitė, L.; Maruška, A.; Zacchini, M. Investigation on metal tolerance and phytoremoval activity in the poplar hybrid clone "Monviso" under Cu-spiked water: Potential use for wastewater treatment. *Sci. Total Environ.* **2017**, *592*, 412–418. [CrossRef] [PubMed]
12. Poschenrieder, C.; Bech, J.; Llugany, M.; Pace, A.; Fenés, E.; Barceló, J. Copper in plant species in a copper gradient in Catalonia (North East Spain) and their potential for phytoremediation. *Plant Soil* **2001**, *230*, 247–256. [CrossRef]
13. Zacchini, M.; Pietrini, F.; Mugnozza, G.S.; Iori, V.; Pietrosanti, L.; Massacci, A. Metal Tolerance, Accumulation and Translocation in Poplar and Willow Clones Treated with Cadmium in Hydroponics. *Water Air Soil Pollut.* **2009**, *197*, 23–34. [CrossRef]
14. Pietrini, F.; Iori, V.; Pietropaoli, S.; Mughini, G.; Beni, C.; Massacci, A.; Zacchini, M. Phytoremediation of cadmium polluted waters by a eucalypt hybrid clone: A mesocosm study. In Proceedings of the Sixth European Bioremediation Conference, Chania, Creta, Greece, 29 June–2 July 2015; pp. 238–241.
15. Barbosa, B.; Boléo, S.; Sidella, S.; Costa, J.; Duarte, M.P.; Mendes, B.; Cosentino, S.L.; Fernando, A.L. Phytoremediation of heavy metal-contaminated soils using the perennial energy crops *Miscanthus* spp. and *Arundo donax* L. *Bioenergy Res.* **2015**, *8*, 1500–1511. [CrossRef]
16. Fiorentino, N.; Ventorino, V.; Rocco, C.; Cenvinzo, V.; Agrelli, D.; Gioia, L.; Di Mola, I.; Adamo, P.; Pepe, O.; Fagnano, M. Giant reed growth and effects on soil biological fertility in assisted phytoremediation of an industrial polluted soil. *Sci. Total Environ.* **2017**, *575*, 1375–1383. [CrossRef] [PubMed]
17. Nsanganwimana, F.; Marchand, L.; Douay, F.; Mench, M. *Arundo donax* L., a Candidate for Phytomanaging Water and Soils Contaminated by Trace Elements and Producing Plant-Based Feedstock. A Review. *Int. J. Phytoremediat.* **2014**, *16*, 982–1017. [CrossRef] [PubMed]
18. Oustriere, N.; Marchand, L.; Lottier, N.; Motelica, M.; Mench, M. Long-term Cu stabilization and biomass yields of Giant reed and poplar after adding a biochar, alone or with iron grit, into a contaminated soil from a wood preservation site. *Sci. Total Environ.* **2017**, *579*, 620–627. [CrossRef]
19. Elhawat, N.; Alshaal, T.; Domokos-Szabolcsy, E.; El-Ramady, H.; Antal, G.; Márton, L.; Czakó, M.; Balogh, P.; Fári, M. Copper uptake efficiency and its distribution within bioenergy grass giant reed. *Bull. Environ. Contam. Toxicol.* **2015**, *95*, 452–458. [CrossRef]
20. Marchiol, L. Prospettive e limitazioni del fitorisanamento. *Not. ERSA* **2008**, *4*, 65–72.
21. Ceotto, E.; Di Candilo, M. Shoot cuttings propagation of giant reed (*Arundo donax* L.) in water and moist soil: The path forward? *Biomass Bioenergy* **2010**, *34*, 1614–1623. [CrossRef]
22. Ruyters, S.; Salaets, P.; Oorts, K.; Smolders, E. Copper toxicity in soils under established vineyards in Europe: A survey. *Sci. Total Environ.* **2013**, *443*, 470–477. [CrossRef] [PubMed]
23. Cerovic, Z.G.; Masdoumier, G.; Ghozlen, N.B.; Latouche, G. A new optical leaf-clip meter for simultaneous non-destructive assessment of leaf chlorophyll and epidermal flavonoids. *Physiol. Plant.* **2012**, *146*, 251–260. [CrossRef] [PubMed]

24. Strasser, R.J.; Tsimilli-Michael, M.; Srivastava, A. Analysis of the Chlorophyll a Fluorescence Transient. In *Chlorophyll Fluorescence: A Signature of Photosynthesis. Advances in Photosynthesis and Respiration*; Papageorgiou, G., Govindjee, Eds.; Kluwer Academic Publishers: Dordrecht, The Netherlands, 2004; pp. 321–362.
25. Genty, B.; Briantais, J.-M.; Baker, N.R. The relationship between the quantum yield of photosynthetic electron transport and quenching of chlorophyll fluorescence. *Biochim. Biophys. Acta Gen. Subj.* **1989**, *990*, 87–92. [CrossRef]
26. Maxwell, K.; Johnson, G.N. Chlorophyll fluorescence—A practical guide. *J. Exp. Bot.* **2000**, *51*, 659–668. [CrossRef] [PubMed]
27. Evans, J.R. Photosynthetic acclimation and nitrogen partitioning within a lucerne canopy. 2. Stability through time and comparison with a theoretical optimum. *Aust. J. Plant Physiol.* **1993**, *20*, 69–82. [CrossRef]
28. Adrees, M.; Ali, S.; Rizwan, M.; Ibrahim, M.; Abbas, F.; Farid, M.; Zia-ur-Rehman, M.; Irshad, M.K.; Bharwana, S.A. The effect of excess copper on growth and physiology of important food crops: A review. *Environ. Sci. Pollut. Res.* **2015**, *22*, 8148–8162. [CrossRef]
29. Oustriere, N.; Marchand, L.; Roulet, E.; Mench, M. Rhizofiltration of a Bordeaux mixture effluent in pilot-scale constructed wetland using *Arundo donax* L. coupled with potential Cu-ecocatalyst production. *Ecol. Eng.* **2017**, *105*, 296–305. [CrossRef]
30. Yruela, I. Copper in plants: Acquisition, transport and interactions. *Funct. Plant Biol.* **2009**, *36*, 409–430. [CrossRef]
31. Elhawat, N.; Alshaal, T.; Domokos-Szabolcsy, É.; El-Ramady, H.; Márton, L.; Czakó, M.; Kátai, J.; Balogh, P.; Sztrik, A.; Molnár, M.; et al. Phytoaccumulation potentials of two biotechnologically propagated ecotypes of *Arundo donax* in copper-contaminated synthetic wastewater. *Environ. Sci. Pollut. Res.* **2014**, *21*, 7773–7780. [CrossRef]
32. Angelini, L.G.; Ceccarini, L.; Nassi o Di Nasso, N.; Bonari, E. Comparison of *Arundo donax* L. and Miscanthus x giganteus in a long-term field experiment in Central Italy: Analysis of productive characteristics and energy balance. *Biomass Bioenergy* **2009**, *33*, 635–643. [CrossRef]
33. Li, C.; Xiao, B.; Wang, Q.H.; Yao, S.H.; Wu, J.Y. Phytoremediation of Zn- and Cr-contaminated soil using two promising energy grasses. *Water Air Soil Pollut.* **2014**, *225*, 2027. [CrossRef]
34. Mathialagan, T.; Viraraghavan, T. Adsorption of cadmium from aqueous solutions by perlite. *J. Hazard. Mater.* **2002**, *94*, 291–303. [CrossRef]
35. Vijayaraghavan, K.; Raja, F.D. Experimental characterisation and evaluation of perlite as a sorbent for heavy metal ions in single and quaternary solutions. *J. Water Process. Eng.* **2014**, *4*, 179–184. [CrossRef]
36. Ghassabzadeh, H.; Mohadespour, A.; Torab-Mostaedi, M.; Zaheri, P.; Maragheh, M.G.; Taheri, H. Adsorption of Ag, Cu and Hg from aqueous solutions using expanded perlite. *J. Hazard. Mater.* **2010**, *177*, 950–955. [CrossRef] [PubMed]
37. Sari, A.; Tuzen, M.; Citak, D.; Soylak, M. Adsorption characteristics of Cu(II) and Pb(II) onto expanded perlite from aqueous solution. *J. Hazard. Mater.* **2007**, *148*, 387–394. [CrossRef] [PubMed]
38. Pedescoll, A.; Sidrach-Cardona, R.; Hijosa-Valsero, M.; Bécares, E. Design parameters affecting metals removal in horizontal constructed wetlands for domestic wastewater treatment. *Ecol. Eng.* **2015**, *80*, 92–99. [CrossRef]
39. Marchand, L.; Nsanganwimana, F.; Oustrière, N.; Grebenshchykova, Z.; Lizama-Allende, K.; Mench, M. Copper removal from water using a bio-rack system either unplanted or planted with *Phragmites australis*, *Juncus articulatus* and *Phalaris arundinacea*. *Ecol. Eng.* **2014**, *64*, 291–300. [CrossRef]
40. Galletti, A.; Verlicchi, P.; Ranieri, E. Removal and accumulation of Cu, Ni and Zn in horizontal subsurface flow constructed wetlands: Contribution of vegetation and filling medium. *Sci. Total Environ.* **2010**, *408*, 5097–5105. [CrossRef]
41. Rehman, M.; Maqbool, Z.; Peng, D.; Liu, L. Morpho-physiological traits, antioxidant capacity and phytoextraction of copper by ramie (*Boehmeria nivea* L.) grown as fodder in copper-contaminated soil. *Environ. Sci. Pollut. Res.* **2019**, *26*, 5851–5861. [CrossRef]
42. Brunetto, G.; Bastos de Melo, G.W.; Terzano, R.; Del Buono, D.; Astolfi, S.; Tomasi, N.; Pii, Y.; Mimmo, T.; Cesco, S. Copper accumulation in vineyard soils: Rhizosphere processes and agronomic practices to limit its toxicity. *Chemosphere* **2016**, *162*, 293–307. [CrossRef]

43. Lequeux, H.; Hermans, C.; Lutts, S.; Verbruggen, N. Response to copper excess in *Arabidopsis thaliana*: Impact on the root system architecture, hormone distribution, lignin accumulation and mineral profile. *Plant Physiol. Biochem.* **2010**, *48*, 673–682. [CrossRef] [PubMed]
44. Azeez, M.O.; Adesanwo, O.O.; Adepetu, J. Effect of Copper (Cu) application on soil available nutrients and uptake. *Afr. J. Agric. Res.* **2015**, *10*, 359–364.
45. Feigl, G.; Kumar, D.; Lehotai, N.; Tugyi, N.; Molnár, Á.; Ördög, A.; Szepesi, Á.; Gémes, K.; Laskay, G.; Erdei, L.; et al. Physiological and morphological responses of the root system of Indian mustard (*Brassica juncea* L. Czern.) and rapeseed (*Brassica napus* L.) to copper stress. *Ecotoxicol. Environ. Saf.* **2013**, *94*, 179–189. [CrossRef] [PubMed]
46. Marzilli, M.; Di Santo, P.; Palumbo, G.; Maiuro, L.; Paura, B.; Tognetti, R.; Cocozza, C. Cd and Cu accumulation, translocation and tolerance in *Populus alba* clone (Villafranca) in autotrophic in vitro screening. *Environ. Sci. Pollut. Res.* **2018**, *25*, 10058–10068. [CrossRef] [PubMed]
47. Dai, Y.; Shen, Z.; Liu, Y.; Wang, L.; Hannaway, D.; Lu, H. Effects of shade treatments on the photosynthetic capacity, chlorophyll fluorescence, and chlorophyll content of *Tetrastigma hemsleyanum* Diels et Gilg. *Environ. Exp. Bot.* **2009**, *65*, 177–182. [CrossRef]
48. Cedergreen, N.; Streibig, J.C. Can the choice of endpoint lead to contradictory results of mixture-toxicity experiments? *Environ. Toxicol. Chem.* **2005**, *24*, 1676–1683. [CrossRef] [PubMed]
49. Xu, D.; Li, C.; Chen, H.; Shao, B. Ecotoxicology and Environmental Safety Cellular response of freshwater green algae to perfluorooctanoic acid toxicity. *Ecotoxicol. Environ. Saf.* **2013**, *88*, 103–107. [CrossRef]
50. Pätsikkä, E.; Kairavuo, M.; Frantisek, S.; Aro, E.; Tyystja, E. Excess Copper Predisposes Photosystem II to Photoinhibition in Vivo by Outcompeting Iron and Causing Decrease in Leaf Chlorophyll. *Plant Physiol.* **2002**, *129*, 1359–1367.
51. Srivastava, S.; Mishra, S.; Tripathi, R.D.; Dwivedi, S.; Gupta, D.K. Copper-induced oxidative stress and responses of antioxidants and phytochelatins in *Hydrilla verticillata* (L.f.) Royle. *Aquat. Toxicol.* **2006**, *80*, 405–415. [CrossRef]
52. Yruela, I. Copper in plants. *Braz. J. Plant Physiol.* **2005**, *17*, 145–156. [CrossRef]
53. Baker, N.R. Chlorophyll Fluorescence: A Probe of Photosynthesis In Vivo. *Annu. Rev. Plant Biol.* **2008**, *59*, 89–113. [CrossRef] [PubMed]
54. Bussotti, F.; Desotgiu, R.; Cascio, C.; Pollastrini, M.; Gravano, E.; Gerosa, G.; Marzuoli, R.; Nali, C.; Lorenzini, G.; Salvatori, E.; et al. Ozone stress in woody plants assessed with chlorophyll a fluorescence. A critical reassessment of existing data. *Environ. Exp. Bot.* **2011**, *73*, 19–30. [CrossRef]
55. Gilmore, A.M.; Hazlett, T.L.; Debrunner, P.G. Comparative time-resolved photosystem II chlorophyll a fluorescence analyses reveal distinctive differences between photoinhibitory reaction center damage and xanthophyll cycledependent energy dissipation. *Photochem. Photobiol.* **1996**, *64*, 552–563. [CrossRef] [PubMed]
56. Strasser, R.J.; Stirbet, A.D. Heterogeneity of photosystem II probed by the numerically simulated chlorophyll a fluorescence rise (O–J–I–P). *Math. Comput. Simul.* **1998**, *1*, 3–9. [CrossRef]
57. Strasser, R.J.; Srivastava, A.; Tsimilli-Michael, M. The Fluorescence Transient as a Tool to Characterize and Screen Photosynthetic Samples. In *Probing Photosynthesis: Mechanisms, Regulation and Adaptation*; Mohammad, Y., Pathre, U., Mohanty, P., Eds.; CRC Press: Boca Raton, FL, USA, 2000; pp. 445–483.
58. Sbihi, K.; Cherifi, O.; El Gharmali, A.; Oudra, B.; Aziz, F. Accumulation and toxicological effects of cadmium, copper and zinc on the growth and photosynthesis of the freshwater diatom *Planothidium lanceolatum* (Brébisson) Lange-Bertalot: A laboratory study. *J. Mater. Environ. Sci.* **2012**, *3*, 497–506.
59. Cuchiara, C.C.; Silva, I.M.C.; Martinazzo, E.G.; Braga, E.J.B.; Bacarin, M.A.; Peters, J.A. Chlorophyll Fluorescence Transient Analysis in *Alternanthera tenella* Colla Plants Grown in Nutrient Solution with Different Concentrations of Copper. *J. Agric. Sci.* **2013**, *5*, 8–16. [CrossRef]
60. Cuchiara, C.C.; Silva, I.M.C.; Dalberto, D.S.; Bacarin, M.A.; Peters, J.A. Chlorophyll a fluorescence in sweet potatoes under different copper concentrations. *J. Soil Sci. Plant Nutr.* **2015**, *15*, 179–189. [CrossRef]
61. Perreault, F.; Samadani, M.; Dewez, D. Effect of soluble copper released from copper oxide nanoparticles solubilisation on growth and photosynthetic processes of *Lemna gibba* L. *Nanotoxicology* **2016**, *8*, 374–382. [CrossRef]
62. Vassilev, A.; Lidon, F.C.; Scotti Campos, P.; Ramalho, J.C.; Barreiro, M.G.; Yordanov, I. Cu-induced changes in chloroplast lipids and photosystem 2 activity in barley plants. *Bulg. J. Plant Physiol.* **2003**, *29*, 33–43.

63. Takahashi, S.; Murata, N. How do environmental stresses accelerate photoinhibition? *Trends Plant Sci.* **2008**, *13*, 178–182. [CrossRef]
64. Wilhelm, C.; Selmar, D. Energy dissipation is an essential mechanism to sustain the viability of plants: The physiological limits of improved photosynthesis. *J. Plant Physiol.* **2011**, *168*, 79–87. [CrossRef] [PubMed]
65. Goh, C.; Ko, S.; Koh, S.; Kim, Y.-Y.; Bae, H.-J. Photosynthesis and Environments: Photoinhibition and Repair Mechanisms in Plants. *J. Plant Biol.* **2012**, *55*, 93–101. [CrossRef]

© 2019 by the authors. Licensee MDPI, Basel, Switzerland. This article is an open access article distributed under the terms and conditions of the Creative Commons Attribution (CC BY) license (http://creativecommons.org/licenses/by/4.0/).

Article

Mining Rock Wastes for Water Treatment: Potential Reuse of Fe- and Mn-Rich Materials for Arsenic Removal

Barbara Casentini [1,*], **Marco Lazzazzara** [1], **Stefano Amalfitano** [1], **Rosamaria Salvatori** [2], **Daniela Guglietta** [3], **Daniele Passeri** [3], **Girolamo Belardi** [3] **and Francesca Trapasso** [3]

1. Water Research Institute, National Research Council of Italy (IRSA-CNR), Via Salaria km 29.300, Monterotondo, 00015 Rome, Italy
2. Institute of Atmospheric Pollution, Italian National Research Council (IIA-CNR), Via Salaria km 29.300, Monterotondo, 00015 Rome, Italy
3. Institute of Environmental Geology and Geoengineering, Italian National Research Council (IGAG-CNR), Via Salaria km 29.300, Monterotondo, 00015 Rome, Italy
* Correspondence: casentini@irsa.cnr.it

Received: 30 July 2019; Accepted: 28 August 2019; Published: 11 September 2019

Abstract: The worldwide mining industry produces millions of tons of rock wastes, raising a considerable burden for managing both economic and environmental issues. The possible reuse of Fe/Mn-rich materials for arsenic removal in water filtration units, along with rock properties, was evaluated. By characterizing and testing 47 samples collected from the Joda West Iron and Manganese Mine in India, we found As removal up to 50.1% at 1 mg/L initial As concentration, with a corresponding adsorption capacity of 0.01–0.46 mgAs/g mining waste. The As removal potential was strictly related to spectral, mineralogical, and elemental composition of rock wastes. Unlike rock crystallinity due to quartz and muscovite, the presence of hematite, goethite, and kaolinite, in association with the amorphous fractions of Fe and Al, enhanced the As adsorption. The natural content of arsenic indicated itself the presence of active sorptive sites. The co-occurrence of site-specific competitors (i.e., phosphate) represented a consequent limitation, whereas the content of Ce, Cu, La, and Pb contributed positively to the As adsorption. Finally, we proposed a simplified multiple linear model as predictive tool to select promising rock wastes suitable for As removal by water filtration in similar mining environments: As predicted = 0.241 + 0.00929[As] + 0.000424[La] + 0.000139[Pb] − 0.00022[P].

Keywords: mining wastes; iron and manganese minerals; water filtration; arsenic adsorption

1. Introduction

Millions of tons of waste rock, overburden, and beneficiation wastes are produced by the global mining industry. Due to their limited economic value and the remote location of most mining settings, over 95% of these materials are disposed of, forming enormous stockpiles in the mining area [1–4]. In mining companies, the cost of waste handling and storage can represents a financial loss around 1.5–3.5% of total costs [5]. The transformation of mining wastes is promoted to pursue a zero-waste circular model economy by evaluating solutions for their re-use [6]. The chemical composition and geotechnical properties of the source rock determine which uses are most appropriate and whether reuse is economically feasible. Possible second life pathways of solid mining wastes include the recovery of critical raw materials, the use as backfill materials for open voids, the extraction of valuable minerals and metals from low-grade resources, their application as landscaping materials and capping materials for waste repositories, substrates for mine revegetation, and civil engineering constructions [1,2,7]. Among mining materials, Fe-, Mn- and Al-rich rock wastes could be recovered as end of life products and converted into adsorbents for water treatment.

In recent years, a range of inexpensive water clean-up technologies have been developed to address the major problem of arsenic contamination in water sources. The adsorption onto filtration units filled with Fe, Mn, and Al (hydr-)oxides phases represents the prominent technological treatment [8–12]. Surface complexation accounts for the high selectivity of the adsorption of arsenic onto iron, aluminum, and manganese (hydr-)oxides [13–16]. Close to point of zero charge, arsenate adsorption through anion exchange could also occur [17]. Iron hydroxide is usually considered to be a superior arsenic adsorbent when compared to aluminum and manganese (hydr-)oxides, due to its highest efficiency at natural pH range [8,10,18]. A large body of the literature is focused on As adsorption studies based on synthetic minerals, such as hematite [19–21], magnetite [19–22], goethite [12,21,22], activated alumina [23,24], gibbsite [16], kaolinite and other clays [25,26], zeolites, and modified zeolites [27–30]. Arsenic adsorption up to 50 mg/g adsorbent were reported, with enhanced adsorption capacity, relying on the homogeneity and activity of adsorption sites [8]. Naturally occurring minerals are more attractive for arsenic water treatment due to their large availability and cost effectiveness. Unlike synthetic iron minerals, the naturally occurring iron ores contain a variety of mineral phases and other elements. Hence, final As adsorption is expected to be lower due to the reduced number of available and accessible sorption sites and interfering and competing ions.

Nevertheless, the need for effective, robust, and low-cost devices for widespread small-scale application (i.e., at the scale of an individual household) increased the interest in testing low-cost waste materials as arsenic adsorbents [8]. Even if their adsorption capacity can be a few mgAs/g, their performance to treat As-rich waters could be satisfactory, especially if applied to drinking water treatment targeted to groundwaters with As concentration below 200 µg/L. Nguyen et al. [31] used a purified and enriched magnetite waste from iron ore mine to treat arsenic-rich waters. This material showed arsenic maximum adsorption capacity of 0.74 mg/g. Zhang et al. [32] tested waste rock from natural iron ores, with hematite and goethite as prevailing mineralogical phases, and maximum adsorption capacity by Langmuir was estimated to be 0.4 mgAs/g. A low-cost material (76% pyrolusite with <10% goethite and quartz) from ferruginous manganese ore efficiently removed As at pH 2–8 from six groundwaters with As concentration in the range 40–180 µg/L [33]. Different tools for the characterization of mining wastes are based on either conventional methods, such as X-ray diffraction and scanning electron microscopy, or advanced approaches, such as synchrotron-based microanalysis and automated mineralogy [34].

Previous studies on mining waste reuse for arsenic removal were based on a limited number of samples with homogeneous mineral distribution. However, since rock wastes in mining stockpiles are highly heterogeneous in terms of mineralogical and chemical composition, the correct identification and selection of suitable materials for the re-use in water treatments will require cross-disciplinary approaches, primarily based on field measurements and sampling site selection.

In this study, we explored the suitability of various mining rock wastes to realize water filters for As removal from contaminated waters. More specifically, we aimed (i) to evaluate if spectral information based on field measurements could help in discriminating materials with different As adsorption potential, (ii) to assess how and to what extent the mineralogical composition and element content of rock wastes can contribute to As removal processes, (iii) to elaborate a pre-screening statistical procedure to identify and select promising materials to be potentially reused in water reclamation practices.

2. Materials and Methods

2.1. Study Area and Sampling

Joda West Iron and Manganese Mine (JWIMM) is located at about 20 km from Barbil town in Keonjhar, Odisha district, Eastern India (Figure 1). The iron ores belong to the Iron Ore Group

(IOG) and manganese ore deposits. They are confined to shale formation of the Precambrian IOG. In particular, manganese ore bodies are associated with shales, laterite, chert, and quartzite of the IOG and are distributed within the horseshoe-shaped synclinorium, plunging towards NNE over folded towards SW. The shale formation occurs as a core of the synclinorium along Jamda-Koira valley overlying the banded iron formation (See geological details in Supplementary Material, Figure S1). From 1933 onward, the mining lease was granted in favor of Tata Steel. An enormous amount of solid waste is produced each year by mining activities. Valuable material possibly interesting for reuse or recovery, or rock waste not suitable for steel production, is all disposed of in stockpiles. Local workers accumulate wastes into stock deposits as big as mountains (see details in Figure S2), following the color/weight identification of rocks.

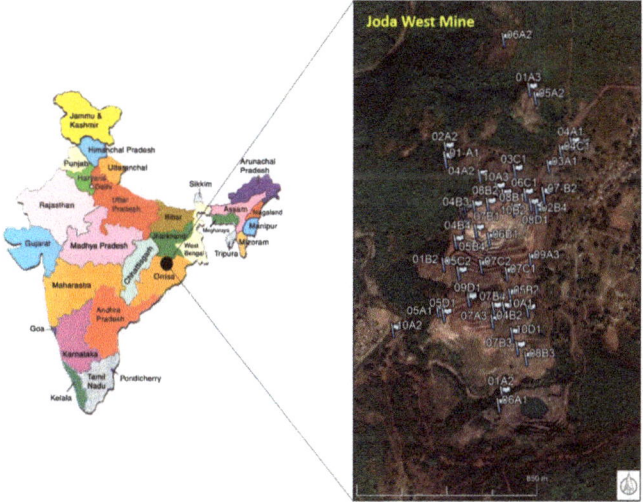

Figure 1. Study area with location and codes of sampled rock wastes at Joda West Iron and Manganese Mine.

During November 2018, a field sampling campaign at the JWIMM was conducted by National Research Council (CNR, Rome, Italy) in collaboration with Tata Steel and National Environmental Engineering Research Institute (NEERI, Nagpur, India). In particular, 47 waste rocks were collected from stock deposits and dumps (Figure 1). They were taken in different areas of the mine in order to ensure the heterogeneity of waste materials to be further tested for arsenic removal. Collected samples were micronized under 70 μm by vibrating rotary cup mill at 900 rpm motor speed for further tests.

2.2. Rock Waste Characterization

2.2.1. X-Ray Diffraction (XRD)

Mineralogical characterization was carried out by XRD analysis on micronized samples with a fully automated AXS D8 Advance diffractometer (Bruker, Billerica, MA, USA) operating in reflection mode with θ-θ geometry, equipped with high-resolution energy dispersive 1-D Lynxeye XE detector opening 3° in 2θ. Measurement parameters used were CuKα, 40 kV, 30 mA, 2.5° Soller collimators, 0.6 mm divergence slit, anti-scatter screen, scan angle (2θ) = 0–70°; step width (2θ) = 0.02°; counting time 0.3 sec per step. Diffraction data were elaborated with DIFFRAC.EVA software and identified using Crystallography Open-Access Database (COD, www.crystallography.net/cod/).

After fitting the major peaks, a semi-quantitative analysis was performed based on the XRD peak relative heights and reference intensity ratio (RIR) values. RIR is the ratio between the intensities of the strongest line of the compound of interest and the strongest line of corundum in a 1:1 mixture (by weight). The quality of the results depends on the graphic adjustment of the *y*-scale values of each XRD peak. Moreover, the method assumes that the peak height is proportional to the net area of the peak, which may be different for different minerals. An approximate crystallinity index was given as maximum counts in XRD spectra. In addition, the percentage of amorphous phases, as semi-quantitative indicator, was calculated for Al, Mn, and Fe, major adsorption phases, as %Amorphous = %TotalXRF − %CrystallineXRD. The crystalline contribution was calculated based on the sum of all percentage of element contained in each mineral form containing it (Supplementary Materials, Table S1).

2.2.2. X-Ray Fluorescence (XRF)

The chemical composition of samples was assessed by an X-ray EDS fluorescence analysis carried out by XEPOS HE spectrometer (AMETEK, Berwin, PA, USA), optimized for heavy elements with max power of 50 W and max voltage of 50 kV. The calibration curves were constructed using certified materials (OREAS, https://www.ore.com.au/) and the common linear model developed by Lucas-Tooth and Price [35].

2.2.3. Spectral Characterization

The spectral signature of each undisturbed rock waste sample was recorded using a field hyperspectral spectrometer (FieldSpec FR3 PRO, Analytical Spectral Devices-ASD, Boulder, CO, USA) operating in visible, near-infrared (NIR) and short-wave infrared (SWIR) domain (0.35–2.5 µm). We intentionally selected only the reflectance values related to red band resampled according to band 4 in Sentinel image (range 0.645–0.683 µm), since it was more representative of the target mineral phases (i.e., Fe minerals). A white Spectralon®panel (regarded as a Lambertian reflector) was used as reference to calculate the reflectance of the sample, expressed as ratio to reference (unitless).

2.3. Batch Tests for Arsenic Removal

Batch tests were carried out to evaluate arsenic removal capacity of all the 47 sampled rock wastes. Arsenic(V) stock solution (1000 mg/L) was prepared using $Na_2HAsO_4 \cdot 7H_2O$ (Fluka). Standards in the range 1–100 µg/L were prepared by dilution. To test arsenic adsorption properties of samples mining waste, 20 mL Milli-Q spiked to initial concentration of 1000 µg/L As(V) was placed in contact with 20 ± 2 mg of sample (liquid/solid ratio of 1 g/L). Solution pH was 7.0 ± 0.5. Arsenic adsorption capacity was expressed as mg of adsorbed As per grams of mining waste (mgAs/g). Initial As concentration of 1000 µg/L was selected to keep concentration sufficiently high and not far from arsenic levels typically found in groundwaters (20–200 µg/L). Samples were mildly shaken onto orbital shaker at 160 oscillations/min for 5 h. Samples were filtered on 0.2 µm acetate cellulose filters. Arsenic in solution was measured, following appropriate dilution, by Atomic Absorption Spectrometry AAnalyst 800 (Perkin Elmer, MA, USA) equipped with Ir-coated THGA furnace (range of linearity 0–100 µg/L). Duplicate samples were carried out on 10% of entire dataset. According to As adsorption capacity, mining rock wastes were classified into two groups (i.e., "not suitable (-)", "suitable (+)") by using the median as discriminatory value (i.e., 0.249 mgAs/g). Samples with significantly higher removal efficiency (i.e., 0.35–0.5 mgAs/g) were classified as "promising (++)".

2.4. Statistical Analysis

Descriptive statistic and multivariate analyses were performed on the dataset using the freeware software PAST [36]. Non-parametric statistics were applied because the normal distribution was rejected for many of the measured variables. The Kruskal–Wallis test was used to verify the equality of medians per single variable between the three groups of sampled materials with different As adsorption potential. Spearman's correlation coefficients (r_s) were calculated between all pairs of variables.

When required, the min-max data normalization was applied $y = (x - \min)/(\max - \min)$, where min and max are the minimum and maximum values of selected parameter.

Considering the multivariate dataset, the significance difference between sample groups was tested by the non-parametric multivariate analysis of variance (PERMANOVA), based on the Euclidean distance measure of normalized data. The similarity percentage analysis (SIMPER), based on the Bray–Curtis dissimilarity matrix, was used to calculate the percentage contribution of each variable (i.e., among the mineralogical phases, the major and trace elements) to the overall dissimilarity between the sample groups [37].

The factor analysis (CABFAC) was used to verify whether all information conveyed by the analyzed variables ($n = 53$) could be used to consistently predict the As adsorption potential in comparison to the measured As adsorption values [38]. Data were also modeled using multiple linear forward stepwise regression through SigmaPlot (v. 11.0)(Dundas Software LTD, GmbH, Germany). The goodness of fit was then evaluated in terms of coefficient of determination (R^2) and root mean square error (RMSE), where R^2 represents the relative measure of fit (trend prediction), while RMSE is an absolute measure of fit (model accuracy).

Coefficient of Determination

$$R^2 = \frac{\sum (x_m - \overline{x_e})^2}{\sum (x_m - \overline{x_e})^2 + \sum (x_m - x_e)^2} \tag{1}$$

Root Mean Square Error

$$\text{RMSE} = \sqrt{\frac{1}{n} \sum_{i=1}^{i=n} (x_m - x_e)^2} \tag{2}$$

where x_m is the value given by the model, x_e is the experimental data, and $\overline{x_e}$ is the mean value of experimental dataset. To have an estimate of the overall prediction error, the value of RMSE was then divided by mean of predicted values.

Predicted data were divided into correctly assigned (high and low) and mistakenly classified (false high and false low).

3. Results

3.1. Mineralogical and Chemical Composition of Rock Wastes

The entire samples dataset is shown in Table S2a–c, while a summary is reported in Table 1a,b. Among mineralogical phases, hematite, goethite, kaolinite, pyrolusite, and quartz were the most frequently found (>30% of sampled materials). Hematite was the dominant Fe mineral with samples showing more than 80% content. Goethite contribution was on average 11.2%. Samples with high crystallinity (i.e., 4000 counts in XRD spectra with sharp high peaks), were characterized by the presence of quartz (54.3–87.8%). Radiometer signal (in red band) ranges from 0.07 (dark minerals rich in Mn or hematite) to 0.38 (higher reflectance characterized by whitish minerals, like quartz, muscovite, or kaolinite). Mining rock wastes showed a heterogeneous composition with Fe, Mn, and Al as main constituents (51.3%, 14.3%, and 6% average content, respectively), with rocks having Fe and Mn above 70%. Arsenic was also naturally present in selected samples in the range 2.8–139.8 mg/kg, with a mean value of 36.2 mg/kg. Phosphorus concentration was one order of magnitude higher than As (mean = 350.1 mg/kg), due to the presence of phosphate minerals, such as berlinite, zanazziite and hopeite (Table S2). Sulfide concentration was low (mean = 140.9 mg/kg).

Table 1. Major properties of sampled rock wastes. Mineralogical properties (**a**), including major mineral phases (%) occurring in >30% of total samples are shown along with the content of major (**b**) and trace (**c**) elements.

Mineral Properties	Red Band Reflectance	Crystallinity Index	Hematite (α-Fe$_2$O$_3$)	Goethite (α-FeOOH)	Kaolinite (Al$_2$Si$_2$O$_5$(OH)$_4$)	Pyrolusite (MnO$_2$)	Quartz (SiO$_2$)
Mean	0.14	3015	37.4	11.2	10.7	4.5	19.0
Median	0.13	2000	41.4	8.0	0.0	0.0	10.7
Std Dev.	0.07	3219	20.1	9.7	15.3	10.4	23.2
Min	0.07	800	0.0	0.0	0.0	0.0	0.0
Max	0.38	13000	81.5	34.1	53.4	57.6	87.8

(**a**)

Major Elements (%)	Al	Fe	Mn	Ca	K	Mg	Si	Ti
Mean	6.0	51.3	14.3	0.10	1.1	0.15	9.7	0.42
Median	5.3	54.8	5.1	0.10	0.7	0.07	6.4	0.34
Std Dev.	3.5	22.2	18.1	0.05	1.2	0.28	9.6	0.29
Min	0.5	8.9	0.4	0.02	0.1	0.01	0.8	0.04
Max	17.2	87.4	75.8	0.24	5.5	1.39	40.5	1.33

(**b**)

Minor Elements (mg/kg)	As	Ce	Cr	Cu	La	Mo	Ni	P	Pb	Rb	S	Y	Zn
Mean	36.2	49.6	290.5	28.5	46.6	65.8	221.2	350.1	272.7	42.8	140.9	44.3	162.8
Median	32.7	38.0	241.9	24.6	34.1	33.1	171.0	350.7	231.0	41.8	109.8	32.8	150.4
Std Dev.	34.7	41.5	191.3	14.9	51.9	65.1	293.2	114.5	260.1	14.2	101.5	38.6	90.9
Min	2.8	1.5	52.6	8.1	1.5	0.2	0.8	93.3	0.3	18.4	26.6	1.1	29.4
Max	139.8	175.9	846.2	82.4	284.2	263.0	1914	678.8	1360	112.6	473.5	192.7	432.6

(**c**)

3.2. Arsenic Removal Capacity

Arsenic removal ranged from 1.2% to 50.1%, with a calculated adsorption capacity of 0.01–0.456 mgAs/g. Variation in duplicate samples were in the range 5.4–14.3%. Adsorption capacity distribution showed a mean value of 0.255 and median of 0.249 mgAs/g. Sampled rock wastes with a potential As adsorption lower than the median were classified as "not suitable" (45% of total samples), while 55% of samples are classified suitable, including 13% of them showing promising capacity for As removal application (Figure 2).

3.3. Influence of Spectral, Chemical, and Mineralogical Parameters on As Adsorption

The three identified groups were significantly different in terms of As adsorption and reflectance (Kruskal–Wallis test, $p < 0.05$). Reflectance values above 0.2 were only found in the group with lower As adsorption capacity (Figure 2). Samples with >0.2 reflectance in red band were characterized by the presence of muscovite (13.7–35.1%), and quartz (35.6–78%).

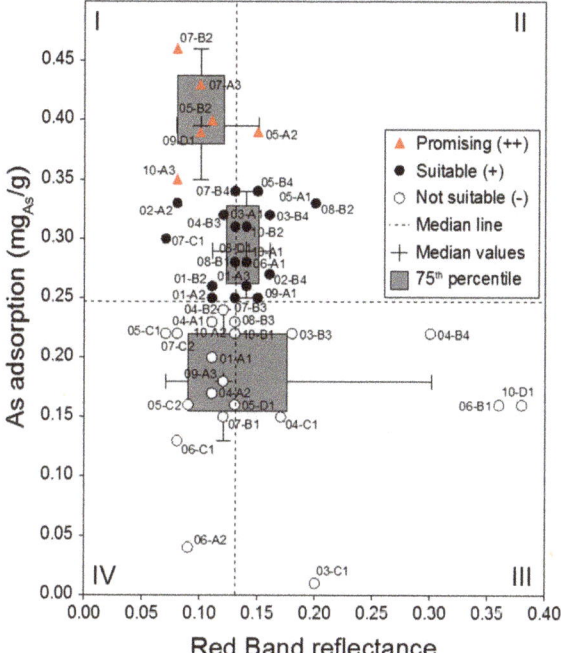

Figure 2. Bi-directional box plots of Red Band reflectance and As adsorption (mgAs/g) of sampled rock wastes, coded as in Figure 1. The overall median values of the two parameters (dashed lines) were used to discriminate samples with relatively lower and higher transmittance and adsorption potential (i.e., not suitable (-), suitable (+), and promising (++) for As removal). Samples were divided into four quadrants according to their characteristics. Quadrant I: high As adsorption and low Red Band reflectance (26% of samples); Quadrant II: High As adsorption and high Red Band reflectance (30% of samples): Quadrant III: low As adsorption and high Red Band reflectance (19% of samples) and Quadrant IV: low As adsorption and low Red Band reflectance (26% of samples).

The rock waste groups were also proven to be significantly different considering the entire normalized dataset of spectral, mineralogical, and chemical parameters (PERMANOVA, $p = 0.011$).

Samples characterized by lower As adsorption and classified as "not suitable (-)", showed a relatively lower content of hematite, goethite, and kaolinite, along with the prevalence of quartz. Samples 03A1 and 06A2 showed extremely low adsorption capacity (<0.05 mgAs/g) and crystalline hematite above 60% (Figure 3). On the contrary, the concomitant presence of iron minerals, with high content of kaolinite and low contribution of quartz led to the relatively higher arsenic adsorption measured for samples classified as "suitable (+)" and "promising (++)".

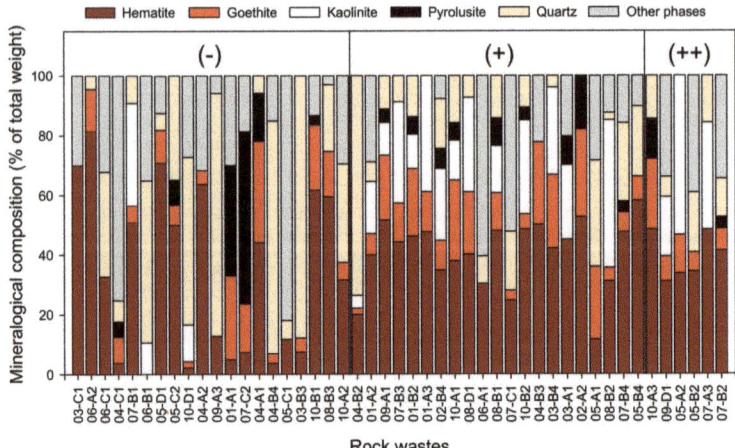

Figure 3. Mineralogical composition of sampled rock wastes (i.e., not suitable (-), suitable (+), and promising (++) for As removal), ordered by increasing As adsorption potential (from left to right). Mineralogical phases occurring in less than 30% of the samples were grouped and plotted as "other phases".

The SIMPER tests, carried out on mineralogical phases, major, and trace elements (Table 3a–c), showed that the mineral phases kaolinite, goethite, quartz, and hematite explained most of the overall dissimilarity in the dataset (>30% of cumulative contribution). Notably, the less represented mineral phases with contribution close to 5% were silico-aluminates (staurolite and clinochlore) and phosphate rocks (zanazziite). Among major elements, including % of amorphous (am-) and crystalline (cryst-), Fe, and Al, especially the forms crystalline/amorphous Al and amorphous Fe, were higher in suitable and promising groups, indicating their predominant role in As adsorption (Figure S3). Moreover, the presence of Mn and Si did not promote or inhibit As adsorption.

Table 2. Outputs of the similarity percentage analysis (SIMPER) test performed on normalized data on the three sample groups with different As adsorption potential, classified as "not suitable (-)", "suitable (+)", and "promising (++)". The mineralogical phases (a), major (b) and trace (c) elements were tested separately. The parameters were sorted in descending order of percentage contribution (Contrib %) to the observed difference between sample groups. Mean values for each variable and sample group are also reported.

Mineral Phases % of Total Weight	Contrib %	Mean Values		
		(-)	(+)	(++)
Kaolinite	10.1	2.9	16.6	18.2
Goethite	8.4	8.8	14.2	9.7
Quartz	7.3	28.3	11.3	11.7
Muscovite	6.3	5.9	2.1	0.0
Hematite	5.9	32.8	41.6	39.6
Zanazziite	5.5	1.1	0.5	0.0
Staurolite	4.7	4.3	1.6	0.0
Clinochlore	4.7	2.9	1.1	0.0
Gjerdingenite-Fe	4.4	1.6	0.8	0.0
Gibbsite	3.7	0.8	4.1	7.9
Birnessite	3.6	1.4	0.9	1.3
Krettnichite	3.5	0.0	0.2	1.6
Ellenbergerite	3.2	0.0	0.0	3.4
Ferrierite-Na	3.2	0.0	0.0	2.5
Hopeite	3.2	0.0	0.0	1.3
Pyrolusite	3.1	6.1	3.4	2.9
Siderite	2.8	0.0	0.6	0.0
Gehlenite	2.1	0.0	0.2	0.0
Inesite	2.1	0.0	0.4	0.0
Kogarkoite	2.1	0.0	0.4	0.0
Berlinite	2.0	0.3	0.0	0.0
Chalcophanite	2.0	0.2	0.0	0.0
Lazurite	2.0	0.8	0.0	0.0
Magnesiochromite	2.0	0.3	0.0	0.0
Pyroxene-ideal	2.0	1.3	0.0	0.0

(a)

Major Elements %	Contrib %	(-)	(+)	(++)
cryst-Al	10.8	2.4	4.2	5.7
am-Fe	10.2	17.4	17.7	18.6
am-Al	9.7	2.8	2.4	1.3
Fe	9.4	46.6	56.0	52.4
Mn	6.8	17.8	10.3	15.2
Si	6.8	11.8	8.3	7.0
cryst-Fe	6.8	21.8	15.2	21.0
K	6.7	1.6	0.8	0.6
am-Mn	6.7	13.4	7.7	12.4
Ca	6.3	0.1	0.1	0.1
Ti	5.9	0.4	0.4	0.4
Al	5.4	5.1	6.5	6.9
Mg	5.0	0.2	0.1	0.1
cryst-Mn	3.6	4.7	2.8	2.7

(b)

Table 3. Cont.

Minor Elements mg/kg	Contrib %	(-)	(+)	(++)
As	10.7	26.1	39.5	60.6
Ce	10.5	38.4	52.8	78.1
Mo	9.7	72.5	54.0	81.9
Cr	9.2	312.0	258.0	321.0
Zn	8.8	155.0	159.0	205.0
S	8.7	130.0	151.0	143.0
Cu	7.1	24.3	30.7	36.1
P	6.8	394.0	305.0	347.0
La	6.8	34.8	46.8	86.8
Y	6.6	54.6	41.0	19.6
Pb	6.4	187.0	291.0	510.0
Ni	5.2	187.0	175.0	494.0
Rb	3.6	43.0	42.8	41.9

(c)

Bivariate correlation plots evidenced that arsenic adsorption (mgAs/g) was significantly correlated with As, Ce, Cu, P, Pb, and Y naturally occurring in the sampled materials (mg/g, XRF measurements). The presence of Ce, Cu, and Pb led to an increase of As adsorption, while the presence of P and Y was inversely correlated. Despite data of As, Pb and Ce corresponding to their LOD value (2.8, 0.3, 1.5 mg/kg, respectively) being close to the x-axis (Figure 4), the As adsorption was measurable since the adsorption driving forces in rock waste were dependent on a combination of factors.

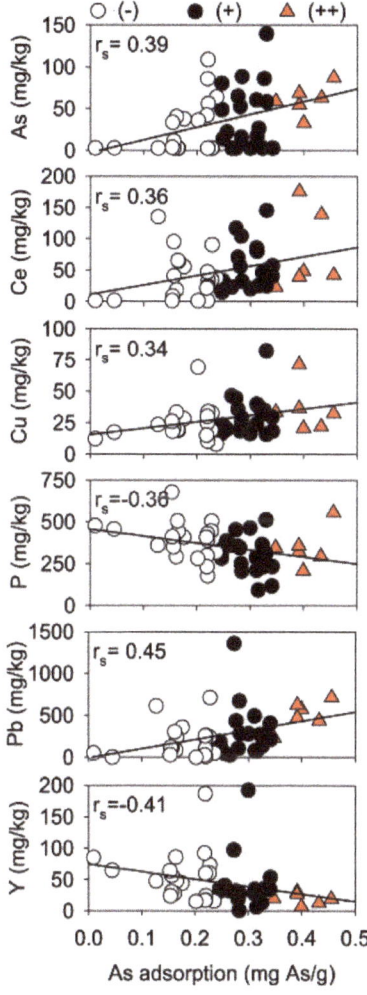

Figure 4. Bivariate plots and significant Spearman correlations (r_s) between the As adsorption potential and the content of selected trace elements ($p < 0.02$, always). Symbols indicate samples not suitable (-), suitable (+), and promising (++) for As removal.

3.4. Predicting Arsenic Removal by Wastes Characteristics of Fe- and Mn-Rich Ores

Factor analysis, based on the entire dataset, was tested to formulate an arsenic adsorption predictive model. Arsenic adsorption could be predicted by all variables, with and $R^2 = 0.6$, RMSE of 0.06 corresponding to 22.8% prediction error. To our scope, the error that more affected materials selection and further filter performance was the one represented by false high, that is those samples predicted as suitable adsorbents that ended up not being suitable. In the case of false low materials, the error represents an underestimation of our materials and leads only to non-inclusion of wastes that are possibly good adsorbents. False high error corresponded to 9%.

Factor analysis proved the possibility of building a predictive model. Thus, we developed a simplified predictive tool by extrapolating a multiple linear model based on forward stepwise regression (Figure 5). The resultant equation is:

$$\text{As predicted} = 0.241 + 0.00929[\text{As}] + 0.000424[\text{La}] + 0.000139[\text{Pb}] - 0.00022[\text{P}] \tag{3}$$

where As predicted is given in mg/g, and chemical concentrations of single elements, measured by XRF, are in mg/kg.

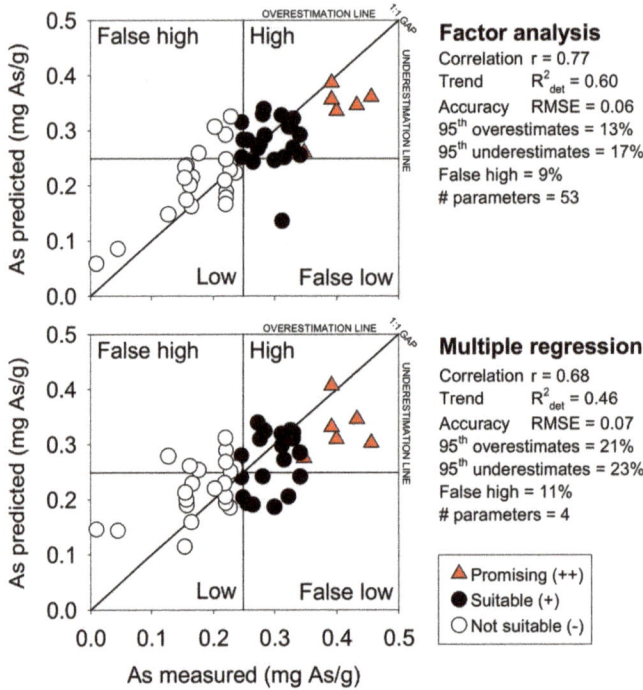

Figure 5. Multiparameter linear regression results based on (**a**) factor analysis performed on 53 parameters and (**b**) multiple linear stepwise forward regression, based on four selected parameters (As, La, Pb, and P). Goodness of fit are expressed through R^2 and root mean square error (RMSE). Percentiles lines (95%) of prediction are plotted (dashed lines). Symbols indicate samples not suitable (-), suitable (+), and promising (++) for As removal.

Among heavy metals, the one mostly affecting As adsorption ability is Pb, while among rare earth elements, lanthanum contribution was dominant. The presence of P in the materials limited the adsorption potential. The presence of natural arsenic enrichment in the sample was the major predictor for As adsorption potential capacity. The prediction using a multiple linear regression model showed $R^2 = 0.46$, RMSE of 0.067, and error of prediction 26.3%, while false high corresponded to 11%. Overestimation of 21% and underestimation of 23% was calculated using multiple linear model based on 95th percentile interval.

4. Discussion

Mining activities, together with construction/demolition and manufacturing, contributed to nearly 74% of all wastes disposed of in the European Union [39]. North America produces more than 10 times as much solid mine waste as municipal solid waste per capita [34]. In India, more than 200 million tons of non-hazardous inorganic solid wastes are being generated every year, out of which about 80 million tons are mine tailings/ores [40].

At Joda West mine, Fe, Mn and Al (hydr-)oxides, clay mineral, and quartz were the predominant mineral phases of collected rock wastes. The natural arsenic content reached up to 140 mg/kg, (higher than soil baseline concentration generally 5–10 mg/kg). In different mining environments (e.g., gold mines), arsenic content in rocks was reported in the order of g/kg, due to the presence of primary and secondary minerals [41]. Among transition metals, Cr, Ni, Pb, and Zn showed higher concentrations than average values in soils and sediments reported by FOREGS European geochemical Atlas [42] but much lower than their reported maximum values (Table S3).

First exploratory adsorption tests allowed us to test and select materials that could be potentially interesting for the realization of filters for treating As-rich waters. In heterogeneous natural iron oxides, adsorption of 0.3–0.5 mg/g of arsenate were commonly found [12,19,20]. Some of the tested samples showed a satisfactory As adsorption capacity (>0.25 mgAs/g at Ci = 1 mgAs/L). At this concentration, treated magnetite waste from iron ore showed adsorption below 0.2 mg/g [31]. Chakravarty et al. [33] tested a ferruginous manganese ore material, mainly constituted by pyrolusite and goethite, and concentration in condition similar to this study resulted in 0.2 mgAs/g adsorption according to Langmuir As(V) equation. The As adsorption capacity of three hematite-rich iron ore samples was 0.17–0.48 mgAs/g [32]. The presence of quartzite and less reactive clays at high crystallinity (e.g., muscovite) were the major limiting factors for adsorption.

Since the reflectance spectrum of rocks depends on their mineralogical composition [43,44], we noted that the reflectance in red band range, measured by hyperspectral field radiometer on undisturbed rock wastes, provided valuable indications for the on-site pre-selection of materials with lower As removal potential. The rock wastes with a limited availability of adsorption sites could be discarded, with no need to carry out further measurements. A variety of reflectance spectroscopy-based applications, relying on the spectral signatures of minerals able to bind/sorb metal(loids), have been developed to promote indirect detection and avoid expensive laboratory measurements. Pallottino et al. [45] realized a predictive model for As contamination in calcareous soil surrounding thermal springs based on the diffuse VIS-NIR spectral reflectance. In that study, As content was largely associated (>46%) with the sole $CaCO_3$ phase. According to our outcomes, a first pre-screening step could be used to exclude spotted materials containing high amounts of quartzite and muscovite, but the presence of dark-red minerals (amorphous and crystalline) cannot be directly discriminated through the spectral signals in the red band, given the observed mineralogical complexity and heterogeneity of As adsorption phases. Appropriate spectral information should be collected by better refining band selection in order to exclude the less adsorptive materials (i.e., quartz and muscovite) and to identify good adsorbents (i.e., kaolinite).

The presence of iron minerals (goethite and hematite), together with Al-rich kaolinite, contributed the most to As adsorption. At natural pH range of 6–8, the adsorption onto iron (hydr-)oxides is the most competitive, since Fe-based materials have a favorable surface charge (pHpzc 7-9) for oxyanion adsorption, while Mn oxides are mostly negatively charged (pHpzc 2–3). Adsorption of aluminum (hydr-)oxides is known to be maximum at pH 4–5 [10]. On the contrary, Fe-Al binary oxides showed to be attractive adsorbents for both As(V) and As(III) removal from contaminated waters [46]. The ability of Mn dioxides to sorb As(III) and As(V) appeared to be related also to materials with highly ordered pyrolusite having low specific surface (7.9 m^2/g). Conversely, poorly crystalline birnessite has higher specific surface area of 27.7 m^2/g [47].

The presence of Mn minerals, either amorphous or crystalline, lowered As adsorption. Arsenic adsorption onto hydroxides was correlated to Fe and amorphous phases, characterized by edge structures more efficient in hosting arsenate ions than crystalline minerals. For example, the transformation of amorphous FeOOH to crystalline FeOOH would reduce sorption sites and surface area, thus lowering the number of ions that can be adsorbed [15,41,48]. We found that amorphous fraction of Fe and Al were important in promoting As adsorption. Pedersen et al. [49] observed a decrease in adsorbed arsenic clearly correlated with the transformation of ferrihydrite and lepidocrocite into more crystalline phases as goethite and hematite. Pigna et al. [50] showed better As removal capacity of non-crystalline Al(OH)x than gibbsite. Fine-grained and poorly crystalline Mn oxides showed good adsorption properties, even if in Mn-ores pyrolusite (most stable and abundant) and birnessite minerals are often encountered [51]. Fe, Al, and Mn minerals with medium grade crystallinity are responsible for As adsorption, since crystallization process kinetic and environmental conditions might induce defects in crystalline structures, which are suitable as adsorption sites.

In line with literature reports, the correlation of As adsorption with the presence of rare earth elements (Ce and La) and transition metals (Pb, Cu) suggested that their variation mostly explained changes in As adsorption capacity [10,52–55]. These elements could be incorporated in mineral structures or adsorbed on specific sites. The adsorption of heavy metals onto clay and oxides surface might cause a pHpzc shift towards higher values, thus rendering surfaces more positive at higher pH and promoting adsorption of oxyanions [8,56–58]. Fe (hydr-)oxides structure may incorporate metals cations and adsorb As-ions more effectively, due to a better matching of ion size and orientation, also by shortening the atom-to-atom distances between adsorbent and adsorbate [59]. Mohapatra et al. [60] modified goethite surface by doping Cu(II), Ni(II), or Co(II) to enhance arsenate uptake capacities. Lu et al. [61] observed that the presence of Pb during the process of ferrihydrite transformation to hematite induced the formation of nanoparticles with a loose and porous structure in comparison with the compact structure of pure hematite nanoparticles.

The presence of bivalent cations (namely, Ni(II), Co(II), Mg(II)) were reported to enhance As adsorption capacity [33]. Natural and modified enriched clays with exchangeable cations and anions have been widely tested as adsorbents for water treatment [62–65]. In pillared or intercalated clays with transition positively charged metals, clay sheets increase each other's distances. Adsorption increased due to change in surface area and charge, with positively charged surface enhancing penetrability of As oxyanions [66]. Na et al. [67] demonstrated that Ti-pillared montmorillonite was an efficient material for the removal of arsenate and arsenite from aqueous solutions. Doušová et al. [68] proved that pre-treatment of low-grade clay materials Fe (Al, Mn) salts can significantly improve their sorption affinity to As oxyanions. A simplified multiple linear model was proposed, based on XRF measurements, as a predictive tool to guide mining wastes selection to realize removal filters for As contaminated water. As predicted = 0.241 + 0.00929[As] + 0.000424[La] + 0.000139[Pb] − 0.00022[P]. This model applicability is site-specific and strictly related to the mineralogical, geological and chemical context encountered at JWIMM. The natural presence of arsenic in the sampled materials was one of the best predictors for As adsorption, thus indicating that rock wastes kept their original and natural As-adsorbing affinity. On the contrary, higher concentrations of Y and Rb were found at low As adsorption levels only. These two elements are found in association with phosphate rocks [69,70]. Due to chemical similarity, phosphate is also known to be a competitor for the As adsorption sites and, together with silicate, a major interfering ion for As removal processes onto oxides [71–74]. Furthermore, waste materials active for arsenate could be also successfully tested for phosphate removal to reduce P-load from surface water and promote its recovery as critical raw material [75,76].

Overall, due to the elevated presence of iron oxides and the co-occurring kaolinite in most suitable samples, the factors that turned out to be more significant to differentiate potential adsorption capacity were elements adsorbed onto major phases, which were able to positively modify surrounding adsorption site structure. The possible release of As together with other potentially toxic metals from reused rock wastes should be consciously investigated in the long term to promote safer applications, especially if intended for human consumption purposes.

5. Conclusions

Mining rock wastes, accumulated in stockpiles at Fe-Mn ores, showed a good potential to be reused for water treatment, due to the presence of iron minerals and kaolinite clays. The adsorption capacity of suitable materials was not exceptionally high (0.25–0.46 mgAs/g), but satisfactory for treating As-rich groundwaters. The use of spectral, mineralogical and chemical information proved useful to select heterogeneous materials and promisingly suitable to remove arsenic from contaminated waters. Given the significant correlation with As adsorption, the role of positively charged ions in the structure of clays and oxides should be considered specifically, also evaluating thermodynamic and kinetic properties which may affect the efficiency of filtration-based As removal by reused mining rock wastes.

Supplementary Materials: The following are available online at http://www.mdpi.com/2073-4441/11/9/1897/s1, Figure S1: Geological map of Joda West Mine area, Odisha, India; Figure S2: Joda West mining site and waste stockpiles; Figure S3: Amorphous (dark bars) and crystalline (light bars) contribution to overall presence of Fe (blue), Mn (green) and Al (grey) divided into As adsorption groups: (-) "not suitable", (+) "suitable" and (++) "promising". Samples were ordered by increasing As adsorption capacity (from left to right). Semi-quantitative contribution (%) of amorphous fraction for Fe, Mn and Al is given by subtracting by XRD data (% crystalline phases) from XRF measurements (% Total element).Table S1: Part I: Mineral formula and contribution (%) of Fe, Al, Mn to each phase for the calculation of Crystalline phase used to derive semi-quantitative amorphous value. Part II: Other mineralogical phases; Table S2: (a) All samples codes, site and color description, As removal efficiency (% and mg Arsenic/g material), spectral information and contribution of Crystalline and amorphous fraction of selected major elements (Fe, Al, Mn); (b) All samples codes with major (%) and minor (mg/kg) elements measured by XRF; (c) All samples codes with mineralogical phases major (PartI) and minor (PartII) (%) measured by XRD. Other phase is the sum of all minor phases. n.d. = not detectable phases; Table S3: European soil and sediments major and minor elements content (mg/kg) (FOREGS geochemical Atlas.

Author Contributions: Conceptualization: B.C. and S.A.; Data curation: B.C., D.G. and F.T.; Formal analysis, B.C., S.A and G.B.; Investigation: M.L., R.S., D.G., D.P., G.B., and F.T.; Resources: D.G., F.T. and D.P; Supervision:, B.C.; Validation: S.A.; Visualization: B.C. and S.A.; Writing—original draft: B.C.; Writing—review & editing, S.A., R.S. and F.T.

Funding: The authors are grateful to TECO Project ICI+/2014/342-817-"Technological Eco - Innovations for the Quality Control and the Decontamination of Polluted Waters and Soils", for partly funding sampling campaign.

Acknowledgments: The authors are grateful to the staff of Tata Steel for providing general support during fieldwork.

Conflicts of Interest: The authors declare no conflict of interest.

References

1. Bian, Z.; Miao, X.; Lei, S.; Chen, S.; Wang, W.; Struthers, S. The Challenges of Reusing Mining and Mineral-Processing Wastes. *Science* **2012**, *337*, 702–703. [CrossRef]
2. Lottermoser, B.G. Recycling, reuse and rehabilitation of mine wastes. *Elements* **2011**, *7*, 405–410. [CrossRef]
3. Hudson-Edwards, K.A.; Jamieson, H.E.; Lottermoser, B.G. Mine wastes: Past, present, future. *Elements* **2011**, *7*, 375–380. [CrossRef]
4. Ndlovu, S.; Simate, G.S.; Matinde, E.; Simate, G.S.; Matinde, E. *Waste Production and Utilization in the Metal Extraction Industry*; CRC Press: Boca Raton, FL, USA; Taylor & Francis: Abington, Thames, UK, 2017; ISBN 9781315153896.
5. *Symonds Group Report to DG Environment, European Commission: A Study on the Costs of Improving the Management of Mining Waste*; Symonds Group Ltd: East Grinstead, UK, 2001.
6. Matinde, E.; Simate, G.S.; Ndlovu, S. Mining and metallurgical wastes: A review of recycling and re-use practices. *J. South. Afr. Inst. Min. Metall.* **2018**, *118*, 825–844. [CrossRef]
7. Yellishetty, M.; Karpe, V.; Reddy, E.H.; Subhash, K.N.; Ranjith, P.G. Reuse of iron ore mineral wastes in civil engineering constructions: A case study. *Resour. Conserv. Recycl.* **2008**, *52*, 1283–1289. [CrossRef]
8. Mohan, D.; Pittman, C.U. Arsenic removal from water/wastewater using adsorbents-A critical review. *J. Hazard. Mater.* **2007**, *142*, 1–53. [CrossRef] [PubMed]
9. Mondal, P.; Bhowmick, S.; Chatterjee, D.; Figoli, A.; Van der Bruggen, B. Remediation of inorganic arsenic in groundwater for safe water supply: A critical assessment of technological solutions. *Chemosphere* **2013**, *92*, 157–170. [CrossRef] [PubMed]
10. Giles, D.E.; Mohapatra, M.; Issa, T.B.; Anand, S.; Singh, P. Iron and aluminium based adsorption strategies for removing arsenic from water. *J. Environ. Manag.* **2011**, *92*, 3011–3022. [CrossRef] [PubMed]
11. Cundy, A.B.; Hopkinson, L.; Whitby, R.L.D. Use of iron-based technologies in contaminated land and groundwater remediation: A review. *Sci. Total Environ.* **2008**, *400*, 42–51. [CrossRef] [PubMed]

12. Ladeira, A.C.; Ciminelli, V.S. Adsorption and desorption of arsenic on an oxisol and its constituents. *Water Res.* **2004**, *38*, 2087–2094. [CrossRef] [PubMed]
13. Goldberg, S.; Johnston, C.T. Mechanisms of Arsenic Adsorption on Amorphous Oxides Evaluated Using Macroscopic Measurements, Vibrational Spectroscopy, and Surface Complexation Modeling. *J. Colloid Interface Sci.* **2001**, *234*, 204–216. [CrossRef] [PubMed]
14. Zhang, G.; Liu, F.; Liu, H.; Qu, J.; Liu, R. Respective Role of Fe and Mn Oxide Contents for Arsenic Sorption in Iron and Manganese Binary Oxide: An X-ray Absorption Spectroscopy Investigation. *Environ. Sci. Technol.* **2014**, *48*, 10316–10322. [CrossRef] [PubMed]
15. Dixit, S.; Hering, J. Comparison of arsenic (V) and arsenic (III) sorption onto iron oxide minerals: Implications for arsenic mobility. *Environ. Sci. Technol.* **2003**, *37*, 4182–4189. [CrossRef] [PubMed]
16. Ladeira, A.C.Q.; Ciminelli, V.S.T.; Duarte, H.A.; Alves, M.C.M.; Ramos, A.Y. Mechanism of anion retention from EXAFS and density functional calculations: arsenic (V) adsorbed on gibbsite. *Geochim. Cosmochim. Acta* **2001**, *65*, 1211–1217. [CrossRef]
17. Ouvrard, S.; Simonnot, M.O.; Sardin, M. Reactive Behavior of Natural Manganese Oxides toward the Adsorption of Phosphate and Arsenate. *Ind. Eng. Chem. Res.* **2002**, *41*, 2785–2791. [CrossRef]
18. Jeong, Y.; Fan, M.; Singh, S.; Chuang, C.L.; Saha, B.; Hans van Leeuwen, J. Evaluation of iron oxide and aluminum oxide as potential arsenic(V) adsorbents. *Chem. Eng. Process. Process Intensif.* **2007**, *46*, 1030–1039. [CrossRef]
19. Giménez, J.; Martínez, M.; de Pablo, J.; Rovira, M.; Duro, L. Arsenic sorption onto natural hematite, magnetite, and goethite. *J. Hazard. Mater.* **2007**, *141*, 575–580. [CrossRef]
20. Aredes, S.; Klein, B.; Pawlik, M. The removal of arsenic from water using natural iron oxide minerals. *J. Clean. Prod.* **2013**, *60*, 71–76. [CrossRef]
21. Mamindy-Pajany, Y.; Hurel, C.; Marmier, N.; Roméo, M. Arsenic adsorption onto hematite and goethite. *C. R. Chim.* **2009**, *12*, 876–881. [CrossRef]
22. Mamindy-Pajany, Y.; Hurel, C.; Marmier, N.; Roméo, M. Arsenic (V) adsorption from aqueous solution onto goethite, hematite, magnetite and zero-valent iron: Effects of pH, concentration and reversibility. *Desalination* **2011**, *281*, 93–99. [CrossRef]
23. Singh, T.S.; Pant, K. Equilibrium, kinetics and thermodynamic studies for adsorption of As(III) on activated alumina. *Sep. Purif. Technol.* **2004**, *36*, 139–147. [CrossRef]
24. Lin, T.-F.; Wu, J.-K. Adsorption of Arsenite and Arsenate within Activated Alumina Grains: Equilibrium and Kinetics. *Water Res.* **2001**, *35*, 2049–2057. [CrossRef]
25. Saada, A.; Breeze, D.; Crouzet, C.; Cornu, S.; Baranger, P. Adsorption of arsenic (V) on kaolinite and on kaolinite—Humic acid complexes. *Chemosphere* **2003**, *51*, 757–763. [CrossRef]
26. Mohapatra, D.; Mishra, D.; Chaudhury, G.R.; Das, R.P. Arsenic(V) adsorption mechanism using kaolinite, montmorillonite and illite from aqueous medium. *J. Environ. Sci. Health Part A* **2007**, *42*, 463–469. [CrossRef] [PubMed]
27. Simsek, E.B.; Özdemir, E.; Beker, U. Zeolite supported mono- and bimetallic oxides: Promising adsorbents for removal of As(V) in aqueous solutions. *Chem. Eng. J.* **2013**, *220*, 402–411. [CrossRef]
28. Shevade, S.; Ford, R.G. Zeolite Performance as an Anion Exchanger for Arsenic Sequestration in Water. In *Advances in Arsenic Research*; ACS Symposium Series; American Chemical Society: Washington, DC, USA, 2005; Volume 915, pp. 306–320.
29. Chutia, P.; Kato, S.; Kojima, T.; Satokawa, S. Arsenic adsorption from aqueous solution on synthetic zeolites. *J. Hazard. Mater.* **2009**, *162*, 440–447. [CrossRef] [PubMed]
30. Jeon, C.S.; Baek, K.; Park, J.-K.; Oh, Y.-K.; Lee, S.-D. Adsorption characteristics of As(V) on iron-coated zeolite. *J. Hazard. Mater.* **2009**, *163*, 804–808. [CrossRef] [PubMed]
31. Nguyen, T.V.; Nguyen, T.V.T.; Pham, T.L.; Vigneswaran, S.; Ngo, H.H.; Kandasamy, J.; Nguyen, H.K.; Nguyen, D.T. Adsorption and removal of arsenic from water by iron ore mining waste. *Water Sci. Technol.* **2009**, *60*, 2301–2308. [CrossRef] [PubMed]
32. Zhang, W.; Singh, P.; Paling, E.; Delides, S. Arsenic removal from contaminated water by natural iron ores. *Miner. Eng.* **2004**, *17*, 517–524. [CrossRef]
33. Chakravarty, S.; Dureja, V.; Bhattacharyya, G.; Maity, S.; Bhattacharjee, S. Removal of arsenic from groundwater using low cost ferruginous manganese ore. *Water Res.* **2002**, *36*, 625–632. [CrossRef]

34. Jamieson, H.E. Geochemistry and mineralogy of solid mine waste: Essential knowledge for predicting environmental impact. *Elements* **2011**, *7*, 381–386. [CrossRef]
35. Lucas-Tooth, H.J.; Price, B.J. A Mathematical Method for the Investigation of Interelement Effects in X-Ray Fluorescence Analysis. *Metallurgia* **1961**, *64*, 149–152.
36. Hammer, Ø.; HARPER, D.A.T.; Ryan, P.D. PAST: Paleontological statistics software package. *Palaeontol. Electron.* **2001**, *4*, 9.
37. Amalfitano, S.; Del Bon, A.; Zoppini, A.; Ghergo, S.; Fazi, S.; Parrone, D.; Casella, P.; Stano, F.; Preziosi, E. Groundwater geochemistry and microbial community structure in the aquifer transition from volcanic to alluvial areas. *Water Res.* **2014**, *65*, 384–394. [CrossRef] [PubMed]
38. Klovan, J.E.; Imbrie, J. An algorithm andFortran-iv program for large-scaleQ-mode factor analysis and calculation of factor scores. *J. Int. Assoc. Math. Geol.* **1971**, *3*, 61–77. [CrossRef]
39. Hering, J.G. An end to waste? *Science* **2012**, *337*, 623. [CrossRef]
40. Pappu, A.; Saxena, M.; Asolekar, S.R. Solid wastes generation in India and their recycling potential in building materials. *Build. Environ.* **2007**, *42*, 2311–2320. [CrossRef]
41. Smedley, P.L.; Kinniburgh, D.G. A review of the source, behaviour and distribution of arsenic in natural waters. *Appl. Geochem.* **2002**, *17*, 517–568. [CrossRef]
42. Geochemical Atlas of Europe. Available online: http://www.gtk.fi/publ/foregsatlas/ (accessed on 29 August 2019).
43. Clark, R.N. Chapter 1: Spectroscopy of Rocks and Minerals, and Principles of Spectroscopy. In *Manual of Remote Sensing, Volume 3, Remote Sensing for the Earth Sciences*; Sons, J.W., Ed.; John Wiley & Sons, Inc.: New York, NY, USA, 1999; pp. 3–58.
44. Shi, C.; Ding, X.; Liu, Y.; Zhou, X. Reflectance Spectral Features and Significant Minerals in Kaishantun Ophiolite Suite, Jilin Province, NE China. *Minerals* **2018**, *8*, 100. [CrossRef]
45. Pallottino, F.; Stazi, S.R.; D'Annibale, A.; Marabottini, R.; Allevato, E.; Antonucci, F.; Costa, C.; Moscatelli, M.C.; Menesatti, P. Rapid assessment of As and other elements in naturally-contaminated calcareous soil through hyperspectral VIS-NIR analysis. *Talanta* **2018**, *190*, 167–173. [CrossRef]
46. Zhang, G.; Liu, H.; Qu, J.; Jefferson, W. Arsenate uptake and arsenite simultaneous sorption and oxidation by Fe–Mn binary oxides: Influence of Mn/Fe ratio, pH, Ca^{2+}, and humic acid. *J. Colloid Interface Sci.* **2012**, *366*, 141–146. [CrossRef] [PubMed]
47. Oscarson, D.W.; Huang, P.M.; Liaw, W.K.; Hammer, U.T. Kinetics of Oxidation of Arsenite by Various Manganese Dioxides1. *Soil Sci. Soc. Am. J.* **1983**, *47*, 644. [CrossRef]
48. Tufano, K.J.; Fendorf, S. Confounding Impacts of Iron Reduction on Arsenic Retention. *Environ. Sci. Technol.* **2008**, *42*, 4777–4783. [CrossRef] [PubMed]
49. Pedersen, H.D.; Postma, D.; Jakobsen, R. Release of arsenic associated with the reduction and transformation of iron oxides. *Geochim. Cosmochim. Acta* **2006**, *70*, 4116–4129. [CrossRef]
50. Pigna, M.; Krishnamurti, G.S.R.; Violante, A. Kinetics of Arsenate Sorption–Desorption from Metal Oxides. *Soil Sci. Soc. Am. J.* **2006**, *70*, 2017. [CrossRef]
51. Post, J.E. Manganese oxide minerals: Crystal structures and economic and environmental significance. *Proc. Natl. Acad. Sci. USA* **1999**, *96*, 3447–3454. [CrossRef] [PubMed]
52. Hunter, R.J. Charge Reversal of Kaolinite by Hydrolyzable Metal Ions: An Electroacoustic Study. *Clays Clay Miner.* **1992**, *40*, 644–649. [CrossRef]
53. Mary Ugwu, I.; Anthony Igbokwe, O. Sorption of Heavy Metals on Clay Minerals and Oxides: A Review. In *Advanced Sorption Process Applications*; IntechOpen: Rijeka, Croatia, 2019.
54. Ding, M.; de Jong, B.H.W.S.; Roosendaal, S.J.; Vredenberg, A. XPS studies on the electronic structure of bonding between solid and solutes: Adsorption of arsenate, chromate, phosphate, Pb^{2+}, and Zn^{2+} ions on amorphous black ferric oxyhydroxide. *Geochim. Cosmochim. Acta* **2000**, *64*, 1209–1219. [CrossRef]
55. Sun, X.; Doner, H.E. Adsorption and oxidation of arsenic on goethite. *Soil Sci.* **1998**, *163*, 278–287. [CrossRef]
56. Mohapatra, M.; Sahoo, S.K.; Anand, S.; Das, R.P. Removal of As(V) by Cu(II)-, Ni(II)-, or Co(II)-doped goethite samples. *J. Colloid Interface Sci.* **2006**, *298*, 6–12. [CrossRef] [PubMed]
57. Lu, Y.; Hu, S.; Wang, Z.; Ding, Y.; Lu, G.; Lin, Z.; Dang, Z.; Shi, Z. Ferrihydrite transformation under the impact of humic acid and Pb: Kinetics, nanoscale mechanisms, and implications for C and Pb dynamics. *Environ. Sci. Nano* **2019**, *6*, 747–762. [CrossRef]

58. Srinivasan, R. Advances in Application of Natural Clay and Its Composites in Removal of Biological, Organic, and Inorganic Contaminants from Drinking Water. *Adv. Mater. Sci. Eng.* **2011**, *2011*, 1–17. [CrossRef]
59. Uddin, M.K. A review on the adsorption of heavy metals by clay minerals, with special focus on the past decade. *Chem. Eng. J.* **2017**, *308*, 438–462. [CrossRef]
60. Bhattacharyya, K.G.; Gupta, S. Sen Adsorption of a few heavy metals on natural and modified kaolinite and montmorillonite: A review. *Adv. Colloid Interface Sci.* **2008**, *140*, 114–131. [CrossRef] [PubMed]
61. Mukhopadhyay, R.; Manjaiah, K.M.; Datta, S.C.; Yadav, R.K.; Sarkar, B. Inorganically modified clay minerals: Preparation, characterization, and arsenic adsorption in contaminated water and soil. *Appl. Clay Sci.* **2017**, *147*, 1–10. [CrossRef]
62. Lenoble, V.; Bouras, O.; Deluchat, V.; Serpaud, B.; Bollinger, J.-C. Arsenic Adsorption onto Pillared Clays and Iron Oxides. *J. Colloid Interface Sci.* **2002**, *255*, 52–58. [CrossRef]
63. Na, P.; Jia, X.; Yuan, B.; Li, Y.; Na, J.; Chen, Y.; Wang, L. Arsenic adsorption on Ti-pillared montmorillonite. *J. Chem. Technol. Biotechnol.* **2010**, *85*, 708–714. [CrossRef]
64. Doušová, B.; Fuitová, L.; Grygar, T.; Machovič, V.; Koloušek, D.; Herzogová, L.; Lhotka, M. Modified aluminosilicates as low-cost sorbents of As(III) from anoxic groundwater. *J. Hazard. Mater.* **2009**, *165*, 134–140. [CrossRef]
65. Raichur, A.M.; Panvekar, V. Removal of As(V) by adsorption onto mixed rare earth oxides. *Sep. Sci. Technol.* **2002**, *37*, 1095–1108. [CrossRef]
66. Zhang, Y.; Yang, M.; Dou, X.; He, H.; Wang, D.-S. Arsenate Adsorption on an Fe–Ce Bimetal Oxide Adsorbent: Role of Surface Properties. *Environ. Sci. Technol.* **2005**, *39*, 7246–7253. [CrossRef]
67. Dong, S.; Wang, Y. Characterization and adsorption properties of a lanthanum-loaded magnetic cationic hydrogel composite for fluoride removal. *Water Res.* **2016**, *88*, 852–860. [CrossRef] [PubMed]
68. Jais, F.M.; Ibrahim, S.; Yoon, Y.; Jang, M. Enhanced arsenate removal by lanthanum and nano—magnetite composite incorporated palm shell waste—based activated carbon. *Sep. Purif. Technol.* **2016**, *169*, 93–102. [CrossRef]
69. Emsbo, P.; McLaughlin, P.I.; Breit, G.N.; du Bray, E.A.; Koenig, A.E. Rare earth elements in sedimentary phosphate deposits: Solution to the global REE crisis? *Gondwana Res.* **2015**, *27*, 776–785. [CrossRef]
70. Innocenzi, V.; De Michelis, I.; Kopacek, B.; Vegliò, F. Yttrium recovery from primary and secondary sources: A review of main hydrometallurgical processes. *Waste Manag.* **2014**, *34*, 1237–1250. [CrossRef] [PubMed]
71. Gupta, K.; Ghosh, U.C. Arsenic removal using hydrous nanostructure iron(III)-titanium(IV) binary mixed oxide from aqueous solution. *J. Hazard. Mater.* **2009**, *161*, 884–892. [CrossRef] [PubMed]
72. Brunsting, J.H.; McBean, E.A. Phosphate interference during in situ treatment for arsenic in groundwater. *J. Environ. Sci. Health Part A* **2014**, *49*, 671–678. [CrossRef] [PubMed]
73. Violante, A.; Pigna, M. Competitive Sorption of Arsenate and Phosphate on Different Clay Minerals and Soils. *Soil Sci. Soc. Am. J.* **2002**, *66*, 1788. [CrossRef]
74. O'Reilly, S.E.; Strawn, D.G.; Sparks, D.L. Residence Time Effects on Arsenate Adsorption/Desorption Mechanisms on Goethite. *Soil Sci. Soc. Am. J.* **2001**, *65*, 67. [CrossRef]
75. Lalley, J.; Han, C.; Li, X.; Dionysiou, D.D.; Nadagouda, M.N. Phosphate adsorption using modified iron oxide-based sorbents in lake water: Kinetics, equilibrium, and column tests. *Chem. Eng. J.* **2016**, *284*, 1386–1396. [CrossRef]
76. De-Bashan, L.E.; Bashan, Y. Recent advances in removing phosphorus from wastewater and its future use as fertilizer (1997–2003). *Water Res.* **2004**, *38*, 4222–4246. [CrossRef]

© 2019 by the authors. Licensee MDPI, Basel, Switzerland. This article is an open access article distributed under the terms and conditions of the Creative Commons Attribution (CC BY) license (http://creativecommons.org/licenses/by/4.0/).

Article

Groundwater Autochthonous Microbial Communities as Tracers of Anthropogenic Pressure Impacts: Example from a Municipal Waste Treatment Plant (Latium, Italy)

David Rossi [1], Anna Barra Caracciolo [1,*], Paola Grenni [1], Flavia Cattena [1], Martina Di Lenola [1], Luisa Patrolecco [1], Nicoletta Ademollo [1], Ruggiero Ciannarella [2], Giuseppe Mascolo [2] and Stefano Ghergo [1]

1 National Research Council, Water Research Institute (IRSA-CNR), Area della Ricerca RM1, 00015 Rome, Italy; david.rossi@irsa.cnr.it (D.R.); grenni@irsa.cnr.it (P.G.); flavia.cattena@hotmail.it (F.C.); dilenola@irsa.cnr.it (M.D.L.); patrolecco@irsa.cnr.it (L.P.); ademollo@irsa.cnr.it (N.A.); ghergo@irsa.cnr.it (S.G.)
2 National Research Council, Water Research Institute (IRSA-CNR), Viale Francesco de Blasio, 5, 70132 Bari, Italy; ruggero.ciannarella@ba.irsa.cnr.it (R.C.); giuseppe.mascolo@ba.irsa.cnr.it (G.M.)
* Correspondence: barracaracciolo@irsa.cnr.it; Tel.: +39-06-9067-2786

Received: 11 July 2019; Accepted: 10 September 2019; Published: 16 September 2019

Abstract: The groundwater behavior at a municipal solid waste disposal dump, located in Central Italy, was studied using a multi-parameter monitoring over 1 year consisting of 4 seasonal samples. The hydrological and hydrogeological dynamics of water circulation, microbiological parameters (microbial abundance and cell viability of the autochthonous microbial community), dissolved organic carbon, and several contaminants were evaluated and related to the geological structures in both two and three dimensions and used for geostatistical analysis in order to obtain 3D maps. Close relationships between geological heterogeneity, water circulation, pollutant diffusion, dissolved organic carbon, and cell viability were revealed. The highest cell viability values were found with dissolved organic carbon (DOC) values ≤ 0.5 mg/L; above this value, DOC negatively affected the microbial community. The highest DOC values were detected in groundwater at some sampling points within the site indicating its probable origin from the waste disposal dump. Although legislation limits for the parameters measured were not exceeded (except for a contaminant in one piezometer), the 1-year multi-parameter monitoring approach made it possible to depict both the dynamics and the complexity of the groundwater flux and, with "non-legislative parameters" such as microbial cell viability and DOC, identify the points with the highest vulnerability and their origin. This approach is useful for identifying the most vulnerable sites in a groundwater body.

Keywords: geostatistical analysis; geological heterogeneity; dissolved organic carbon; autochthonous microbial community

1. Introduction

Contamination of groundwater is a serious environmental problem and a risk to human health, especially because many communities depend on groundwater as the sole or major source of drinking water. In fact, more than 65% of the drinking water produced in Europe is sourced from groundwater [1,2]. Most subsurface systems consist of rock fissures and caves or pores in sand, silt, and gravel where the water flows and connects the aquifer with the surface water [3–5]. Soil pollution can influence groundwater quality because groundwater bodies are the final recipient of pollutants of both a point and nonpoint nature due to anthropic pressure. Groundwater is increasingly threatened by contaminants from human activities, like agriculture, industrial plants, livestock farming, septic tank leaking, and subsurface storage of household, municipal solid and industrial wastes [6–9].

The attenuation capacity of the unsaturated zone is crucial in order to reduce the contamination of the water reservoir below. In this context, persistence and mobility in soil of a specific chemical are key aspects for evaluating groundwater vulnerability [10,11]. An organic compound persistence is strictly dependent on its degradation route, which is a complex phenomenon dependent on both abiotic and biotic processes occurring mainly in the surface environment. However, the role of natural microbial communities in degrading or removing contaminants in situ [12,13] or in groundwater microcosm experiments has recently been discovered [1,14–16].

Leaching of a contaminant from surface to groundwater depends on various chemical and physical processes which take place in the underground profile and are linked to the chemodynamic properties of a substance (e.g., sorption to soil organic matter, reactivity of the solid matrix) and to specific hydrogeological settings. The process occurs through matrix and macropore flow. In the latter case, large and discontinuous macropores can operate as preferential flow pathways causing a rapid movement of chemicals through the unsaturated zone [17,18]. In other cases, pollutant persistence is related to the water travel times, which determine their concentration and impact on biota [19,20].

Volcanic areas are generally considered protected from groundwater contamination due to their intrinsically low permeability [21–24]; however, volcanic substrata, considered to be un-deformed or blandly deformed, could be intensely fractured due to brittle deformation processes [25]. In particular, meso-scale brittle deformation elements such as faults and extensional fractures, define a potential gradient of water circulation related to the strain processes [26]. In this context, geological heterogeneity is likely to be a problem in planning management programmes (e.g., sampling design). The linkage of brittle elements and their frequency, orientation, and spatial distribution influence the mechanical stratigraphy [27,28] and the environmental conditions [29,30]. Consequently, the understanding of the mechanisms underlying the reservoir shape and water travel time facilitates the optimization and planning of sustainable use of natural resources [31] in areas characterized by intense anthropogenic activities. The concept of geological heterogeneity is key in creating a well-defined model able to predict directions, flux and time-space movement of water, including possible pollutants, within a reservoir. Accordingly, the understanding of mechanisms underlying the rate of diffusion of water-borne pollutants will facilitate the optimization of sampling efforts (e.g., number of measurements in living organisms).

The EU Directive 2000/60/EC (Water Framework Directive, WFD) [32] defines a general objective of reaching a good quality status in all surface and groundwater bodies and introduces the principle of preventing any further deterioration in status. Analyses of ecological and chemical parameters are foreseen for surface water, while only quantitative and chemical parameters are considered for groundwater bodies. Consequently, for assessing a groundwater body status, WFD does not take into consideration biota and considers that subsurface environments are without life. The key role of microbial communities in biochemical processes and ecosystem services [33], including water purification and organic contaminant degradation in groundwater, is nowadays recognized [1,34,35]. Nevertheless, there has been little research into these communities owing to the lack for a long time of consistent and suitable methods for identifying and characterizing their structure and functioning. The application to microbial ecology of molecular methods has made it possible to study groundwater microbial communities in their natural environment [1,34].

Microorganisms constitute the major group of organisms in groundwater systems, both in terms of biomass and activity [36]. Community composition is key for all the ecosystem processes [37]. Where groundwater is in a near-natural state and far from the surface, one may assume a steady state without much development over time, but spatial gradients are important [38]. The microbial community response may prove an indicator of residual toxicity linked to the disappearance or sequestration of pollutants and their degradation products [34,39,40].

The factors that regulate the occurrence and abundance of microorganisms, their biotic interactions and their community structure are relatively difficult to discern, due to their large numbers, small size, and difficulty in evaluating the spatial or temporal scale of the environmental gradients influencing microbial communities [41,42]. The organic material load, in particular the dissolved organic carbon (DOC), has proved useful for the study of factors governing microbial activity in groundwater [43,44].

The current study aims to show the importance of taking into consideration the autochthonous microbial communities in groundwater in addition to other ecological parameters (DOC) in quality monitoring control because of their sensitivity to anthropogenic pressures. We illustrate this by a study of a municipal solid waste dump located in Central Italy, which collects daily waste from the city of Rome.

The overall results showed that legislation limits were exceeded only for a contaminant in one piezometer, while the 1-year multi-parameter monitoring approach made it possible to depict the dynamicity and complexity of the groundwater flux and identify where contaminant movement to groundwater was concentrated. Close relationships between geological heterogeneity, water circulation, pollutant diffusion, dissolved organic carbon, and microbial viability were found. This approach can be extremely useful for improving the monitoring plans and to identifying the most vulnerable sampling points in groundwater.

2. Materials and Methods

2.1. Geological and Hydrogeological Setting

The investigated area is about 52 ha of which 11 ha comprise waste disposal dumps and the rest treatment facilities (Figure 1). It is located in the South-Western sector of the Colli Albani volcanic district (Albano Laziale, Rome), which is represented by Quaternary volcanisms belonging to the ultrapotassic Roman province [45]. The treatment facility manages waste with a mechanical biological treatment aimed at the production of secondary solid fuel and stabilized organic fraction, and metal recovery. It also manages the biogas collection system and related electricity production facility. The plant is able to treat 500 tons of municipal solid waste per day.

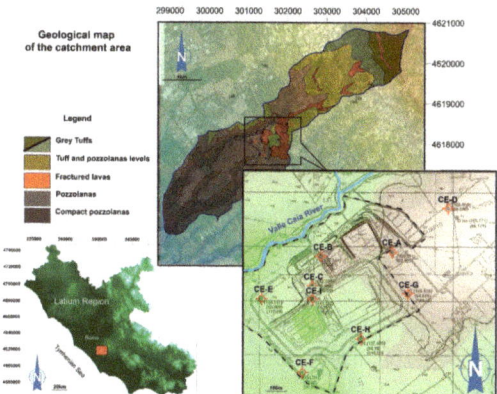

Figure 1. Study area and sampling site location map: the catchment basin is highlighted with a blue line (intermediate map); the dashed black line shows the site (detail map) where each red circle represents a sampled piezometer (CE-A, CE-B, CE-C, CE-D, CE-F, CE-G, CE-H, and CE-I).

The area is located at the top of the Tuscolano-Artemisio volcanic stage [46–48]. Moreover, the outcropping formations are characterized by grey tuffs, fractured lava, and compact pozzolanas. The last, from a hydrological point of view, represents the impermeable layer of the aquifer hosted in the fractured lavas. The groundwater circulation is linked to the extensive Colli Albani main hydrogeological unit [21], characterized by a centrifugal drainage trend around the Albano Lake volcanic complex [49,50]. The whole area is also characterized by several minor perched aquifers maintained by thin waterproof layers represented by compacted tuffs and paleosols. It has long been recognized in literature that the permeability of rocks is extremely variable, from low to high [51–53]. This is mainly due to a secondary process, represented by fracturation, which completely changes the hydrogeological behavior of the un-deformed rocks.

2.2. Digital Elevation Model

With the integration of elevation data gathered from topographic maps, aerial photo interpretation and field measurements, a Digital Elevation Model (DEM) was created (Figure 1) using the Surfer10 software.

A rectangular grid (numerical matrix) was realized using the intersection of rows and columns to obtain a three-dimensional representation of the entire drainage basin.

A highly detailed Digital Elevation Model (DEM) was used to automatically map the stream channel and divide the networks of a watershed located in the study area. The construction of a code describing the network topology, each drainage area and the associated stream link represents the basis for an efficient watershed information system. A detailed DEM was superimposed on a thematic slope map and on drainage direction/intensity ones to obtain dynamic surface maps and to identify the local-scale run-off potentially affecting the municipal solid waste disposal site [54–57].

Meso and macrostructural surveys together with surface and sub-surface geological analyses were carried out to spatially identify the geological structures, geometry, and when possible the fracturation network. The latter represents a fundamental aspect that directly affects the water circulation and facilitates infiltration processes and groundwater circulation.

The DEM was realized with a mesh of 5 meters in order to have a reproduction as close as possible to reality. When necessary, in order to increase the ground resolution of the altimetric variations, the mesh was thickened up to 2 m ground resolution. A geological cross section and block diagram were realized by processing data from the stratigraphical logs of the piezometer used as a monitoring network.

2.3. Groundwater Sampling and on Site Characterization

It was first necessary to depict the dynamicity and complexity of the groundwater flux and identify where contaminant movement to groundwater was concentrated. Therefore, water samples were collected from eight piezometers four times and the main chemical (inorganic and organic contaminants) and quantitative parameters (groundwater regime level) required in both the Italian [58] and European (Water Frame Directive) [32] legislation were measured. An ecological evaluation was performed in addition. It includes the analyses of the groundwater natural microbial community (cell abundance and viability) and the dissolved organic carbon (DOC). Finally, we integrated the overall data with the geological structures in both two and three dimensions.

A preliminary campaign made it possible to select 8 internal piezometers and an external one (Figure 1, Table 1). Water samplings for chemical and microbiological analyses and hydrogeological measurements were performed in all piezometers within the municipal solid waste disposal site.

Table 1. Maximum depth, piezometer-top/bottom and diameter, and static water level of the CE-A, CE-B, CE-C, CE-D, CE-E, CE-F, CE-G, CE-H, and CE-I piezometers in the four sampling campaigns (I, II, III and IV). Static level was not measured for CE-E piezometer. a.s.l. = above sea level.

Piezometer	Maximum Depth from Surface (m)	Piezometer-Top/Bottom (m) (a.s.l.)	Diameter (mm)	Static Level (m) (a.s.l.)				Range
				I September	II February	III July	IV October	
CE-A	153	156.39/3.39	150	63.5	63.8	63.8	64.2	0.7
CE-B	100	140.74/40.74	150	59.6	59.7	60	60	0.4
CE-C	142	142.58/0.58	250	57.1	57.5	57.8	57.6	0.7
CE-D	158	166.31/8.31	180	68.8	68.9	69.4	69.3	0.6
CE-E	15	134.52/119.52	75	-	-	-	-	-
CE-F	90	134.04/44.04	75	54.2	54.8	56.1	56.1	1.9
CE-G	120	155.82/35.82	75	55.5	53.8	62.8	57.2	7.3
CE-H	115	137.31/20.31	75	54.9	54.2	58.9	56.4	4.7
CE-I	125	143.96/16.96	170	56.2	56.6	56.8	56.6	0.4

Each single water point was sampled in four campaigns during one year (between 2012 and 2013: I September, II February, III July, IV October) in order to have a multi-temporal analysis of the parameters investigated. Owing to accessibility problems, the CE-G piezometer was sampled only in the third and fourth campaigns. Hydrogeological (piezometric depth-average: 124 m), chemical (main inorganic and organic contaminants), DOC and microbiological data (total microbial abundance and cell viability) were analyzed inside and outside the area studied in order to assess if the water body might be negatively affected by the waste disposal plant.

For this purpose, data from the eight internal piezometers (named CE-A, CE-B, CE-C, CE-E, CE-F, CE-G, CE-H, CE-I) and the external one (CE-D, which is the most upstream one) were compared during 1 year and processed with a geostatistical approach [59–61].

For each piezometer the depth-average piezometric level and main physico-chemical parameters (pH, redox potential, electrical conductivity, dissolved oxygen, temperature, and alkalinity) were measured on site using a pre-calibrated multiparameter probe (WTW, Germany) [53,62]. Water samples were collected from each single piezometer after 30 minutes of purging and placed in sterile polyethylene bottles to avoid any contamination. Subsamples were immediately processed for different purposes. For trace elements, dissolved organic carbon (DOC) and main cations and anions, water samples were filtered (0.45 µm) and then put in polyethylene bottles (previously cleaned with HNO_3 and then washed with milliQ water). A pre-acidification of the sample (HNO_3 1%) was performed before metal and DOC analyses. For organic contaminants, water samples were collected in pre-cleaned 2.5 L dark glass bottles. For microbiological analyses, water samples were immediately fixed with formaldehyde (2% final concentration) for microbial abundance or were kept fresh for microbial cell viability. All the groundwater samples were stored (24 h maximum) at 4 °C before use.

The reconstruction of the piezometric level of the aquifer was performed through the analysis of the static groundwater level using several measuring points consisting of springs, piezometers and equipped piezometers within the municipal solid waste disposal site. These data were processed to obtain a three-dimensional surface, suitable for obtaining a 3D volumetric geometry of the aquifer. All the data were analyzed using geostatistical programs.

2.4. Chemical Analyses

The main inorganic and organic pollutants considered in the WFD [32] and Italian legislation (Decree law 152/06) [58] for evaluating the water quality state were searched for.

Analyses of thirty-two inorganic elements (anions, cations, and metals) were performed in accordance with the Italian Environment Protection Agency Guidelines [63] and APHA, AWWA, WEF methods [64]. Inorganic Anions (Fluoride, Chloride, Nitrite, Nitrate, and Sulfate) were determined by ion chromatography using a Dionex DX-120 Ion Chromatograph. Inductively coupled plasma-mass spectrometry (ICP-MS-Agilent technologies 7500c, Santa Clara, CA, USA) with an Octopole Reaction System (ORS) was used for the determination of Boron, Barium, Antimony, Arsenic, Cadmium,

Total Chromium, Copper, Lead, Mercury, Nickel, Selenium, Vanadium, Iron, Zinc, Manganese, Aluminum, Strontium, Lithium, Cesium, Uranium, and Cobalt.

Calcium, Magnesium, Sodium, and Potassium were determined by Inductively Coupled Plasma Optical Emission Spectroscopy (ICP-OES-Perkin Elmer, Waltham, MA, USA) using a Perkin Elmer P400 spectrometer.

Analysis of ninety-two organic contaminants was performed. Polycyclic Aromatic Hydrocarbons (PAHs) were extracted by SPE (solid-phase extraction) and then determined using a high-performance liquid chromatography (HPLC-Varian 9010-Agilent technologies, Santa Clara, CA, USA) with fluorescence detection (FL-*Perkin Elmer* LS30-Perkin Elmer, Waltham, MA, USA) [65]. The identity of each analyte was confirmed by liquid chromatography-mass spectrometry (LC-MS) in accordance with Patrolecco et al. [65].

Volatile organic compounds (VOCs) were determined by the dynamic purge-and-trap technique using a Tekmar dynamic purge-and-trap system (Teledyne Tekmar, Mason, OH USA), equipped with a 25 mL glass vessel, coupled to a gas Saturn 2200 (Varian) chromatograph/mass spectrometric system (GC/MS-Agilent technologies, Santa Clara, CA, USA). An aliquot of 25 mL of the water sample was manually introduced into the vessel using a syringe. The VOCs purged by a helium flow were trapped onto a Tenax cartridge and then thermally desorbed at 210 °C and swept to the GC column were they were separated and detected by the MS system. LODs obtained were between 0.005 and 0.1 µg/L depending on the VOC selected.

Chlorinated pesticides were extracted applying solid-phase extraction (SPE) and then determined by gas chromatography (GC) with an electron capture detector (ECD) using *Perkin Elmer, Clarus* 480 and a fused silica capillary column [66].

The list of overall organic compounds analyzed in the groundwater samples is reported in Table 2.

Table 2. List of organic contaminants examined in the nine piezometers sampled.

Group of Organic Contaminants	Chemicals
Polycyclic Aromatic Hydrocarbons (PAHs)	Naphthalene, Acenaphthene, Fluorene, Phenanthrene, Anthracene, Fluoranthene, Pyrene, Benzo(a)Anthracene, Chrysene, Benzo(b)fluoranthene, Benzo(k)fluoranthene, Benzo(a)pyrene, Dibenz(a,h)anthracene, Benzo(g,h,i)perylene, Indeno(1,2,3-c,d)pyrene, Dibenzo(a,e) pyrene, Dibenzo(a,i) pyrene, Dibenzo(a,h) pyrene, Dibenzo(a,l) pyrene
Volatile Organic Compounds (VOCs)	Trichloromethane, Tribromomethane, 1,2-Dibromoethane, Dibromochloromethane, Bromodichloromethane, 1,2-Dichloroethane, 1,1-Dichloroethane, 1,2-Dichloropropane, 1,1,2-Trichloroethane, 1,2,3-Trichloropropane, 1,1,2,2-Tetrachloroethane, Chloroethene, Chloromethane, 1,2-Dichloroethene, 1,1-Dichloroethene, Trichloroethylene, Tetrachloroethylene, Hexachloro-1,3-butadiene, Chlorobenzene, 1,2-Dichlorobenzene, 1,3-Dichlorobenzene, 1,4-Dichlorobenzene, 1,2,4-Trichlorobenzene, 1,2,3-Trichlorobenzene, n-Hexane, Benzene, Ethylbenzene, Styrene, Toluene, ortho-Xylene, meta-Xylene, para-Xylene, (1-methylethyl)benzene, Propylbenzene, Methylbenzene, Bromobenzene, Phenylamine, Diphenylamine, o-Toluidine, prop-2-enamide, pyridine-3-carboxylic acid, 1,2-Dinitrobenzene, 1,3-Dinitrobenzene, 1-methyl-2,4-dinitro benzene, n-Nitrotoluene, 1-Chloro-2-nitrobenzene, 1-Chloro-3-nitrobenzene, 1-Chloro-4-nitrobenzene, 1,2,3-Trichloropropane, Dichloromethane, Bromomethane, Tetrachloromethane, Chloroethane, Dichlorodifluoromethane, Trichlorofluoromethane, Bromoethene, 1-bromo-2-chloroethane, Dichlorobutane, 1,1,1-Trichloroethane, 1,1,1,2-Tetrachloroethane, Bromodichloroethane, 4-Bromochlorobenzene, 1,1-Dichlorocycloutane, 1,4-Dichlorobut-2-ene, 2-Methoxy-2-methylpropane, Diiodomethane, n- Propylbenzene
Chlorinated Pesticides	Dichlorodiphenyldichloroethylene (DDE) Gamma-hexachlorocyclohexane (γ HCH)

2.5. Dissolved Organic Carbon (DOC)

Aliquots of water samples (20–500 µL) were analyzed for dissolved organic carbon content with high temperature catalytic oxidation using a Shimadzu TOC-5000A Total Organic Carbon Analyzer, with a detection limit of 0.15 mg/L.

2.6. Microbiological Analysis

The microbial abundance (No. cells/mL) was determined in at least four replicates in each piezometer. Water aliquots (20 mL) fixed in formaldehyde (2%) were filtered on black polycarbonate filters (pore size 0.22 µm, 25 mm diameter) and stained for 20 minutes at 4 °C in the dark with the DNA intercalant DAPI (at the final concentration of 1 µg/L). Finally, the filters were observed under an epifluorescence microscope (Leica DM 4000B). The DAPI method was used because it is able to detect all the microbial cells in a sample whatever their physiological state and metabolic activity and for this reason is suitable for the total microbial counts [67,68].

Cell viability (% live cells/live + dead) was measured in four fresh replicate subsamples (20 mL each) using two fluorescent dyes. SYBR Green II and propidium iodide (Sigma-Aldrich, Germany) were used for distinguishing between viable (green) and dead or damaged (red) cells under a fluorescence microscope [34]. The cells were counted in both analyses with the epifluorescence microscope at 1000 magnification, counting a minimum of 300 cells per filter. Each microbiological measure was the average of four replicate filters comprising at least 1200 microbial cells counted.

2.7. Statistical Analysis

The statistical analysis was performed, using the R software version 2.15 (downloaded from the Comprehensive R Archive Network (CRAN), which is located at http://cran.r-project.org/). The t test was used to compare the overall organic contaminant concentrations among the different piezometers. Moreover, any correlations between DOC and abiotic (conductivity, dissolved oxygen, bicarbonates, $\sum$ organic contaminants, PAHs, VOCs) and biotic (microbial abundance and cell viability) parameters, whatever the piezometer location and the sampling time, were also evaluated.

2.8. Geostatistical Analysis

In order to investigate the distribution and evolution patterns of pollutants in the groundwater an ordinary kriging [69–71] was performed taking into consideration the overall data obtained from the sampling, using the Surfer10 software.

This method consists of a spatial interpolation technique, which makes it possible to obtain contour distribution maps. Among the numerous geostatistical gridding methods that are useful in many fields of applications, kriging provides visually appealing maps from irregularly spaced data and incorporates anisotropy and underlying trends in an efficient and natural manner [72]. Moreover, kriging is very flexible and makes it possible to obtain spatial autocorrelation graphs, allowing prediction, prediction standard errors, and probability maps, minimizing at the same time errors in predicted values. To measure the mean variability between two sampled points x and $x + h$, as a function of their distance h semi-variograms were produced. The semi-variogram is an autocorrelation statistic defined as follows:

$$\hat{y}(h) = \frac{1}{2n} \sum_{i=1}^{n} (z(x_i) - z(x_i + h))^2$$

where n is the number of pairs of sample points separated by distance h, $z(x_i)$ is the value of the variable z at point i, and $z(x_i + h)$ is the value of the variable z at point $i + h$. The assumptions of kriging are a stationary difference between x and $x + h$ and variance of the differences, which define the requirements for the intrinsic hypothesis. This means that semi-variance does not depend on the location of samples and only depends on the distance between samples, so that the semi-variance is isotropic. Consequently, several geostatistic thematic maps of the site were made using the most suitable model with the least errors.

3. Results

3.1. Geological and Hydrogeological Characterization

The reconstruction of a detailed geological cross section (Figure 2) allows us to identify the limit of the aquiclude represented by the "Pozzolane inferiori" formation. Using the surface geological and geometrical data acquired directly in the field and well log stratigraphy it was possible to correlate and model the complete volcanic sequences observed in the area investigated. Furthermore, each single geologic unit was geometrically identified and superimposed on the others to obtain a complete 3D picture of the whole volcanic template succession using the Voxler3 software. In fact, the overall volcanic sequence lies in an unconformity above a thick sedimentary succession mainly represented by clay, sand, and silt. Heterogeneity of the geological substrata is represented by the alternation between tuff levels and lava lenses in the upper half portion of the cross section (between 0 and 50 m), while the lower half one is characterized by thick fractured lava levels. The latter are characterized by constant thickness in the NE–SW direction and by important lateral thickness variations in the W–E direction. This aspect represents an important factor that strongly limits water circulation towards West and East.

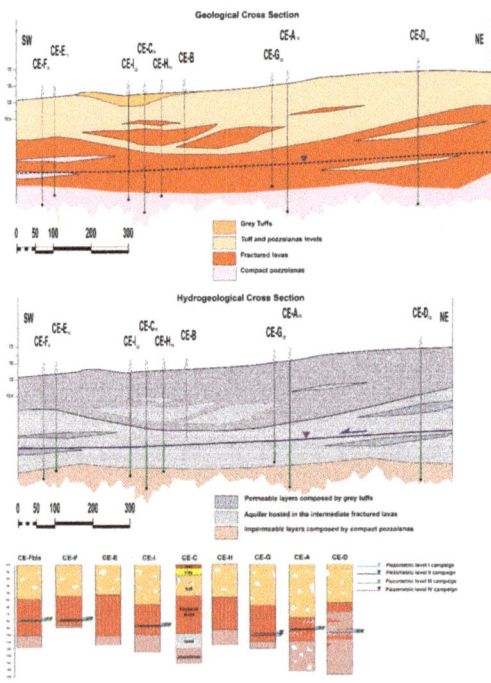

Figure 2. Geological and hydrogeological cross sections within the municipal solid waste disposal dump and stratigraphic correlation of the piezometers investigated.

With a total thickness of 30–35 m and a widespread areal distribution, the aquifer located in the thick fractured lavas and supported by compact pozzolanas, which act as an aquiclude (Figures 2 and 3), represents a fundamental element able to rapidly convey pollutants toward SSW, driven by the main drainage direction (Figure 4). The 3D geological and hydrogeological model [73–76] allows us to better define the vertical and lateral variations of each single rock body [73–78]. Because both the fractured lavas and the tuff layers do not show a constant thickness in the whole area studied, it is fundamental to build up a detailed geometry of the volcanic deposits, showing a progressive reduction of the thickness towards both West and South. The latter aspect is key to a better understanding of the hydraulic behavior of the aquifer and a better identification of the relationships between the dump activities, water circulation and bacterial behavior.

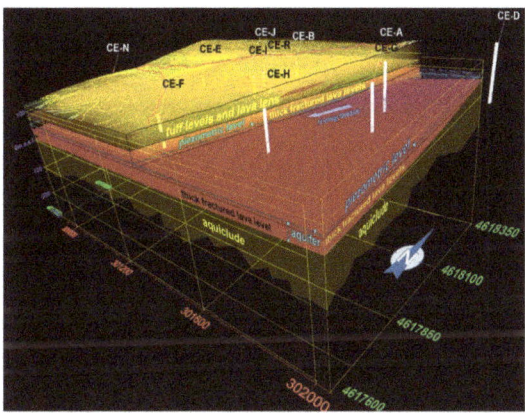

Figure 3. Detailed 3D sketch of the area investigated: groundwater (blue line), aquiclude (green), lavas (red) and tuff formations (yellow). Each white vertical pin corresponds to a piezometer (CE-A, CE-B, CE-C, CE-D, CE-F, CE-G, CE-H, and CE-I). The geological cross-section is oriented from North-East to South-West (NE–SW). The surface red line corresponds to the contour of the waste treatment plant.

The reconstruction of the piezometric level shows a time-space variation with a progressive increase in the water level from 68.5 m (I campaign) to 69.3 m (IV campaign) at piezometer CE-D, which is the upstream of the overall piezometers sampled (Table 1).

The piezometric variation shows a trend that does not seem influenced by seasonal variation, as highlighted by the high level of the piezometric level measured during the February 2013 campaign in the upper stream sector. Despite the vertical variation observed, the drainage direction remained constant during the investigation period showing a main drainage direction from N-NE to S-SW. During the last campaign, the southern sector was characterized by a counterclockwise rotation of the drainage direction toward South. This process was mainly observable within the area comprising the CE-F, CE-H, and CE-I piezometers, as shown in Figure 4.

It is important to underline that the groundwater drainage direction did not have a uniform trend; the gradient in the northern sector (CE-D, CE-A, CE-B, CE-C, CE-I piezometers) was higher than that in the southern sector (CE-F, CE-H, CE-G piezometers) (Figure 4). This result can be ascribed to the inhomogeneities in the permeability profile affecting the groundwater level. Based on the subsurface geological analysis, the CE-B piezometer was the most vulnerable to the waste disposal occurrence owing to the presence of several fractured lava layers between the surface and groundwater levels. This makes possible a passive transport of both organic matter and contaminants from surface to groundwater. On the contrary, the CE-D piezometer was intrinsically less vulnerable because it was located upstream and outside the waste disposal area and was characterized by a vertical geological homogeneity with a low fracturation process. For this reason, it was considered a control piezometer.

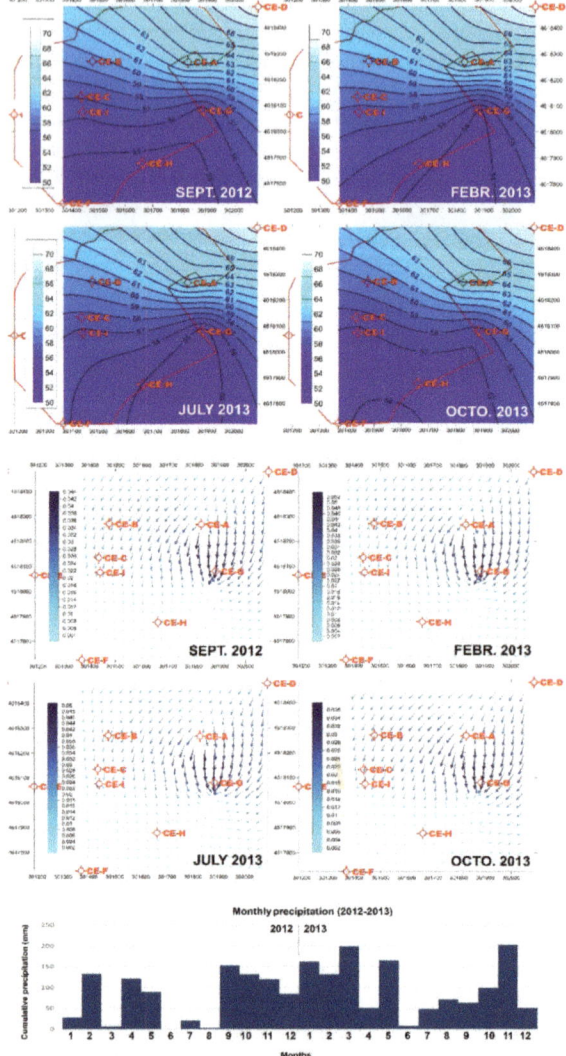

Figure 4. Reconstruction of the piezometric level variations (meters above sea level) in the different sampling campaigns: I September 2012; II February 2013; III July 2013; IV October 2013. Each red circle represents a piezometer sampled (CE-A, CE-B, CE-C, CE-D, CE-F, CE-G, CE-H, and CE-I). The blue arrows show the drainage direction and magnitude. The red line corresponds to the contour of the waste treatment plant. In the bottom part of the Figure, the monthly cumulative precipitation are reported.

3.2. Main Chemical–Physical Parameters on Site

Temperature and pH in all piezometers were quite constant (Supplementary Materials Table S1), with no significant differences between different sampling times and piezometers, with average values of 17.54 ± 0.09 °C and 7 ± 0.03, respectively. Conductivity and dissolved oxygen values are reported in Table 3 and Table S1 (Supplementary Materials).

Table 3. Conductivity (μS/cm) and dissolved oxygen (mg/L) found in each piezometer (CE-A, CE-B, CE-C, CE-D, CE-E, CE-F, CE-G, CE-H, and CE-I) at each sampling campaign.

Piezometer	Conductivity (μS/cm)				Dissolved Oxygen (mg/L)			
	I September	II February	III July	IV October	I September	II February	III July	IV October
CE-A	704	1404	789	758	9.14	8.30	8.8	4.3
CE-B	1029	1022	1028	1039	3.3	3.50	3.2	3.3
CE-C	855	905	847	840	8.1	5.2	4.0	7.5
CE-D	718	1412	713	710	8.87	8.9	8.9	8.8
CE-E	713	707	709	718	6.8	6.58	5.1	6.4
CE-F	1039	1040	1035	1049	4.4	3.64	3.7	3.7
CE-G	-	850	862	-	-	6.97	5.1	-
CE-H	850	843	790	798	3.54	6.28	7.5	7.4
CE-I	769	792	770	774	6.5	6.73	6.7	7.3

The highest values for conductivity and the lowest for oxygen were in the CE-B and CE-F piezometers. All the other parameters are reported in Supplementary Materials Table S1.

3.3. Chemical Analysis

Anion and cation concentrations found in the four samplings rarely exceeded the corresponding threshold limits required by the Italian legislation (D.Lgs 152/2006 Annex 5; D.Lgs 30/2009, Annex 3), as can be seen in Table 4 and in detail in Supplementary Materials (Table S1).

Table 4. Minimum (Min) and maximum (Max) concentrations of anions and cations found in the four sampling campaigns (second and third column); threshold limits reported in the Italian legislations (fourth column); Numbers of exceeding values and corresponding piezometers (fifth and sixth column).

Ions	Min	Max	Legal limits *	>Limits	Piezometers
Fluorides	0.86	6.64	1.5 mg/L	11	CE-C; CE-E; CE-H; CE-I
Nitrites	0	0	0.5 mg/L	0	
Sulphates	19.1	40.39	250 mg/L	0	
Be	0.1	0.9	4 μg/L	0	
B	114	803.54	1000 μg/L	0	
Al	3.1	1718	200 μg/L	8	CE-E; CE-F; CE-G; CE-H
Cr Tot	0.8	3.7	50 μg/L	0	
Cr (VI)	<0.5	<0.5	5 μg/L	0	
Mn	0.2	46.2	50 μg/L	0	
Fe	7.5	543	200 μg/L	3	CE-G; CE-H
As	8.3	26.58	10 μg/L	28	CE-A; CE-B; CE-C; CE-E; CE-F; CE-G; CE-H; CE-I
Ni	0.3	7.8	20 μg/L	0	
Cu	0.21	4.53	1000 μg/L	0	
Co	0.3	2.5	50 μg/L	0	
Zn	2.5	838	3000 μg/L	0	
Se	0.1	0.6	10 μg/L	0	
Ag	<0.1	<0.1	10 μg/L	0	
V	30.2	49.8	50 μg/L	0	
Cd	0.11	0.15	5 μg/L	0	
Sb	0.2	0.4	5 μg/L	0	
Hg		<0.1	1 μg/L	0	
Pb	0.1	1.5	10 μg/L	0	

* D.Lgs 152/06; D.Lgs 30/2009, Annex 3.

Nitrites were never found and the legislation limits for inorganic elements were exceeded only in the case of fluorides (>1.5 mg/L), aluminum, iron and arsenic in several piezometers

(Supplementary Materials Table S1). However, these values were in line with those naturally found in volcanic areas of Central Italy and can therefore be ascribed to a geogenic origin [53,79].

The overall chemical analysis results (see Supplementary Materials, Tables S2, S3, and S4) showed that although 92 contaminants were searched for, only 31 were found in residual concentrations. The legislation limit (0.15 µg/L) for the volatile organic compound 1,2 Dichloropropane was exceeded just in the case of CE-B (0.17 µg/l) and CE-F (0.55 µg/L) piezometers in the third campaign (July), (Table S4).

However, in order to assess a possible impact from the waste dump and the groundwater vulnerability to residual multiple contamination, at each single piezometer sampled, the residual concentrations of each single compound were added together and a "sum value" is reported for each single piezometer in Table 5.

Table 5. Sum value of all organic contaminants (µg/L) found in each piezometer (CE-A, CE-B, CE-C, CE-D, CE-E, CE-F, CE-G, CE-H, and CE-I) and at each sampling.

Piezometer	I September (µg/L)	II February (µg/L)	III July (µg/L)	IV October (µg/L)
CE-A	0.398	0.122	0.434	0.287
CE-B	0.813	0.238	1.08	0.597
CE-C	0.508	0.129	0.92	0.353
CE-D	0.312	0.049	0.674	0.259
CE-E	0.291	0.061	0.465	0.236
CE-F	0.667	0.529	1.579	-
CE-G	0.646	0.163	0.347	0.179
CE-H	0.651	0.121	-	0.546
CE-I	0.398	0.219	1.12	0.288

In the first sampling, 15 micro-contaminants were found (trichloroethylene and tetrachloroethylene were in all piezometers sampled); however, the legislation limits were never exceeded. The highest number of contaminants and the highest sum value were found at the CE-B piezometer (0.8 µg/L: VOCs 0.78 µg/L + PAHs 0.021 µg/L), followed by the CE-F (0.7 µg/L: VOCs 0.66 µg/L + PAHs 0.04 µg/L), CE-H (0.65 µg/L: VOCs 0.633 µg/L + PAHs 0.016 µg/L) and CE-I (0.4 µg/L: VOCs 0.38 µg/L + PAHs 0.022 µg/L).

In the second sampling, 21 micro-contaminants were found, although their concentrations were very low. The highest sum values were found at the CE-F piezometer (0.53 µg/L: 0.13 µg/L VOCs + 0.4 µg/L PAHs) and at the CE-B one (0.24 µg/L: 0.09 µg/L VOCs + 0.15 µg/L PAHs), respectively.

In the third sampling (July) 22 micro-contaminants were found and their sum values were significantly higher (t test, $p < 0.01$) than those in the two previous samplings. In particular, at the CE-F piezometer the sum value of all contaminants was 1.58 µg/L (1.56 µg/L for VOCs and 0.016 µg/L for PAHs) and at the CE-B piezometer 1.08 µg/L (only VOCs were found). In this sampling, beyond the most frequently found Trichloroethylene, Tetrachloroethylene, 1,2-Dichloropropane, Etilbenzene, o/p-Xilene, and Phenantrene were also found (Tables S3 and S3). However, the legislation limit (0.15 µg/L) was exceeded just for the volatile organic compound 1,2-Dichloropropane at both the CE-B (0.17 µg/L) and CE-F (0.55 µg/L) piezometers.

In the fourth sampling (October) 15 micro-contaminants were found; 1,2-Dichloropropane was found at the CE-B, CE-C, CE-F, and CE-I piezometers, however its concentrations were always in trace amounts (Tables S2 and S3), below the legislation limit.

3.4. Dissolved Organic Carbon

The average value of DOC was 0.6 mg/L ± 0.1 at the CE-D piezometer located upstream from the dump and at the CE-A and CE-E piezometers located on the boundary of the dump; the DOC values were significantly higher ($p < 0.05$) (ranging from 0.7 to 1.1 mg/L) at the other piezometers (CE-B, CE-C, and CE-F). No relationships were found between the DOC values and sum values of all organic contaminants; while a positive relationship was found between DOC and dissolved oxygen and conductivity ($p < 0.01$) (Table 6).

Table 6. Dissolved Organic Carbon (DOC) values determined in each piezometer (CE-A, CE-B, CE-C, CE-D, CE-E, CE-F, CE-G, CE-H, and CE-I) at each sampling campaign.

Piezometer	DOC (mg/L)					
	I September	II February	III July	IV October	Average	±se
CE-A	0.4	0.9	0.5	0.5	0.6	±0.1
CE-B	1.4	1.3	0.9	0.7	1.1	±0.2
CE-C	0.5	1.2	0.6	0.4	0.7	±0.2
CE-D	0.5	0.7	0.6	0.4	0.6	±0.1
CE-E	0.3	0.7	0.7	0.6	0.6	±0.1
CE-F	1.0	1.2	1.2	1.0	1.1	±0.0
CE-G	0.4	0.8	1.4	0.6	0.8	±0.3
CE-H	0.5	1.1	2.0	0.7	1.1	±0.3
CE-I	0.4	0.9	0.7	0.6	0.7	±0.1

3.5. Microbial Abundance and Cell Viability

The microbial abundance values (No. cells/mL) ranged from a minimum of $5.3 \times 10^3 \pm 1.3 \times 10^3$ to a maximum of $1.2 \times 10^6 \pm 7.7 \times 10^4$ (Table 7). The CE-D piezometer had the lowest average value if compared with the other ones.

Table 7. Microbial abundance (No. Cells/mL) values determined in each piezometer (CE-A, CE-B, CE-C, CE-D, CE-E, CE-F, CE-G, CE-H, and CE-I) at each sampling campaign.

Piezometer	Microbial Abundance (No. Cells/mL)					
	I September	II February	III July	IV October	Average	±se
CE-A	6.9×10^3	8.8×10^4	5.2×10^4	1.4×10^5	7.2×10^4	$\pm 2.9 \times 10^4$
CE-B	1.7×10^4	3.1×10^4	2.9×10^4	6.6×10^4	3.6×10^4	$\pm 1.1 \times 10^4$
CE-C	9.2×10^3	3.1×10^4	9.0×10^3	2.6×10^4	1.9×10^4	$\pm 5.7 \times 10^3$
CE-D	1.0×10^4	1.4×10^4	1.2×10^4	2.2×10^4	1.4×10^4	$\pm 2.5 \times 10^3$
CE-E	1.5×10^4	2.6×10^4	1.9×10^4	1.1×10^5	4.2×10^4	$\pm 2.2 \times 10^4$
CE-F	5.3×10^3	2.6×10^4	6.6×10^4	2.5×10^5	8.8×10^4	$\pm 5.7 \times 10^4$
CE-G	-	1.7×10^4	1.7×10^5	-	9.4×10^4	$\pm 7.7 \times 10^4$
CE-H	2.5×10^4	6.5×10^4	1.5×10^4	1.2×10^6	3.3×10^5	$\pm 2.9 \times 10^5$
CE-I	1.2×10^4	5.8×10^4	1.5×10^4	3.5×10^4	3.0×10^4	$\pm 1.1 \times 10^4$

The cell viability values (expressed as % live cells/live + dead) ranged from 10.7% to 69.5% (Table 8). The data for the CE-H piezometer regarding the third sampling are not reported, owing to the interference of suspended particles, which hampered cell observations under the microscope.

Table 8. Cell viability (% live cells/live + dead) values determined in each piezometer at each sampling campaign.

Piezometer	Cell Viability (%)					
	I September	II February	III July	IV October	Average	±se
CE-A	33.8	12.3	51.5	47.2	36.2	±8.8
CE-B	11.3	11.2	60.3	50.2	33.2	±12.9
CE-C	61.4	10.7	58.6	30.6	40.3	±12.1
CE-D	57.3	44.3	23.5	44.3	42.3	±7.0
CE-E	50.0	38.9	64.5	39.5	48.2	±6.0
CE-F	18.0	19.0	37.1	32.7	26.7	±4.8
CE-G	-	19.6	16.8	-	18.2	±1.4
CE-H	40.4	43.8	-	18	34.1	±8.1
CE-I	34.9	25.7	69.5	55.5	46.4	±12.2

A significant negative correlation was found ($p < 0.01$) between the cell viability and DOC values (Figure 5).

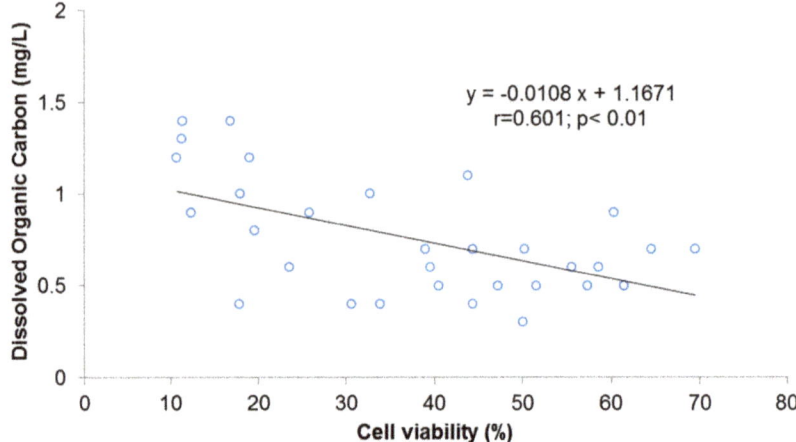

Figure 5. Linear regression (n = 34) which shows the relationship between the cell viability and dissolved organic carbon (DOC).

3.6. Spatial and Multi-Parameter Monitoring over 1 Year with 4 Season Analysis

Data collected were used for the geostatistical analysis. Close relationships among the geological heterogeneity, water circulation, pollutant diffusion, dissolved organic carbon, and cell viability were found. The thematic maps regarding microbial abundance, cell viability, and DOC are reported with a chromatic scale in Figure 6.

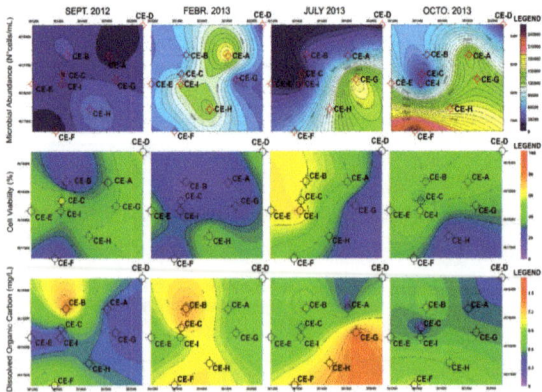

Figure 6. Geostatistical maps of microbial abundance (No. cells/mL), cell viability (%), dissolved organic carbon (mg/L) for each sampling campaign. The higher values are reported in red and the lower ones in blue.

Microbial abundance was not directly related either to cell viability or to the DOC values (Figure S1). On the contrary, a close relationship ($p < 0.01$) was found between cell viability and dissolved organic carbon. The highest values for cell viability were found with DOC values ≤ 0.5 mg/L; above these values DOC seems to negatively affect the microbial community (Figure 5).

4. Discussion

The study carried out allowed us to reconstruct in detail the surface and subsurface geological structure of the investigated area. Using field analyses, piezometer stratigraphy (when available) and computer modelling, a new and up to date 3D geological model of the area was performed. The latter, mainly composed of a layer cake template succession of tuffs, lavas and compacted pozzolanas, with a total thickness up to 300 m, was characterized by a shallow aquifer comprised within the lava layers. A detailed structural analysis carried out on the outcropping formation showed an intense fracturation process, which substantially modified, as a second order process, the first order mechanical and hydraulic characters of the rocks (very low permeability). This last aspect is key to a better understanding of water circulation and drainage directions. The brittle deformation processes involving the most rigid formations of the entire stratigraphic succession are characterized by piezometer-organized and sub-parallel extensional fracture lineaments, showing different overlapped generations. In particular, the intersection among different fracture systems represents a preferential water vertical direction flux from the surface through the volcanic succession to the aquiclude levels, composed of the compact pozzolanas layers. However, the fracturation processes, which are not evenly distributed either planimetrically or vertically, favor the development of areas with higher induced permeability (secondary permeability) values such as those at CE-B and CE-F piezometers. Using the computer modelling software, it was possible to model the geological structures in the area investigated in 3D and evaluate the direction, geometry and volumes of both groundwater and volcanic formations. In this configuration, the surface successions of tuffs, unevenly fractured, represent the unsaturated layers, which should perform the function of protecting aquifers from contamination processes coming from the surface. Despite this, the pervasive fracturation processes, which also affects the tuffs, can make it possible for chemicals to directly reach the aquifer.

The overall data analysis regarding the 6 piezometers belonging to this volcanic aquifer showed that the threshold values for evaluating the chemical and quantitative status established by the European WFD and entered in force in the Italian legislation [32,58] were not exceeded, except for one chemical at two points (CE-B and CE-F) in the July sampling campaign. Using the current legislation approach, we can state that the waste dump did not negatively affect the groundwater quality.

However, by considering unconventional parameters such as DOC and cell viability and applying a geostatistical analysis we were able to identify some points where the groundwater was unexpectedly vulnerable to inputs from the surface. In this context, the high DOC value observed for the most vulnerable piezometer CE-B, whatever the season, can be ascribed to the dump's presence. Using our approach, we identified the initial carbon input peak, at the CE-B piezometer, (September and February) and its transfer with the subsurface water flow in the following months toward the Southern sector of the area investigated though groundwater circulation. In fact, a progressive increase in DOC values during time at the CE-G (February and July), CE-H (February–July) and, downstream, CE-F piezometers (July and October) was observed (Figure 6).

An opposite trend was found for the cell viability (highest DOC, lowest viability), showing that DOC values above 0.6 mg/L had a detrimental effect on the groundwater microbial community (Figure 6). In fact, in the external control piezometer CE-D, DOC always displayed relatively low values, in accordance with those commonly found in pristine groundwater [1,34]. A low DOC content is known to be a factor limiting the growth of bacterial communities; on the contrary, in our samples we observed that the lowest cell viability was associated with the highest DOC values. This suggests that the additional organic carbon comes from the dump and that the microbial communities are sensitive to its origin and quality. This result, showing the negative effects on the microbial community of higher values of DOC in contaminated ecosystems, is in line with those found in in previous works [80–82]. Finally, it is interesting to note that in the third sampling campaign the lowest oxygen concentrations were found and the volatile organic compound 1,2 Dichloropropane detected above the legislation limits in the most vulnerable piezometers (CE-B and CE-F). The presence of volatile organic compounds is an indicator of anthropogenic activities [83]. Negative effects on the microbial community due to high DOC values and trace contamination of chlorinated organics were found in previous studies in groundwater [80]. These results show that a contaminant threshold compliance alone does not exclude possible effects from the trace contaminant mixture on the groundwater ecosystem. Once the most vulnerable groundwater sectors are identified, monitoring surveys can focus on these points and also take into consideration the microbiological parameters.

5. Conclusions

The overall results show that the chemical analysis of single contaminants and quantitative parameters required by the current regulations is not always able to assess the vulnerability of a groundwater and all the possible impacts of a waste deposit area. On the contrary, a detailed geological and hydrogeological analysis made it possible to understand the surface and subsurface geological structures of the area investigated. The new 3D geological model made it possible to identify the geometry and distribution of the different volcanic formations in the area studied. The meso-structural analysis showed different extensional fracture generations affecting both the tuffs and lavas. The intersection among different fracture lineament generations represents a preferential pathway for chemicals and organic carbon to reach and affect groundwater. This process was evident in the most vulnerable CE-B and CE-F piezometers. This study also shows that the geostatistical analysis makes it possible to understand the relationships between parameters and spatial heterogeneity in a better way than the classic statistical analysis. Finally, unconventional ecological parameters such as cell viability and DOC can be more sensitive than the chemical determinations alone for assessing the anthropogenic pressure on an aquifer and can be good indicators of water quality to be added to those required by the current legislation.

Supplementary Materials: The following are available online at http://www.mdpi.com/2073-4441/11/9/1933/s1, Table S1. List of parameters and inorganic ions examined in the nine piezometers samples for different sampling campaigns (I-II-III-IV). Table S2. Organic contaminants (Polycyclic Aromatic Hydrocarbons, PAHs, µg/L) in the water samples of the four campaigns: I September, II February, III July, IV October. Table S3. Organic contaminants (PCB, DDE and γHCH µg/L) in the water samples of the four campaigns: I September, II February, III July, IV October. Table S4. Organic contaminants (VOCs, µg/L) in the water samples of the four campaigns: I September, II February, III July, IV October. Figure S1: Variograms for Microbial abundance.

Author Contributions: Conceptualization, D.R. and A.B.C.; Methodology, D.R., A.B.C., P.G., F.C., M.D.L., L.P., N.A., R.C., G.M.; Software, D.R.; Formal Analysis, D.R., A.B.C., P.G., F.C., M.D.L., L.P., N.A., R.C., G.M. and S.G.; Data curation, D.R., A.B.C., P.G.; Writing—Original Draft Preparation, D.R. and A.B.C.; Writing—Review and Editing, D.R., A.B.C., P.G.; Project Administration, S.G.

Funding: This work was funded by the CNR-IRSA/ Pontina Ambiente S.r.l. Agreement, Project IRSA-CNR N. 0001461/2012.

Acknowledgments: We thank Giuseppe Mininni for funding acquisition. We also thank Domenico Mastroianni and Francesca Falconi for the analysis of inorganic ions, metals, and DOC.

Conflicts of Interest: The Authors declare no conflict of interest.

References

1. Barra Caracciolo, A.; Fajardo, C.; Grenni, P.; Sacca', M.L.; Amalfitano, S.; Ciccoli, R.; Martin, M.; Gibello, A. The role of a groundwater bacterial community in the degradation of the herbicide terbuthylazine. *FEMS Microbiol. Ecol.* **2010**, *71*, 127–136. [CrossRef] [PubMed]
2. European Commission Water Statistics-Share of External Inflow from Neighbouring Territories in Renewable Freshwater Resources-Long-Term Annual Average. Available online: https://ec.europa.eu/eurostat/statistics-explained/index.php/Water_statistics (accessed on 9 May 2019).
3. Ducci, D.; de Melo, M.T.C.; Preziosi, E.; Sellerino, M.; Parrone, D.; Ribeiro, L. Combining natural background levels (NBLs) assessment with indicator kriging analysis to improve groundwater quality data interpretation and management. *Sci. Total Environ.* **2016**, *569–570*, 569–584. [CrossRef] [PubMed]
4. Preziosi, E.; Parrone, D.; Del Bon, A.; Ghergo, S. Natural background level assessment in groundwaters: Probability plot versus pre-selection method. *J. Geochem. Explor.* **2014**, *143*, 43–53. [CrossRef]

5. Looser, M.; Parriaux, A.; Bensimon, M. Landfill underground pollution detection and characterization using inorganic traces. *Water Res.* **1999**, *33*, 3609–3616. [CrossRef]
6. Lapworth, D.J.; Baran, N.; Stuart, M.E.; Ward, R.S. Emerging organic contaminants in groundwater: A review of sources, fate and occurrence. *Environ. Pollut.* **2012**, *163*, 287–303. [CrossRef] [PubMed]
7. Castañeda, S.S.; Sucgang, R.J.; Almoneda, R.V.; Mendoza, N.D.S.; David, C.P.C. Environmental isotopes and major ions for tracing leachate contamination from a municipal landfill in Metro Manila, Philippines. *J. Environ. Radioact.* **2012**, *110*, 30–37. [CrossRef] [PubMed]
8. Samadder, S.R.; Prabhakar, R.; Khan, D.; Kishan, D.; Chauhan, M.S. Analysis of the contaminants released from municipal solid waste landfill site: A case study. *Sci. Total Environ.* **2017**, *580*, 593–601. [CrossRef] [PubMed]
9. Preziosi, E.; Frollini, E.; Zoppini, A.; Ghergo, S.; Melita, M.; Parrone, D.; Rossi, D.; Amalfitano, S. Disentangling natural and anthropogenic impacts on groundwater by hydrogeochemical, isotopic and microbiological data: Hints from a municipal solid waste landfill. *Waste Manag.* **2019**, *84*, 245–255. [CrossRef] [PubMed]
10. Gogu, R.C.; Dassargues, A. Current trends and future challenges in groundwater vulnerability assessment using overlay and index methods. *Environ. Geol.* **2000**, *39*, 549–559. [CrossRef]
11. Civita, M.; De Maio, M. Assessing and mapping groundwater vulnerability to contamination: The Italian "combined" approach. *Geofis. Int.* **2004**, *43*, 513–532. [CrossRef]
12. Tuxen, N.; Albrechtsen, H.-J.; Bjerg, P.L. Identification of a reactive degradation zone at a landfill leachate plume fringe using high resolution sampling and incubation techniques. *J. Contam. Hydrol.* **2006**, *85*, 179–194. [CrossRef] [PubMed]
13. de Lipthay, J.R.; Tuxen, N.; Johnsen, K.; Hansen, L.H.; Albrechtsen, H.-J.; Bjerg, P.L.; Aamand, J. In Situ Exposure to Low Herbicide Concentrations Affects Microbial Population Composition and Catabolic Gene Frequency in an Aerobic Shallow Aquifer. *Appl. Environ. Microbiol.* **2002**, *69*, 461–467. [CrossRef] [PubMed]
14. Naidu, R. Recent Advances in Contaminated Site Remediation. *Water Air Soil Pollut.* **2013**, *224*, 1573–2932. [CrossRef]
15. Safonov, A.V.; Babich, T.L.; Sokolova, D.S.; Grouzdev, D.S.; Tourova, T.P.; Poltaraus, A.B.; Zakharova, E.V.; Merkel, A.Y.; Novikov, A.P.; Nazina, T.N. Microbial Community and in situ Bioremediation of Groundwater by Nitrate Removal in the Zone of a Radioactive Waste Surface Repository. *Front. Microbiol.* **2018**, *9*, 1985. [CrossRef] [PubMed]
16. Taş, N.; Brandt, B.W.; Braster, M.; van Breukelen, B.M.; Röling, W.F.M. Subsurface landfill leachate contamination affects microbial metabolic potential and gene expression in the Banisveld aquifer. *FEMS Microbiol. Ecol.* **2018**, *94*, fiy156. [CrossRef] [PubMed]
17. Guzzella, L.; Capri, E.; Di Corcia, A.; Barra Caracciolo, A.; Giuliano, G. Fate of Diuron and Linuron in a Field Lysimeter Experiment. *J. Environ. Qual.* **2006**, *35*, 312–323. [CrossRef]
18. Alaoui, A.; Lipiec, J.; Gerke, H.H. A review of the changes in the soil pore system due to soil deformation: A hydrodynamic perspective. *Soil Tillage Res.* **2011**, *115–116*, 1–15. [CrossRef]
19. Beeby, A. *Applying Ecology*; Springer: New York, NY, USA, 1994; ISBN 978-0-412-44470-8.
20. Calow, P. Evolution, ecology and environmental stress. *Biol. J. Linn. Soc.* **1989**, *37*, 1. [CrossRef]
21. Capelli, G.; Mazza, R. Water criticality in the Colli Albani (Rome, Italy). *G. Geol. Appl.* **2005**, *1*, 261–271.
22. Hansen, J.E.; Lacis, A.A.; Lee, P.; Wang, W.-C. Climatic effects of atmospheric aerosols. *Ann. N. Y. Acad. Sci.* **1980**, *338*, 575–587. [CrossRef]
23. Capelli, G.; Mazza, R.; Gazzetti, C. *Strumenti e Strategie per la Tutela e uso Compatibile Della Risorsa Idrica nel Lazio. Gli Acquiferi Vulcanici*; Pitagora: Bologna, Italy, 2005; ISBN 978-8837114503.
24. Tatti, F.; Petrangeli Papini, M.; Torretta, V.; Mancini, G.; Boni, M.R.; Viotti, P. Experimental and numerical evaluation of Groundwater Circulation Wells as a remediation technology for persistent, low permeability contaminant source zones. *J. Contam. Hydrol.* **2019**, *222*, 89–100. [CrossRef]
25. Dagan, G.; Neuman, S.P. *Subsurface Flow and Transport: A Stochastic Approach-International Hydrology Series*; Cambridge University Press: Cambridge, UK, 1997; ISBN 9780521020091.
26. Sibson, R.H. Crustal stress, faulting and fluid flow. *Geol. Soc. Lond. Spec. Publ.* **1994**, *78*, 69–84. [CrossRef]
27. Fischer, M.P.; Jackson, P.B. Stratigraphic controls on eformation patterns in fault-related folds: A detachment fold example from the Sierra Madre Oriental, northeast Mexico. *J. Struct. Geol.* **1999**, *21*, 613–633. [CrossRef]
28. Chester, J.S. Mechanical stratigraphy and fault–fold interaction, Absaroka thrust sheet, Salt River Range, Wyoming. *J. Struct. Geol.* **2003**, *25*, 1171–1192. [CrossRef]

29. Engelder, T.; Marshak, S. Disjunctive cleavage formed at shallow depths in sedimentary rocks. *J. Struct. Geol.* **1985**, *7*, 327–343. [CrossRef]
30. Jamison, W.R. Stress controls on fold thrust style. In *Thrust Tectonics*; Springer: Dordrecht, The Netherlands, 1992; pp. 155–164.
31. United Nations World Commission on Environment and Development-Brundtland Commission. *Our Common Future, Report of the World Commission on Environment and Development, World Commission on Environment and Development*; Oxford University Press: Oxford, UK, 1987; Volume 1, Doc. 149; ISBN 019282080X.
32. Parliament, E. Directive 2000/60/EC of the European Parliament and of the Council of 23 October 2000 establishing a framework for Community action in the field of water policy. *Off. J. Eur. Union* **2000**, *12*, 1–51.
33. Gilbert, J.A.; Steele, J.A.; Caporaso, J.G.; Steinbrück, L.; Reeder, J.; Temperton, B.; Huse, S.; McHardy, A.C.; Knight, R.; Joint, I.; et al. Defining seasonal marine microbial community dynamics. *ISME J.* **2012**, *6*, 298–308. [CrossRef]
34. Grenni, P.; Gibello, A.; Barra Caracciolo, A.; Fajardo, C.; Nande, M.; Vargas, R.; Saccà, M.L.; Martinez-Iñigo, M.J.; Ciccoli, R.; Martín, M. A new fluorescent oligonucleotide probe for in situ detection of s-triazine-degrading Rhodococcus wratislaviensis in contaminated groundwater and soil samples. *Water Res.* **2009**, *43*, 2999–3008. [CrossRef]
35. Amalfitano, S.; Del Bon, A.; Zoppini, A.; Ghergo, S.; Fazi, S.; Parrone, D.; Casella, P.; Stano, F.; Preziosi, E. Groundwater geochemistry and microbial community structure in the aquifer transition from volcanic to alluvial areas. *Water Res.* **2014**, *65*, 384–389. [CrossRef]
36. Danielopol, D.L.; Griebler, C. Changing Paradigms in Groundwater Ecology-from the "Living Fossils" Tradition to the "New Groundwater Ecology". *Int. Rev. Hydrobiol.* **2008**, *93*, 565–577. [CrossRef]
37. Allison, S.D.; Martiny, J.B.H. Resistance, resilience, and redundancy in microbial communities. *Proc. Natl. Acad. Sci. USA* **2008**, *105*, 11512–11519. [CrossRef]
38. Seibert, J.; Bishop, K.; Rodhe, A.; McDonnell, J.J. Groundwater dynamics along a hillslope: A test of the steady state hypothesis. *Water Resour. Res.* **2003**, *39*, 1014. [CrossRef]
39. Henriksen, H.J.; Rasmussen, P.; Brandt, G.; von Bülow, D.; Jensen, F.V. Public participation modelling using Bayesian networks in management of groundwater contamination. *Environ. Model. Softw.* **2007**, *22*, 1101–1113. [CrossRef]
40. Yagi, J.M.; Madsen, E.L. Diversity, Abundance, and Consistency of Microbial Oxygenase Expression and Biodegradation in a Shallow Contaminated Aquifer. *Appl. Environ. Microbiol.* **2009**, *75*, 6478–6487. [CrossRef]
41. Haack, S.K.; Fogarty, L.R.; West, T.G.; Alm, E.W.; McGuire, J.T.; Long, D.T.; Hyndman, D.W.; Forney, L.J. Spatial and temporal changes in microbial community structure associated with recharge-influenced chemical gradients in a contaminated aquifer. *Environ. Microbiol.* **2004**, *6*, 438–448. [CrossRef]
42. Griebler, C.; Lueders, T. Microbial biodiversity in groundwater ecosystems. *Freshw. Biol.* **2009**, *54*, 649–677. [CrossRef]
43. Hancock, P.J.; Boulton, A.J.; Humphreys, W.F. Aquifers and hyporheic zones: Towards an ecological understanding of groundwater. *Hydrogeol. J.* **2005**, *13*, 98–111. [CrossRef]
44. Humphreys, W.F. Hydrogeology and groundwater ecology: Does each inform the other? *Hydrogeol. J.* **2009**, *17*, 5–21. [CrossRef]
45. Peccerillo, A. *Plio-Quaternary Volcanism in Italy: Petrology, Geochemistry, Geodynamics*; Springer: Berlin/Heidelberg, Germany; New York, NY, USA, 2005; p. 370. ISBN 13 978-3-540-25885-8.
46. De Rita, D.; Funiciello, R.; Parotto, M. *Geological Map of the Colli Albani Volcanic Complex (Carta geologica del Complesso vulcanico dei Colli Albani)*; Consiglio Nazionale delle Ricerche: Rome, Italy, 1988; Available online: http://repositories.dst.unipi.it/index.php/carte/item/29-carta-geologica-del-complesso-vulcanico-dei-colli-albani-vulcano-laziale (accessed on 10 May 2019).
47. Marra, F.; Karner, D.B.; Freda, C.; Gaeta, M.; Renne, P. Large mafic eruptions at Alban Hills Volcanic District (Central Italy): Chronostratigraphy, petrography and eruptive behavior. *J. Volcanol. Geotherm. Res.* **2009**, *179*, 217–232. [CrossRef]
48. Gaeta, M.; Freda, C.; Marra, F.; Di Rocco, T.; Gozzi, F.; Arienzo, I.; Giaccio, B.; Scarlato, P. Petrology of the most recent ultrapotassic magmas from the Roman Province (Central Italy). *Lithos* **2011**, *127*, 298–308. [CrossRef]

49. Capelli, G.; Mastrorillo, L.; Mazza, R.; Petitta, M. *Map of the Hydrogeological Units of the Lazio Region (Carta delle Unità Idrogeologiche della Regione Lazio, scala 1:250.000*; Regione Lazio, S.EL.C.A.: Firenze, Italy, 2012; Available online: http://www.regione.lazio.it/binary/rl_main/tbl_documenti/AMB_PBL_SIRDIS_Carta_Unita_in_scala_1a250.000_LD.pdf (accessed on 10 May 2019).
50. Capelli, G.; Mastrorillo, L.; Mazza, R.; Petitta, M.; Baldoni, T.; Banzato, F.; Cascone, D.; Di Salvo, C.; La Vigna, F.; Taviani, S.; et al. *Hydrogeological Map of the Lazio Region (Carta Idrogeologica del Territorio della Regione Lazio), (4 sheets)*; Regione Lazio: Rome, Italy, 2016.
51. Cinti, D.; Procesi, M.; Tassi, F.; Montegrossi, G.; Sciarra, A.; Vaselli, O.; Quattrocchi, F. Fluid geochemistry and geothermometry in the western sector of the Sabatini Volcanic District and the Tolfa Mountains (Central Italy). *Chem. Geol.* **2011**, *284*, 160–181. [CrossRef]
52. Cinti, D.; Poncia, P.P.; Brusca, L.; Tassi, F.; Quattrocchi, F.; Vaselli, O. Spatial distribution of arsenic, uranium and vanadium in the volcanic-sedimentary aquifers of the Vicano–Cimino Volcanic District (Central Italy). *J. Geochem. Explor.* **2015**, *152*, 123–133. [CrossRef]
53. Preziosi, E.; Rossi, D.; Parrone, D.; Ghergo, S. Groundwater chemical status assessment considering geochemical background: An example from Northern Latium (Central Italy). *Rend. Lincei* **2016**, *27*, 59–66. [CrossRef]
54. Moore, I.D.; Grayson, R.B.; Ladson, A.R. Digital terrain modelling: A review of hydrological, geomorphological, and biological applications. *Hydrol. Process.* **1991**, *5*, 3–30. [CrossRef]
55. Moore, I.D.; Gessler, P.E.; Nielsen, G.A.; Peterson, G.A. Soil Attribute Prediction Using Terrain Analysis. *Soil Sci. Soc. Am. J.* **1993**, *57*, 443–452. [CrossRef]
56. Wise, S.M. Effect of differing DEM creation methods on the results from a hydrological model. *Comput. Geosci.* **2007**, *33*, 1351–1365. [CrossRef]
57. Vaze, J.; Teng, J.; Spencer, G. Impact of DEM accuracy and resolution on topographic indices. *Environ. Model. Softw.* **2010**, *25*, 1086–1098. [CrossRef]
58. Italian Ministry of the Environment. Legislative Decree no. 152, 2006, Rules in environmental field. *Ital. Off. J.* **2006**, *88*, 425. Available online: https://www.gazzettaufficiale.it/eli/gu/2006/04/14/88/so/96/sg/pdf (accessed on 15 May 2019).
59. Borkowski, A.S.; Kwiatkowska-Malina, J. Geostatistical modelling as an assessment tool of soil pollution based on deposition from atmospheric air. *Geosci. J.* **2017**, *21*, 645–653. [CrossRef]
60. Chang, T.; Shyu, G.; Lin, Y.; Chang, N. Geostatistical analysis of soil arsenic content in taiwan. *J. Environ. Sci. Heal. Part A* **1999**, *34*, 1485–1501. [CrossRef]
61. Brenning, A. Geostatistics without stationarity assumptions within Geographical Information Systems. *Freib. Online Geosci.* **2001**, *102*. Available online: http://www.geo.tu-freiberg.de/fog/FOG_Vol_6.pdf (accessed on 10 March 2019). [CrossRef]
62. Frollini, E.; Rossi, D.; Rainaldi, M.; Parrone, D.; Ghergo, S.; Preziosi, E. A proposal for groundwater sampling guidelines: Application to a case study in southern Latium. *Rend. Online Della Soc. Geol. Ital.* **2019**, *47*, 46–51. [CrossRef]
63. APAT-IRSA CNR (Agenzia per la Protezione Dell'ambiente e per i Servizi Tecnici-Consiglio Nazionale delle Ricerche-Istituto di Ricerca sulle Acque). *Metodi Analitici per le Acque*; Belli, M., Centioli, D., Zorzi, P., Umberto, S., Capri, S., Pagnotta, R., Pettine, M., Eds.; APAT: Rome, Italy, 2003; ISBN 88-448-0083-7.
64. APHA (American Public Health Association). *Standard Methods for the Examination of Water and Waste Waters*, 23rd ed.; Rice, E.W., Baird, R.B., Eaton, A.D., Eds.; American Public Health Associatio; American Water Works Association; Water Environment Federation: Washington, DC, USA, 2017; ISBN 978-0-87553-287-5.
65. Patrolecco, L.; Ademollo, N.; Grenni, P.; Tolomei, A.; Barra Caracciolo, A.; Capri, S. Simultaneous determination of human pharmaceuticals in water samples by solid phase extraction and HPLC with UV-fluorescence detection. *Microchem. J.* **2013**, *107*, 165–171. [CrossRef]
66. Ademollo, N.; Patrolecco, L.; Polesello, S.; Valsecchi, S.; Wollgast, J.; Mariani, G.; Hanke, G. The analytical problem of measuring total concentrations of organic pollutants in whole water. *TrAC Trends Anal. Chem.* **2012**, *36*, 71–81. [CrossRef]
67. Barra Caracciolo, A.; Giuliano, G.; Grenni, P.; Cremisini, C.; Ciccoli, R.; Ubaldi, C. Effect of urea on degradation of terbuthylazine in soil. *Environ. Toxicol. Chem.* **2005**, *24*, 1035–1040. [CrossRef]

68. Barra Caracciolo, A.; Giuliano, G.; Grenni, P.; Guzzella, L.; Pozzoni, F.; Bottoni, P.; Fava, L.; Crobe, A.; Orrù, M.; Funari, E. Degradation and leaching of the herbicides metolachlor and diuron: A case study in an area of Northern Italy. *Environ. Pollut.* **2005**, *134*, 525–534. [CrossRef]
69. Isaaks, E.H.; Srivastava, R.M. *An Introduction to Applied Geostatistics*; Oxford University Press: New York, NY, USA, 1989; Volume 17, ISBN 0-19-505012-6.
70. Cressie, N. The origins of kriging. *Math. Geol.* **1990**, *22*, 239–252. [CrossRef]
71. Cressie, N.A.C. *Statistics for Spatial Data*; Wiley Series in Probability and Statistics; John Wiley & Sons, Inc.: Hoboken, NJ, USA, 1993; ISBN 9781119115151.
72. Burrough, P.A.; McDonnell, R.A. *Principles of Geographical Information Systems*; Oxford University Press: Oxford, UK, 1999; ISBN 019-8233655.
73. Wu, Z.; Guo, F.; Li, J. The 3D modelling techniques of digital geological mapping. *Arab. J. Geosci.* **2019**, *12*, 467. [CrossRef]
74. Kaufmann, O.; Martin, T. 3D geological modelling from boreholes, cross-sections and geological maps, application over former natural gas storages in coal mines. *Comput. Geosci.* **2008**, *34*, 279–290. [CrossRef]
75. Li, X.; Li, P.; Zhu, H.; Liu, J. Geomodeling with integration of multi-source data by Bayesian kriging in underground space. *Tongji Daxue Xuebao/Journal Tongji Univ.* **2014**, *42*, 406–412.
76. Wu, Z.; Guo, F.; Liu, Z.; Hou, M.; Luo, J. Technology and method of multi-data merging in 3D geological modelling. *J. Jilin Univ. Earth Sci. Ed.* **2016**, *6*, 1895–1913.
77. Berg, R.C.; Greenpool, M.R. *Stack-Unit Geologic Mapping: Color-Coded and Computer-Based Methodology*; Department of Energy and Natural Resources, Illinois State Geological Survey Library: Champaign, IL, USA, 1993; pp. 1–11.
78. Berg, R.C.; Kempton, J.P. *Stack-unit Mapping of Geologic Materials in Illinois to a Depth of 15 m*; Department of Energy and Natural Resources, Illinois State Geological Survey Library: Champaign, IL, USA, 1988.
79. Angelone, M.; Cremisini, C.; Piscopo, V.; Proposito, M.; Spaziani, F. Influence of hydrostratigraphy and structural setting on the arsenic occurrence in groundwater of the Cimino-Vico volcanic area (central Italy). *Hydrogeol. J.* **2009**, *17*, 901–914. [CrossRef]
80. Barra Caracciolo, A.; Grenni, P.; Falconi, F.; Caputo, M.C.; Ancona, V.; Uricchio, V.F. Pharmaceutical waste disposal: Assessment of its effects on bacterial communities in soil and groundwater. *Chem. Ecol.* **2011**, *27*, 43–51. [CrossRef]
81. Saccà, M.L.; Ferrero, V.E.V.; Loos, R.; Di Lenola, M.; Tavazzi, S.; Grenni, P.; Ademollo, N.; Patrolecco, L.; Huggett, J.; Barra Caracciolo, A.; et al. Chemical mixtures and fluorescence in situ hybridization analysis of natural microbial community in the Tiber river. *Sci. Total Environ.* **2019**, *673*, 7–19. [CrossRef]
82. Barra Caracciolo, A.; Patrolecco, L.; Grenni, P.; Di Lenola, M.; Ademollo, N.; Rauseo, J.; Rolando, L.; Spataro, F.; Plutzer, J.; Monostory, K.; et al. Chemical mixtures and autochthonous microbial community in an urbanized stretch of the River Danube. *Microchem. J.* **2019**, *147*, 985–994. [CrossRef]
83. Bi, E.; Liu, Y.; He, J.; Wang, Z.; Liu, F. Screening of Emerging Volatile Organic Contaminants in Shallow Groundwater in East China. *Ground Water Monit. Remediat.* **2012**, *32*, 53–58. [CrossRef]

© 2019 by the authors. Licensee MDPI, Basel, Switzerland. This article is an open access article distributed under the terms and conditions of the Creative Commons Attribution (CC BY) license (http://creativecommons.org/licenses/by/4.0/).

Article

The Application of Different Biological Remediation Strategies to PCDDs/PCDFs Contaminated Urban Sediments

Magdalena Urbaniak [1,2,*], Anna Wyrwicka [3], Grzegorz Siebielec [4], Sylwia Siebielec [4], Petra Kidd [5] and Marek Zieliński [6]

1. Department of Applied Ecology, Faculty of Biology and Environmental Protection, University of Lodz, Banacha 12/16, 90-237 Lodz, Poland
2. European Regional Centre for Ecohydrology of the Polish Academy of Sciences, Tylna 3, 90-364 Lodz, Poland
3. Department of Plant Physiology and Biochemistry, Faculty of Biology and Environmental Protection, University of Lodz, Banacha 12/16, 90-237 Lodz, Poland; anna.wyrwicka@biol.uni.lodz.pl
4. Institute of Soil Science and Plant Cultivation—State Research Institute, Czartoryskich 8, 24-100 Puławy, Poland; gs@iung.pulawy.pl (G.S.); ssiebielec@iung.pulawy.pl (S.S.)
5. Instituto de Investigaciones Agrobiológicas de Galicia (IIAG), Consejo Superior de Investigaciones Científicas (CSIC), Santiago de Compostela 15706, Spain; pkidd@iiag.csic.es
6. Nofer Institute of Occupational Medicine, Teresy 8, 91-348 Lodz, Poland; marekz@imp.lodz.pl
* Correspondence: m.urbaniak@unesco.lodz.pl or magdalena.urbaniak@biol.uni.lodz.pl

Received: 5 July 2019; Accepted: 16 September 2019; Published: 20 September 2019

Abstract: Our aim was to assess the efficacy of four different bioremediation strategies applied to soil treated with urban sediments for alleviating soil phytotoxicity (examined using *Lepidium sativum*), by removing polychlorinated dibenzo-*p*-dioxins (PCDDs) and polychlorinated dibenzofurans (PCDFs), and mitigating the toxic effect on plants by the applied sediment: (1) Natural attenuation, (2) phytoremediation with the use of two plants *Tagetes patula* L. and *Festuca arundinacea*, (3) rhizobacterial inoculation with *Massilia niastensis* p87 and *Streptomyces costaricanus* RP92 strains, (4) rhizobacteria-assisted phytoremediation with both plants and strains. The applied sediment had a positive influence on *L. sativum* growth (90% higher than in the unamended soil), mostly due to its high content of nutrients, mainly Ca and Fe, which immobilize pollutants. The positive effect of sediments continued for up to 10-week duration of the experiment; however, the rhizobacterial inoculated samples were characterized by higher growth of *L. sativum*. The application of rhizobacteria-assisted phytoremediation further increased the growth of *L. sativum*, and was also found to improve the efficiency of PCDD/PCDF removal, resulting in a maximum 44% reduction of its content. This strategy also alleviated the negative impact of urban sediments on *T. patula* and *F. arundinacea* biomass, and had a beneficial effect on protein and chlorophyll content in the studied plants.

Keywords: urban sediments; PCDDs/PCDFs; rhizobacterial inoculants; bioremediation; phytoremediation

1. Introduction

The urban water ecosystems, located downstream of city landscapes, often become reservoirs for a variety of pollutants originating from atmospheric deposition, as well as urban runoff, storm water outlets, industrial waste and combined sanitary overflows [1–3]. To minimize the inflow of such pollutants to river ecosystems and their further transport along the river continuum, small dam reservoirs, sedimentation ponds and biofilters might be used. These constructions create ideal conditions for the sedimentation and deposition of particulate matter by decreasing the flow velocity, thus acting as efficient traps for associated compounds of urban origin [4–12]. However, the accelerated accumulation of sediments and associated pollutants leads to rapid siltation of such constructions, and this requires periodical dredging of the accumulated sediments and their further utilization [7]. These dredged urban sediments are usually stored in landfill areas, where they act as potential hazardous material for the surrounding environment. In addition, this solution creates economic burdens for the municipality due to the need to transport the sediments out of the city.

One promising method of managing the increasing amount of urban reservoir sediments is their direct application within the city limits as soil additives on city gardens and lawns. This type of urban sediment utilization is possible due to their richness in nutrients and organic matter, which improve soil properties and promote plant growth. Nevertheless, urban sediments also contain other compounds with harmful properties, such as heavy metals, pesticides washed out from urban green areas, car oils from streets and parking areas, and a variety of other organic compounds of industrial or anthropogenic origin. One of the most toxic groups of compounds, which have carcinogenic, hepatotoxic, immunotoxic and neurotoxic properties, are polychlorinated dibenzo-*p*-dioxins and polychlorinated dibenzofurans (PCDDs/PCDFs) [9,10]. These pollutants are characterized by a wide range of occurrence in the urban water ecosystems, because their main source is the load of domestic and industrial wastewater, atmospheric emission and deposition as well as emission from other sources associated with human activity in the city space such as car traffic [10].

Due to highly hydrophobic characteristics of PCDDs/PCDFs, they undergo rapid deposition in river and reservoir sediments, thus lowering the quality of the urban ecosystem and decreasing its biodiversity. Owing to the toxic properties of PCDDs/PCDFs, their persistence in the environment and their ability to bioaccumulate in aquatic and terrestrial trophic chains, the European Commission classified them as priority hazardous substances in the field of water policy, and imposed a requirement for EU members to monitor and eliminate them from the environment [13].

A potential approach to be applied for inactivation of urban sediments contaminated with PCDDs/PCDFs is phytoremediation—a method aimed at removal or decomposition of the pollutants using plants. Nevertheless, PCDDs/PCDFs due to their hydrophobicity and hence their strong adsorption by sediment and soil particles, transfer to sediment or soil solution to a very small extent. Consequently, the ability of plants to uptake PCDDs/PCDFs from sediments or soil is very limited [14]. In this situation, plants can be used to promote the continued existence of indigenous soil microorganisms, which are able to biodegrade PCDDs/PCDFs (rhizodegradation). In addition, inoculation of these contaminated sediments with appropriately selected specific single strains or consortia of microorganisms may further increase the efficiency of biodegradation processes.

Such remediation approaches as natural bioremediation with indigenous microorganisms (i.e., natural attenuation), inoculation and phytoremediation can be used separately; however, the most promising solution is for their combined use as part of rhizobacteria-assisted phytoremediation, intended to optimize the synergistic effect of selected plants and bacterial strains. Until now, this approach has been successfully used for cleaning soil contaminated with both organic and inorganic compounds [15–18]; however, no such studies have been performed on the remediation of urban sediments used directly within city limits.

The key step in designing an effective rhizobacteria-assisted phytoremediation strategy is the selection of a suitable plant-bacteria partnership to maximize the removal efficiency of a given pollutant [15,19]. Among the plants used in the remediation of contaminated soil, two species are of particular relevance, and are also specific to city gardens and lawns: *Tagetes patula* L. (*Asteraceae*) is commonly used as an ornamental plant, whereas *Festuca arundinacea* Schreb. (*Poaceae*) is often used in reclamation and as a grass species sown on lawns.

Tagetes patula L., commonly known as French marigold, is a robust and non-fussy plant originating from Mexico, used mainly as an edging plant on herbaceous borders. Marigold produces secondary metabolites, which assist in the remediation of combined contaminated sites [20,21]. It also contains bioactive compounds, which are widely employed as insecticides, fungicides and nematicides [22]; while *Festuca arundinacea* Schreb., commonly known as Tall fescue, is an evergreen, tuft-forming grass with a deep root system. It is a cool-season perennial C3 species native to Europe, frequently used in the phytoremediation of soil contaminated with organic compounds [18,20,23,24].

Among the bacterial strains used for bioremediation, *Streptomyces costaricanus* RP92, isolated from the rhizosphere of *Cytisus striatus*, is especially interesting because it can improve the remediation of soil contaminated with chloroorganic compounds such as hexachlorocyclohexane or diesel oils [25–27]. The strain *Massilia niastensis* P87, isolated from the rhizosphere of *Festuca rubra*, found growing on mine tailings with elevated concentrations of Cd, Pb and Zn [27], would further improve the bioremediation efficiency of the mixtures of pollutants, including trace metals which accumulate in urban sediments.

Considering the above, our aim was to assess and compare the efficacy of four different environmentally friendly strategies in the remediation of soil contaminated with PCDDs/PCDFs by urban bottom sediment application: (1) natural attenuation, (2) phytoremediation with two selected plant species *Tagetes patula* L. and *Festuca arundinacea* Schreb., (3) rhizobacterial inoculation with two selected strains *Streptomyces costaricanus* RP92 and *Massilia niastensis* P87, and (4) rhizobacteria-assisted phytoremediation using both sets of plants and bacterial strains given above. At the same time, the effects of urban sediment application and the said remediation strategies on soil phytotoxicity were evaluated. In addition, the plant biomass, total soluble protein content, total chlorophyll content and chlorophyll a/b ratio were measured to assess plant response to the applied urban sediments and remediation strategies.

2. Materials and Methods

2.1. Urban Sediments and Soils

The urban bottom sediments were collected from a sedimentation zone of the sequential sedimentation-biofiltration system located in Lodz, Poland, on the Sokołówka River. The system was constructed to reduce the inflow of particulate matter and a range of organic, inorganic and bacterial pollutants from storm water coming from the most urbanized catchment area of the Sokołówka River.

The system comprises (1) the hydrodynamically intensified sedimentation zone that facilitate the sedimentation and deposition processes, and in this way enable pre-treatment of the inflowing river and storm water; (2) the intensified biogeochemical processes zone, where fine particles are sieved and nutrients are reduced; and (3) intensified biofiltration zone planted with macrophytes (*Phragmites australis*, *Typha latifolia* and *Acorus calamus*) responsible for removal of nutrients and organic compounds [28,29].

Fresh sediments, with a dry matter (DM) content of 35%, were mixed with agricultural uncontaminated Haplic Luvisol type soil, in a 1:10 proportion and transferred into 2 kg pots. Prior to mixing, the soil was sieved using a 2-mm sieve.

2.2. Pot Experiment Design

The experimental design included four variants:

1. Natural attenuation—uncontaminated soil was mixed with fresh urban sediments, and no plants or bacterial inoculants were added;
2. Phytoremediation—uncontaminated soil was mixed with fresh urban sediments and *T. patula* L. or *F. arundinacea* Schreb. were planted;
3. Rhizobacterial inoculation—uncontaminated soil was mixed with fresh urban sediments and *S. costaricanus* RP92 or *M. niastensis* P87 were added, no plants were grown;
4. Rhizobacteria-assisted phytoremediation—using both the studied plants and bacterial strains (Figure 1).

All the variants were performed in triplicate (three separate pots per one remediation variant).

T. patula L. was seeded at a rate of 15 seeds per pot, which was reduced to 10 plants per pot after germination. *F. arundinacea* seeds were seeded at a rate of 2 g per pot.

Fresh cultures of bacterial strains were grown in 869 liquid medium [30] for 24 h. Five mL of this pre-culture was then grown for 12 h in fresh 869 liquid medium. Subsequently, bacterial biomass was collected by centrifugation (6000 rpm, 15 min), washed once and re-suspended in 10 mM $MgSO_4$ solutions to an OD_{660} of 1.0 (about 10^7 cells per mL). Each plant pot was inoculated during the germination phase (four weeks after seeding) with 100 mL of bacterial suspension. The same amount of sterile 10 mM $MgSO_4$ was added to non-inoculated pots. The inoculation was repeated after three weeks using the same procedure.

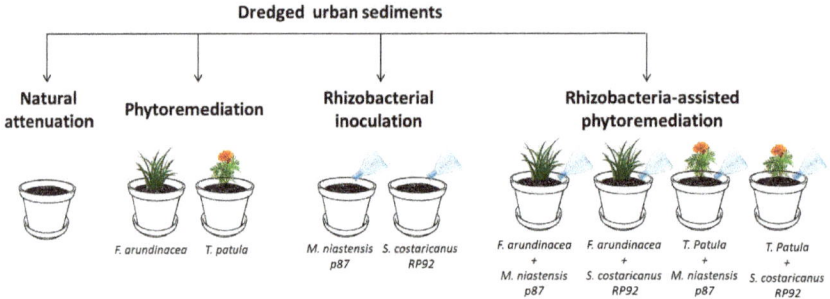

Figure 1. Experiment design.

The pot study was run for a 10-week period in a greenhouse under controlled conditions, i.e., supplemental light and 27 °C/20 °C day/night temperatures. After this time, the plants were harvested, the fresh biomass was weighed and the protein and chlorophyll content, as well as the chlorophyll a/b ratio, were measured in plant green tissues.

After harvesting, soil from each treatment was mixed, sieved to 2-mm sieve to remove the plant residues, and aliquots of soil were collected for further analyses of PCDD/PCDF concentrations and the phytotoxicity bioassay.

2.3. Soil Analyses

2.3.1. Determination of Physico-Chemical Parameters

Soil pH was measured in H_2O using a 1:2.5 soil:solution ratio. Total C was measured by combustion with a CN analyzer (Elementar, vario Macro cube, Langenselbold, Germany). The sediment samples were digested for elemental composition analysis in a 3:1 mixture of concentrated HNO_3:HCl in Teflon PFA vessels in a microwave accelerated reaction system (MarsXpress; CEM Corp., Matthews, NC, USA); total concentrations of elements were analyzed by ICP-MS (Agilent 7500ce, Agilent Technologies Inc., Santa Clara, CA, USA).

2.3.2. Determination of PCDD/PCDF Concentrations

For analysis of the concentrations of 17 toxic congeners of PCDD/PCDF, the PN-EN 1948-3 [31] and US EPA Method 1613 [32] were applied. The analysis was carried out based on isotope dilution and high-resolution gas chromatography (HRGC)/high-resolution mass spectrometry (HRMS) using an HP 6890 N Agilent Technologies GC coupled with a high-resolution mass spectrometer AutoSpec Ultima. The detailed description of the applied analytical procedure is depicted in the work by Urbaniak et al. [8]. The obtained concentrations were calculated using Toxic Equivalency Factor (TEF) and expressed as the Toxic Equivalency (TEQ) [33,34].

Quality assurance/quality control procedure was carried out using certified calibration standards. Each analytical series contained sample blank, control, certified reference material and in-house QC samples. Samples' spikes were used also as an additional check of accuracy. Recoveries of ^{13}C-labeled PCDDs/PCDFs ranged from 74% to 146%. Artefacts were estimated using a reagent blank, and duplicate analysis enabled to verify the precision that ranged from 2% to 11%. The obtained LOD values ranged from 0.070 to 0.143 ng/kg for PCDDs and 0.042 to 0.137 ng/kg for PCDFs.

2.3.3. Phytotoxicity Analysis

The commercially available bioassay Phytotoxkit^{TM} Test (Microbiotest Inc., Nazareth, Belgium) was used to assess the phytotoxicity of the soil samples [35]. The principle of the test is based on measurement of the inhibition of the length of roots of test species after 3 days of exposure to contaminated soil in relation to a reference soil. For the purpose of this experiment, the dicotyledon *Lepidium sativum* (L.) was used as a test plant. Uncontaminated soil with no treatment (control) was used as reference soil to assess the impact of urban sediments and various biological remediation strategies on soil phytotoxicity. The sample was classified as toxic when the root length inhibition exceeded 20% [36].

2.4. Plant Analyses

2.4.1. Determination of Protein Content

The *T. patula* L. and *F. arundinacea* Schreb. leaf extracts were prepared using 50 mM sodium phosphate buffer (pH 7.0) containing 0.5 M NaCl, 1 mM EDTA and 1 mM $C_6H_7NaO_6$. Obtained extracts underwent filtration on Miracloth filters and the obtained filtrates were centrifuged (15,000 g × 15 min). The supernatant was used for protein content determinations according to Bradford method [37]. The measurements were performed using a Helios Gamma spectrophotometer (Thermo Spectronic, Cambridge, UK) based on standard curves with Bovine Serum Albumin. The content of protein was depicted in mg g^{-1} of fresh mass (FM)

2.4.2. Determination of Chlorophyll Content

The *T. patula* L. and *F. arundinacea* Schreb. leaves were homogenized in an ice-cold mortar using sodium phosphate buffer, as it was described in point 2.4.1. The obtained homogenate was filtered and analyzed for chlorophyll content according to Porra et al. [38], using a Helios Gamma spectrophotometer (Thermo Spectronic, Cambridge, UK). The chlorophyll content was shown as mg g^{-1} FM.

2.5. Statistical Analysis

The two-way ANOVA was performed to test the effect of both inoculation and plants on the soil phytotoxicity, and the effect of sediment and inoculation on plant parameters (biomass, protein content, chlorophyll content, chlorophyll a/b ratio). No statistical analyses were conducted for PCDDs/PCDFs as no replicates are available. For the plant analyses statistical significance was tested separately for each plant (*T. patula* and *F. arundinacea*). The post-hoc Duncan test was used to confirm the statistically significant differences. All analyses were performed using STATISTICA 13 software.

3. Results and Discussion

3.1. The Physico-Chemical Properties and PCDD/PCDF Concentrations in Soil, Urban Sediments and Sediment-Amendment Soil

The soil used in the pot study had a loamy sand texture and the soil pH was 6.5. The soil OC content was 11.0 g kg^{-1}, which is lower than the average soil OC content in the climate zone including Poland (sub-oceanic to sub-continental) in the European LUCAS program, which was found to be 15 g kg^{-1} [39]. The concentration of PCDDs/PCDFs in soil was low, amounting to 24.8 ng kg^{-1} and 0.3 ng TEQ kg^{-1}. Similarly, concentrations of potentially toxic trace metals (PTTM) were low and ranged from 0.05 mg kg^{-1} for Cd to 28.5 mg kg^{-1} for Ba (Table 1).

Table 1. The physico-chemical characteristics of the collected urban sediment, the uncontaminated soil used in the experiment and the soil mixed in a proportion of 1:10 with the urban sediments.

Compound	Urban Sediment	Uncontaminated Soil	Urban Sediment Amended Soil
Soil pH	7.15	6.65	7.21
OC (g kg^{-1})	108	11.0	19.9
Sum of 17 PCDDs/PCDFs (ng kg^{-1})	2170	24.8	236
TEQ PCDDs/PCDFs (ng TEQ kg^{-1})	8.8	0.3	2.1
Mg (mg kg^{-1})	7040	403	1030
Ca (mg kg^{-1})	42,300	1020	5440
Fe (mg kg^{-1})	32,000	3170	6140
Zn (mg kg^{-1})	821	16.1	111
Cr (mg kg^{-1})	62.2	5.7	10.7
Cd (mg kg^{-1})	1.2	0.05	0.19
Ba (mg kg^{-1})	282	28.5	51.2
Pb (mg kg^{-1})	90.1	6.5	14.9
Cu (mg kg^{-1})	117	2.7	15.2

The fresh bottom sediments contained 35% of DM, and OC concentration was 108 g kg^{-1} DM. The sediment pH was 7.15. The sediments contained 0.70% magnesium (Mg), 0.52% potassium (K), 4.23% calcium (Ca), 3.20% iron (Fe) and 3.71% aluminum (Al). Among trace elements, only zinc (Zn) concentration exceeded the corresponding Probable Effect Concentration (PEC) value: 821 vs. 459 mg kg^{-1} [40]. Elevated concentrations were recorded for barium (Ba) and copper (Cu), 282 and 117 mg kg^{-1}, respectively, but these values did not exceed threshold values. However, PCDD/PCDF levels were high, exceeding the 0.85 ng TEQ kg^{-1} limit specified in the Sediment Quality Guideline (SQG) (http://ceqg-rcqe.ccme.ca/download/en/245) by more than 10-fold (Table 1).

The application of sediments to soil shifted its pH from neutral (pH 6.65) to slightly alkaline (pH 7.21). Also, OC increased to 19.9 g kg^{-1} after the application of sediment. The concentrations of trace elements also increased; however, they remained below the Probable Effect Concentration (PEC). Only the TEQ concentration of PCDDs/PCDFs grew significantly, exceeding the allowable limit of 0.85 ng TEQ kg^{-1} (SQG) by 2.5-fold as an effect of sediment application (Table 1).

3.2. The Effects of Urban Sediment Amendment and Applied Remediation Strategies on Soil Phytotoxicity and PCDD/PCDF Concentrations

The structure of bottom sediments renders them a perfect geosorbent for the mixture of pollutants introduced to the water environment. Consequently, the assessment of their toxicity based on monitoring of hazardous substances, such as heavy metals and PCDDs/PCDFs, does not encompass all the chemical compounds potentially present therein, nor their interactions. In this situation, the bioindication method (biotests) is a more accurate approach to assessing the toxicity of dredged sediments and may provide more useful information about the phytotoxicity and influence on soil [41–43].

In our bioassay, the application of urban sediments to soil was found to have a positive influence on plant growth, stimulating an 89% increase in *L. sativum* root length, in comparison to control soil. This increased growth can be attributed to the high nutrient content and greater Ca and Fe levels, which are known to be capable of immobilizing pollutants, especially trace elements (Table 1). This positive influence of urban sediments fell to 29% after 10 weeks of the experiment (natural attenuation strategy). However, samples inoculated with rhizobacterial strains were characterized by a better growth of *L. sativum* in comparison to the non-augmented samples, showing an increase of root growth of 38% (p87) and 65% (RP92) in comparison to control samples (Figure 2).The application of inoculation-assisted phytoremediation using both rhizobacterial inoculants and both plants further improves soil quality, with a 97% increase in *L. sativum* root growth compared to untreated samples observed in the case of simultaneous application of *F. arundinacea* and bacterial strain p87. The two-way ANOVA did not confirm, however, the influence of the bacterial inoculation and plants on the soil phytotoxicity.

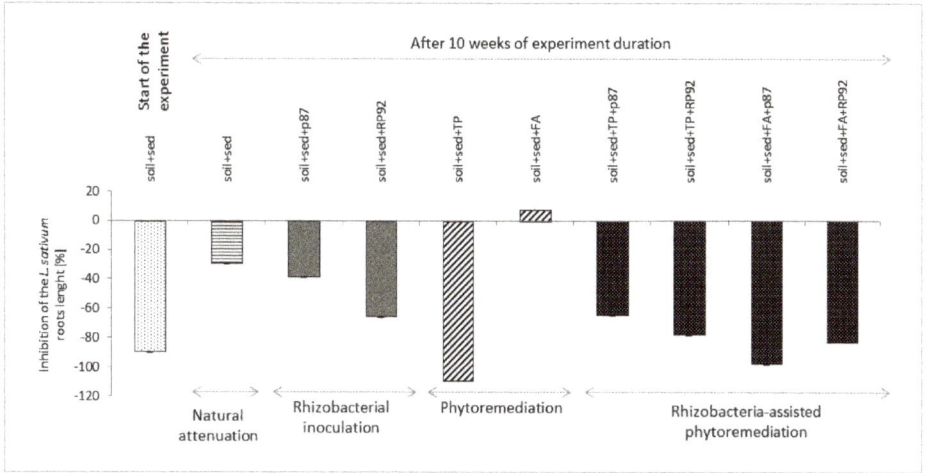

Figure 2. The effect of applied bioremediation strategies on soil phytotoxicity (negative values indicate stimulation of plant root growth).

The application of urban sediment led to nine-fold and seven-fold increases in the concentrations of the sum of 17 PCDDs/PCDFs congeners and TEQ in soil, respectively, in comparison to control soil (Table 1). The application of natural attenuation for 10 weeks increased the total PCDD/PCDF concentration by 47% (Figure 3A); however, TEQ concentration decreased by 14% in comparison to the initial value (Figure 3B). A similar situation was observed when the rhizobacterial strains were applied: Total PCDD/PCDF concentrations increased by 14% and 18% when p87 and RP92 strains were applied (Figure 3A), while TEQ decreased by 23% and 20%, respectively (Figure 3B). The phytoremediation strategy was associated with a 24% (for *T. patula*) and 10% (for *F. arundinacea*) increase of total PCDD/PCDF level (Figure 2A), as well as a 23% (for *T. patula*) and 21% (for *F. arundinacea*) decrease in TEQ value (Figure 3B).

The increasing total concentrations of PCDDs/PCDFs and the lowering of the TEQ values are related to the ongoing degradation processes. PCDDs/PCDFs are subject to both anaerobic and aerobic metabolism. Under anaerobic conditions, dechlorination of higher chlorinated congeners (mostly hexa-, hepta- and octa-chlorinated ones) of lower toxicity, reflected as low TEFs, occurs; while under aerobic conditions, lower chlorinated PCDDs/PCDFs, characterized by higher TEFs contributing in a higher extent to the TEQ, can be removed. Consequently, aerobic bacterial transformation led to the decrease in the content of lower chlorinated and thus more toxic congeners, thus leading to the reduction of the total TEQs in the studied samples.

The most effective approach, however, was the rhizobacteria-assisted phytoremediation strategy based on *F. arundinacea*. This strategy diminished the total concentration by 18% (when used with p87) or 8% (RP92) (Figure 3A), and PCDD/PCDF TEQ value by 44% (p87) or 36% (RP92) (Figure 3B). Literature data also confirms the value of *F. arundinacea* for remediation purposes. Sun et al. [44] reported the degradation of polyaromatic hydrocarbons (PAHs) in soil treated with *F. arundinacea*. The authors demonstrated that 25%, 10% and 30% of 3-ring, 4-ring and 5(+6)-ring PAHs, respectively, were removed by *F. arundinacea*, while this value was 0.6% in unplanted soil. The application of *F. arundinacea* increased the soil PAH-degrading bacterial counts and microbial activity, suggesting that the plant can restore the microbiological functioning of PAH-contaminated soil. However, it must be stated that an essential step toward the biodegradation of a given compound is the expression of the respective degradative genes in bacteria. With this in mind, Siciliano et al. [45] reported greater induction of catabolic genes involved in naphthalene degradation in the rhizosphere soil of *F. arundinacea* than in unplanted soil. This clearly demonstrates the suitability of *F. arundinacea* in the remediation of organic compounds.

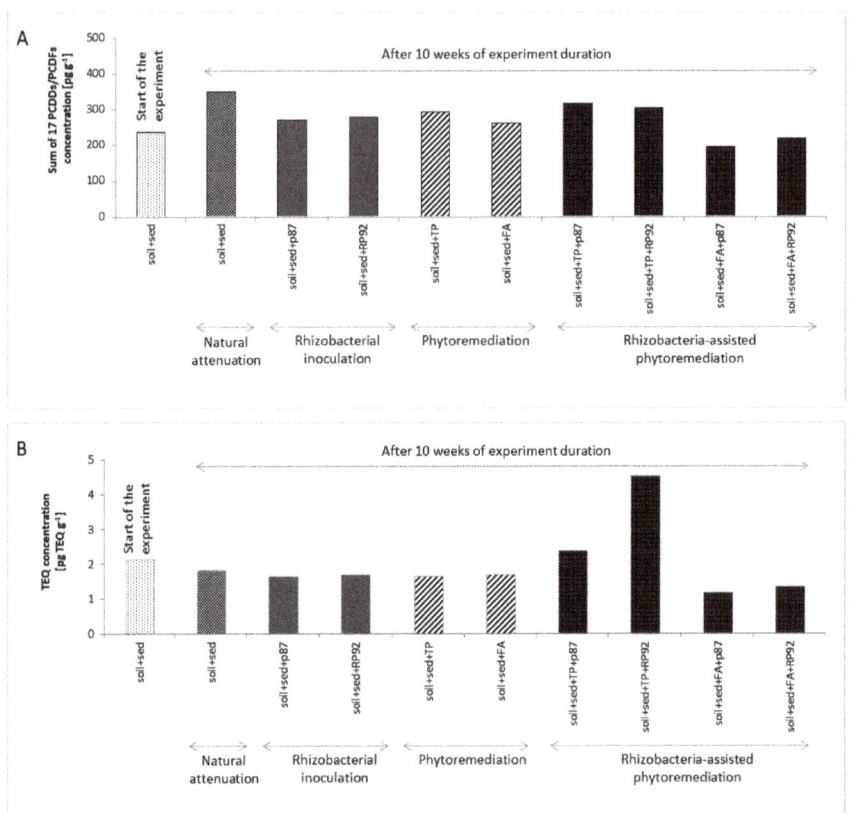

Figure 3. The effect of applied bioremediation strategies on changes in total (**A**) and TEQ (**B**) concentration of PCDDs/PCDFs.

The opposite was observed for the application of *T. patula*: Total values grew by 34% (p87) or 28% (RP92), and PCDD/PCDF TEQ values grew by 13% (p87) or 113% (RP92) (Figure 3A,B). Such increases in the total and TEQ values can be related to the transformation processes, which sometimes led to the production of intermediate compounds of higher toxicity than the parent congeners. In this case, the application of rhizobacteria-assisted phytoremediation strategy with *T. patula*, led to production of penta-chlorinated congeners (data not shown) characterized by higher toxicity in comparison to the hexa-, hepta- and octa-chlorinated compounds being the substrate for dechlorinating processes. The visual inspection of the soil-sediment samples planted with *T. patula* showed possible anoxic conditions therein (compacted, impermeable soil), facilitating dechlorination of higher chlorinated congeners and production of lower chlorinated ones of higher TEFs.

The effectiveness of rhizosphere biodegradation depends on the ability of microorganisms to adapt to a given pollutant concentration and their ability to colonize roots [46]. Kuiper et al. [47] demonstrated that naturally occurring rhizosphere biodegradation may be enhanced by the addition of microorganisms to the rhizosphere. Concluding this, in our case, both rhizobacterial inoculation and phytoremediation strategies gave similar outcomes, resulting in around a 20%–23% decline in the TEQ value. The similar effects of these strategies may be related to the fact that both led to an increase in the activity of soil microbiota: Rhizobacterial inoculation through the artificial addition of selected strains capable of degrading the given pollutant, and phytoremediation/rhizoremediation through the existing interactions between plant exudates, soil and microorganisms. The most promising solution seems to be the rhizobacteria-assisted phytoremediation strategy; however, among the used ornamentals, only F. arundinacea demonstrated the capacity to reduce both PCDD/PCDF total and TEQ levels. T. patula, in turn, despite its positive influence on PCDD/PCDF reduction, when used alone as part of the phytoremediation strategy, increased the total and TEQ concentration when used in combination with the bacterial inoculation.

3.3. The Effects of Urban Sediment Amendments and Applied Remediation Strategies on the Biomass and Physiological Parameters of T. patula L. and F. arundinacea

Pollutant concentration in the soil is certainly a key factor determining plant tolerance or sensitivity. Nevertheless, other factors such as metal speciation, the composition of heterogeneous hydrocarbon fractions, soil-pollutant and pollutant-pollutant interactions, also have to be considered. Another aspect is the protective character of the rhizosphere microbiota, which plays an intrinsic role in the protection of plants against pathogens and stress caused by excessive concentrations of pollutants and eases the uptake of biogenic substances by a given plant [46,48,49].

From the perspective of the proper organization and management of the city space, it is important to select the most resistant plant species which both embellish the environment and resolve the pollution problem in the urban area. Therefore, the inoculation of existing soil microbiota with plant growth-promoting rhizobacteria, may not only improve plant growth, but also enhance phytoremediation rates by assisting in resource acquisition and modulating plant hormone levels, and/or by decreasing the inhibitory effects of pathogens [15,50,51].

In our case, the cultivation of T. patula and F. arundinacea following the application of urban sediments to soil led to different plant responses. T. patula demonstrated 67% lower biomass when grown in soil amended with urban sediments (Figure 4(A1,A2)). The inoculation of soil with bacterial strains alleviated the toxic effect caused by sediment application by 47% for p87 and 54% for RP92, in comparison to plants grown in non-inoculated soil. The addition of p87 and RP92 strains to soil following sediments application led to 2.4-fold and 2.6-fold higher production of plant biomass in comparison to non-inoculated samples (case of T. patula). At the same time, the inoculation did not influence the biomass of plants grown in uncontaminated soil (Figure 4(A1,A2)).

The obtained results demonstrated that soil inoculation had a positive influence on T. patula growth. Two-way ANOVA analysis showed that both the sediment admixture and inoculation have an influence on plant biomass. Also, Duncan post-hoc test confirmed the significantly higher biomass of plants grown in soil with sediments and inoculation, however, no differences were found between used inoculants (p87 vs. RP92). Samples without sediment demonstrated no statistically significant changes (Figure 4(A1)). Although it has been proposed to use T. patula as phytoremediation tool for dyes, tannery solid waste [52], soil co-contaminated by benzo[a]pyrene and metals [20], there is a considerable lack of information regarding the impact of bacterial inoculation on its growth and morphology. However, Agnello et al. [15] reported that bioaugmentation has a positive influence on plant biomass: they noted that bioaugmentation with P. aeruginosa had a positive effect on alfalfa biomass production, resulting in an increase of shoot biomass by as much as 56% and of root biomass by 105%.

In the case of F. arundinacea, the plant grown in amended soil showed only 6% lower biomass than the plants grown in uncontaminated soil. The inoculation with p87 strain did not give any positive or negative response, while inoculation with RP92 strain led to 6% higher plant biomass in comparison to plants growing on untreated soil. In contrast to T. patula, F. arundinacea demonstrated slight decreases in biomass when grown in inoculated unpolluted soil (3% for p87 and 11% for RP92), but no such changes were observed for plants grown in sediment-amended soil (Figure 4(A2)). However, the obtained results were not statistically significant (two-way ANOVA).

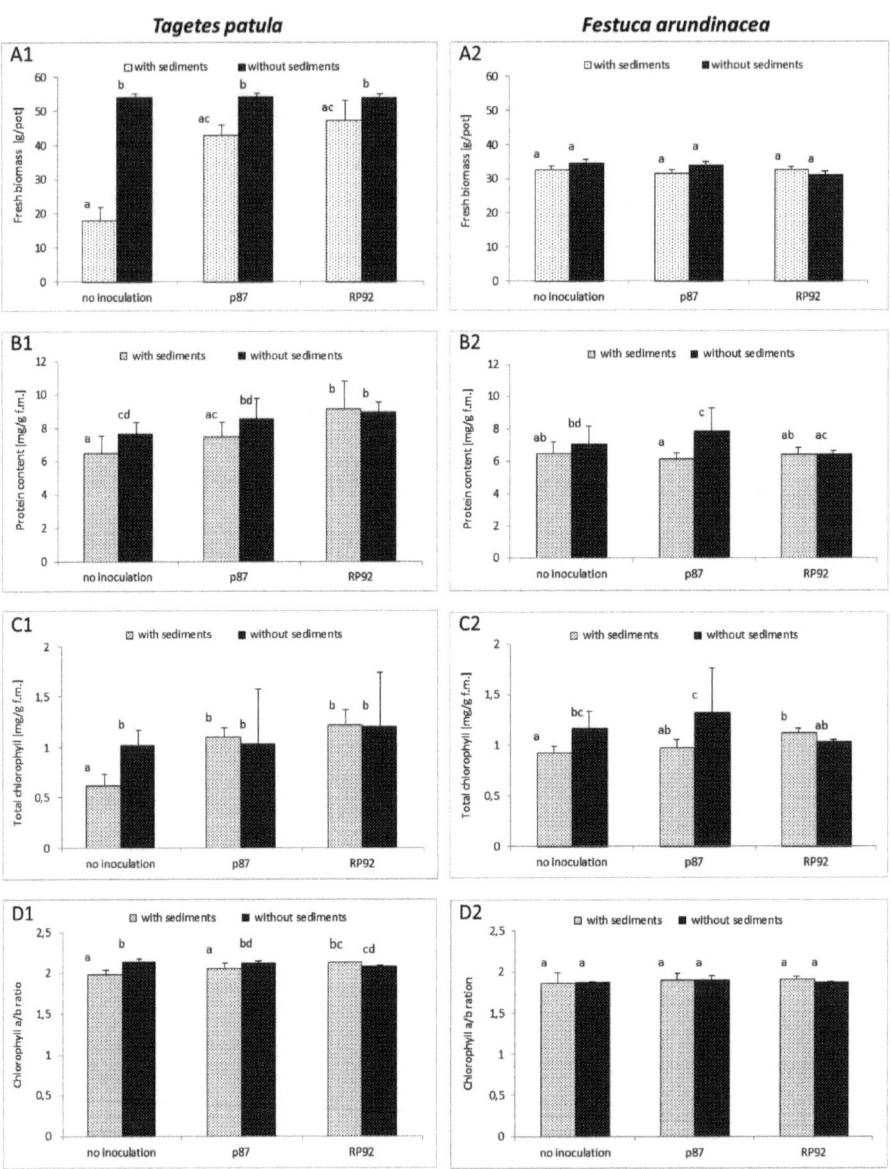

Figure 4. The effect of applied sediments and inoculation of growing medium (soil) with two bacterial strains p87 and RP92 on fresh biomass (**A1,A2**), protein content (**B1,B2**), chlorophyll content (**C1,C2**) and chlorophyll a/b ratio (**D1,D2**) in the leaf tissues of *T. patula* and *F. arundinacea* (at $p < 0.05$, the Duncan post-hoc test, the same letters indicate no statistically significant differences).

While morphological observation of the plant did not reveal the toxic effects related to the use of sediment or bacterial strains, a more diversified response was found when analyzing physiological parameters such as protein content, due to the greater sensitivity and higher degree of response to applied remediation techniques, and the range of compounds present in the sediments.

T. patula grown in sediment-amended soil showed an increase in soluble protein content. Two-way ANOVA confirmed the influence of both sediment amendment and bacterial inoculation on the protein content. The results showed that inoculation with bacterial strains led to a significant increase (Duncan post-hoc test) in protein content being 115% (for p87) and 141% (for RP82) of control values (Figure 4(B1)). In the non-inoculated soil samples, the protein content of *T. patula* was 15% lower in variants fertilized with sediment than in the unfertilized ones. The obtained differences were statistically significant (Duncan post-hoc test).

In contrast, in the case of *F. arundinacea*, sediments had a significant influence on protein content (two-way ANOVA), while inoculation did not affect it. Duncan post-hoc test, however, revealed the statistically significant difference in variant inoculated with p87 and no sediment admixture. In this case, application of p87 strain to the unamended soil led to a 12% increase in the protein content, while simultaneous application of both p87 and sediment resulted in lowering the protein content (Figure 4(B2)).

From the physiological point of view, it is known that soluble protein content decreases as plant senescence-related processes continue [53,54]. This is related to the initiation of the N-remobilization process, during which proteinase activity increases, leading to the degradation of protein to amino acids. The resulting amino acids and/or peptides are then transported through the phloem sap to the growing organs [55]. With this in mind, the increase in soluble protein content observed in *T. patula* tissues grown in inoculated soil may indicate a delay of senescence processes. This increase in soluble protein content is particularly noticeable after sediment application. Considering the toxicity of the sediment used in the experiment, the above described relationship can also indicate that the introduced rhizobacterial strains have a protective effect on studied plants that is reflected in the prevention of plant premature senescence. RP92 proved to have a stronger protective effect in this case.

In the case of total chlorophyll content of *T. patula*, two-way ANOVA demonstrated that sediments have no influence, unlike the type of inoculation that was found to significantly affect the content of chlorophyll. Duncan post-hoc test demonstrated, in turn, that the use of urban sediment significantly lowered the total chlorophyll content in green tissues of *T. patula* grown in non-inoculated variant, being only 60% of that of the value measured in plants grown in non-amended soil (Figure 4(C1)). In this case, the inoculation of sediment-amended soil with p87 and RP92 significantly increased the chlorophyll content to 178% and 198% of non-inoculated and unamended control, respectively (Duncan post-hoc test). Moreover, RP92 increased the chlorophyll content even in non-fertilized plants to 118% of non-amended and non-inoculated control values; however, this increase was not statistically significant (Figure 4(C1)). Regarding the chlorophyll a/b ratio, the two-way ANOVA confirmed the influence of both the sediment admixture and inoculation on the obtained results. Duncan post-hoc test revealed significantly higher chlorophyll a/b ratio in plants grown in soil amended with sediment and inoculated with RP92, in comparison to non-inoculated samples as well as ones inoculated with p87 (both amended with sediments) (Figure 4(C1)). The unamended samples inoculated with RP92 demonstrated a significantly lower chlorophyll a/b ratio, when compared to the unamended and non-inoculated sample (Figure 4(C1)).

For *F. arundinacea*, two-way ANOVA demonstrated a contrary effect to *T. patula*: In this case, the chlorophyll content was dependent on the sediment application, while no such effect was observed for inoculation. Duncan post-hoc test revealed the significant increase in the chlorophyll content in sediment amended samples, in comparison to unamended ones. Moreover, significantly higher chlorophyll content was observed in plants grown in soil amended with sediments and inoculated with RP92, in comparison to the non-inoculated variant (Figure 4(C2)). Inoculation with p87, in turn, led to significantly higher chlorophyll content, but only when compared with samples without sediments and without inoculation. In the case of chlorophyll a/b ratio, the statistical analyses showed neither the influence of sediments application, nor the type of inoculation (Figure 4(D1,D2)).

Our findings indicate that the soil amended with the urban sediment caused a significant reduction of chlorophyll content in both *T. patula* and *F. arundinacea*; while soil inoculation with bacterial strains not only suppressed this effect, but also led to an increase in chlorophyll content in comparison to control samples. The decrease of the chlorophyll content observed in investigated plants due to the application of sediments may be related to the presence of xenobiotics in the sediment. The decrease in chlorophyll content in leaves is one of the first symptoms of plant senescence, during which the breakdown of the thylakoid membranes and degradation of thylakoid-bound proteins occur. Our findings suggest that the use of sediments can exacerbate this process, while the application of bacterial strains protects plants from premature senescence. Furthermore, the reduced chlorophyll a/b ratio may indicate the progress of the aging process of *T. patula* and *F. arundinacea* grown in soil treated with sediments without simultaneous inoculation. Similarly, Nath et al. [56] demonstrated a linear decrease of chlorophyll a/b ratio in *Arabidopsis thaliana* during natural leaf senescence, along with a decrease in total chlorophyll content. The authors noted that with gradual senescence, differential degradation of chlorophyll a and chlorophyll b leads to changes in chlorophyll a/b ratio.

In conclusion, *T. patula* is more susceptible to the range of xenobiotics in the tested sediments, but the application of bacterial strains alleviated the phytotoxicity, especially the RP92 strain. The use of bacteria also gives positive results when the soil is not amended with sediment. In the case of *F. arundinacea*, the use of both sediment application and inoculation with rhizobacterial strains promotes a good plant physiological status; while in sediment-free soil, only strain p87 seems to be beneficial.

Author Contributions: Conceptualization, M.U., A.W., G.S., S.S.; methodology, M.U., A.W., G.S., S.S., P.K., M.Z.; validation, M.U., A.W., G.S., S.S., M.Z.; investigation M.U., A.W., G.S., S.S., M.Z.; resources, M.U., G.S., P.K.; data curation, M.U., G.S.; writing—original draft preparation, M.U., G.S.; writing—review and editing, M.U.; visualization, M.U.; supervision, M.U.; project administration, M.U., G.S.; funding acquisition, M.U., G.S.

Funding: The pot experiment was a part of Greenland project (FP7-KBBE-266124) financed by the European Commission under the Seventh Framework Programme for Research; the research was also conducted as a part of the Project IP2014 049273 financed by the Ministry of Science and Higher Education programme under the name "Iuventus Plus" for the years 2015–2017.

Conflicts of Interest: The authors declare no conflict of interest. The funders had no role in the design of the study; in the collection, analyses, or interpretation of data; in the writing of the manuscript, or in the decision to publish the results.

References

1. Foster, G.D.; Roberts, E.C., Jr.; Gruessner, B.; Velinsky, D.J. Hydrogeochemistry and transport of organic contaminants in an urban watershed of Chesapeake Bay (USA). *Appl. Geochem.* **2000**, *15*, 901–915. [CrossRef]
2. Im, S.H.; Kannan, K.; Matsuda, M.; Giesy, J.P.; Wakimoto, T. Sources and distribution of polychlorinated dibenzo-p-dioxins and dibenzofurans in sediment from Masay Bay, Korea. *Environ. Toxicol. Chem.* **2002**, *21*, 245–252. [CrossRef] [PubMed]
3. Jartun, M.; Ottesen, R.T.; Steinnes, E.; Volden, T. Runoff of particle bound pollutants from urban impervious surfaces studied by analysis of sediments from stormwater traps. *Sci. Total. Environ.* **2008**, *396*, 147–163. [CrossRef] [PubMed]
4. Krishnappan, B.G.; Marsalek, J. Modelling of flocculation and transport of cohesive sediment from an on-stream stormwater detention pond. *Water Res.* **2002**, *36*, 3849–3859. [CrossRef]
5. Urbaniak, M.; Zieliński, M.; Ligocka, D.; Zalewski, M. A comparative analysis of selected Persistent Organic Pollutants (POPs) in reservoirs of different types of anthropopression—Polish and Ethiopian studies. *Fresenius Environ. Bull.* **2010**, *19*, 2710–2718.
6. Urbaniak, M.; Kiedrzyńska, E.; Zalewski, M. The role of a lowland reservoir in the transport of micropollutants, nutrients and the suspended particulate matter along the river continuum. *Hydrol. Res.* **2010**, *43*, 400–411. [CrossRef]
7. Urbaniak, M.; Zieliński, M.; Kaczkowski, Z.; Zalewski, M. Spatial distribution of PCDDs, PCDFs and dl-PCBs along the cascade of urban reservoirs. *Hydrol. Res.* **2012**, *44*, 614–630. [CrossRef]
8. Urbaniak, M.; Kiedrzyńska, E.; Zieliński, M.; Tołoczko, W.; Zalewski, M. Spatial distribution and reduction of PCDD/PCDF Toxic Equivalents along the three shallow lowland reservoirs. *Environ. Sci. Pollut. Res.* **2014**, *21*, 4441–4452. [CrossRef]

9. Urbaniak, M.; Kiedrzyńska, E.; Kiedrzyński, M.; Zieliński, M.; Grochowalski, A. The role of hydrology in the polychlorinated dibenzo-p-dioxin and dibenzofuran distributions in a lowland river. *J. Environ. Qual.* **2015**, *44*, 1171–1182. [CrossRef]
10. Urbaniak, M.; Tygielska, A.; Krauze, K.; Mankiewicz-Boczek, J. Effects of Stormwater and Snowmelt Runoff on ELISA-EQ Concentrations of PCDD/PCDF and Triclosan in an Urban River. *PLoS ONE* **2016**, *11*, 0151756. [CrossRef]
11. Wagner, I.; Zalewski, M. Ecohydrology as a Basis 1 for the Sustainable City Strategic Planning—Focus on Lodz, Poland. *Rev. Environ. Sci. Bio/Technol.* **2009**, *8*, 209–217. [CrossRef]
12. Wagner, I.; Zalewski, M. System solutions in urban water management: The Lodz (Poland) perspective. In *Water Sensitive Cities*; Howe, C., Mitchel, C., Eds.; IWA Publishing: London, UK, 2011; pp. 231–245.
13. European Parliament, Council of the European Union. Directive of the European Parliament and the Council 2013/39/EC of 12 August 2013 Amending Directive 2000/60/EC and 2008/105/EC in Respect of Priority Substances in the Field of Water Policy. 2013. Available online: https://eur-lex.europa.eu/LexUriServ/LexUriServ.do?uri=OJ:L:2013:226:0001:0017:EN:PDF (accessed on 19 September 2019).
14. Briggs, G.G.; Bromilow, R.H.; Evans, A.A. Relationships between lipophilicity and root uptake and translocation of non-ionised chemicals by barley. *Pestic. Sci.* **1982**, *13*, 495–504. [CrossRef]
15. Agnello, A.; Bagard, M.; Van Hullebusch, E.; Esposito, G.; Huguenot, D. Comparative bioremediation of heavy metals and petroleum hydrocarbons co-contaminated soil by natural attenuation, phytoremediation, bioaugmentation and bioaugmentation-assisted phytoremediation. *Sci. Total. Environ.* **2016**, *563*, 693–703. [CrossRef] [PubMed]
16. Huguenot, D.; Bois, P.; Cornu, J.Y.; Jezeguel, K.; Lollier, M.; Lebeau, T. Remediation of sediment and water contaminated by copper in small-scaled constructed wetlands: Effect of bioaugmentation and phytoextraction. *Environ. Sci. Pollut. Res.* **2015**, *22*, 721–732. [CrossRef] [PubMed]
17. Glick, B.R. Using soil bacteria to facilitate phytoremediation. *Biotechnol. Adv.* **2010**, *28*, 367–374. [CrossRef] [PubMed]
18. Lin, X.; Li, X.; Li, P.; Li, F.; Zhang, L.; Zhou, Q. Evaluation of Plant–Microorganism Synergy for the Remediation of Diesel Fuel Contaminated Soil. *Bull. Environ. Contam. Toxicol.* **2008**, *81*, 19–24. [CrossRef] [PubMed]
19. Khan, S.; Afzal, M.; Iqbal, S.; Khan, Q.M. Plant–bacteria partnerships for the remediation of hydrocarbon contaminated soils. *Chemosphere* **2013**, *90*, 1317–1332. [CrossRef] [PubMed]
20. Sun, Y.; Zhou, Q.; Xu, Y.; Wang, L.; Liang, X. Phytoremediation for co-contaminated soils of benzo[a]pyrene (B[a]P) and heavy metals using ornamental plant Tagetes patula. *J. Hazard. Mater.* **2011**, *186*, 2075–2082. [CrossRef] [PubMed]
21. Suresh, B.; Bais, H.; Raghavarao, K.; Ravishankar, G.; Ghildyal, N. Comparative evaluation of bioreactor design using Tagetes patula L. hairy roots as a model system. *Process. Biochem.* **2005**, *40*, 1509–1515. [CrossRef]
22. Vasudevan, P.; Kashyap, S.; Sharma, S. Tagetes: A multipurpose plant. *Bioresour. Technol.* **1997**, *62*, 29–35. [CrossRef]
23. Cruz-Fernandes, A.; Tomasini-Compocosio, A.; Perez-Flores, L.J.; Fernandez-Perrino, F.J.; Gutierrez-Rojas, M. Inoculation of seed-borne fungus in the rhizosphere of Festuca arundinacea promotes hydrocarbon removal and pyrene accumulation in roots. *Plant Soil* **2013**, *362*, 261–270. [CrossRef]
24. Xiao, N.; Liu, R.; Jin, C.; Dai, Y. Efficiency of five ornamental plant species in the phytoremediation of polycyclic aromatic hydrocarbon (PAH)-contaminated soil. *Ecol. Eng.* **2015**, *75*, 384–391. [CrossRef]
25. Balseiro-Romero, M.; Gkorezis, P.; Kidd, P.S.; Vangronsveld, J.; Monterroso, C. Enhanced degradation of diesel in the rhizosphere of Lupinus luteus after inoculation with diesel degrading and PGP bacterial strains. *J. Environ. Qual.* **2016**, *45*, 924–932. [CrossRef] [PubMed]
26. Becerra-Castro, C.; Kidd, P.; Prieto-Fernández, Á.; Weyens, N.; Acea, M.; Vangronsvel, J. Endophytic and rhizoplane bacteria associated with Cytisus striatus growing on hexachlorocyclohexane-contaminated soil: Isolation and characterisation. *Plant Soil* **2011**, *340*, 413–433. [CrossRef]
27. Becerra-Castro, C.; Monterroso, C.; Prieto-Fernandez, A.; Rodríguez-Lamas, L.; Loureiro-Viñas, M.; Acea, M.; Kidd, P. Pseudometallophytes colonising Pb/Zn mine tailings: A description of the plant–microorganism–rhizosphere soil system and isolation of metal-tolerant bacteria. *J. Hazard. Mater.* **2012**, *217*, 350–359. [CrossRef] [PubMed]

28. Szklarek, S.; Wagner, I.; Jurczak, T.; Zalewski, M. Sequential Sedimentation-Biofiltration System for the purification of a small urban river (the Sokolowka, Lodz) supplied by stormwater. *J. Environ. Manag.* **2018**, *205*, 201–208. [CrossRef]
29. Negussie, Y.Z.; Urbaniak, M.; Szklarek, S.; Lont, K.; Gągała, I.; Zalewski, M. Efficiency analysis of two sequential biofiltration systems in Poland and Ethiopia—The pilot study. *Ecohydrol. Hydrobiol.* **2012**, *12*, 271–285. [CrossRef]
30. Mergeay, M.; Nies, D.; Schlegel, H.G.; Gerits, J.; Charles, P.; Van Gijsegem, F. Alcaligenes eutrophus CH34 is a facultative chemolithotroph with plasmid-bound resistance to heavy metals. *J. Bacteriol.* **1985**, *162*, 328–334.
31. Polish Committee for Standardization. *Emission from Stationary Sources. Determination of PCDD/PCDF Mass Concentration. PRT 3: Identification and Quantification*; PN-EN 1948-3; PKN: Warsaw, Poland, 2006. (In Polish)
32. EPA. *Method 1613, Tetra through Octa chlorinated dioxins and furans by Isotope dilution HRGC/HRMS*; Revision B; US EPA: Washington, DC, USA, 1994.
33. Van den Berg, M.; Birnbaum, L.S.; Denison, M.; De Vito, M.; Farland, W.; Feeley, M.; Fiedler, H.; Hakansson, H.; Hanberg, A.; Haws, L.; et al. The 2005 World Health Organization Re-evaluation of Human and Mammalian Toxic Equivalency Factors for Dioxins and Dioxin-like Compounds. *Toxicol. Sci.* **2006**, *93*, 223–241. [CrossRef]
34. Van den Berg, M. The 2005 WHO re-evaluation of toxic equivalency factors for dioxin like compounds—Implications for risk assessment and limitations of the concept. *Toxicol. Letters* **2006**, *164*, S55–S56. [CrossRef]
35. *Phytotoxkit, Seed Germination and Early Growth Microbiotest with Higher Plants*; MicroBioTest Inc.: Nazareth, Belgium, 2004.
36. Persoone, G.; Maršálek, B.; Blinova, I.; Torokne, A.; Zarina, D.; Manusadžianas, L.; Nalecz-Jawecki, G.; Tofan, L.; Stepanova, N.; Tothova, L.; et al. A practical and user-friendly toxicity classification system with microbiotests for natural waters and wastewaters. *Environ. Toxicol.* **2003**, *18*, 395–402. [CrossRef] [PubMed]
37. Bradford, M.M. A rapid and sensitive method for the quantification of microgram quantities of protein utilizing the principle of protein-dye binding. *Anal. Biochem.* **1976**, *72*, 248–254. [CrossRef]
38. Porra, R.; Thompson, W.; Kriedemann, P. Determination of accurate extinction coefficients and simultaneous equations for assaying chlorophylls a and b extracted with four different solvents: Verification of the concentration of chlorophyll standards by atomic absorption spectroscopy. *Biochim. Et Biophys. Acta (Bba)—Bioenerg.* **1989**, *975*, 384–394. [CrossRef]
39. Toth, G.; Jones, A.; Montanarella, L. *LUCAS Topsoil Survey Methodology. Data and Results*; JRC Technical Reports; Publications Office of the European Union: Luxembourg, 2013.
40. MacDonald, D.; Ingersoll, C.; Berger, T. Development and evaluation of consensus-based sediment development and evaluation of consensus-based sediment quality guidelines for freshwater ecosystems. *Arch. Environ. Contam. Toxicol.* **2000**, *39*, 20–31. [CrossRef] [PubMed]
41. Baran, A.; Tarnawski, M. Phytotoxkit/Phytotestkit and Microtox® as tools for toxicity assessment of sediments. *Ecotoxicol. Environ. Saf.* **2013**, *98*, 19–27. [CrossRef]
42. Chen, Y.; Zhu, G.; Tian, G.; Zhou, G.; Luo, Y.; Wu, S. Phytotoxicity of dredged sediment from urban canal as land application. *Environ. Pollut.* **2002**, *117*, 233–241. [CrossRef]
43. Czerniawska-Kusza, I.; Kusza, G. The potential of the Phytotoxkit microbiotest for hazard evaluation of sediments in eutrophic freshwater ecosystems. *Environ. Monit. Assess.* **2011**, *179*, 113–121. [CrossRef]
44. Sun, M.; Fu, D.; Teng, Y.; Shen, Y.; Luo, Y.; Li, Z.; Christie, P. In situ phytoremediation of PAH-contaminated soil by intercropping alfalfa (*Medicago sativa* L.) with tall fescue (Festuca arundinacea Schreb.) and associated soil microbial activity. *J. Soils Sediments* **2011**, *11*, 980–989. [CrossRef]
45. Siciliano, S.D.; Germida, J.J.; Banks, K.; Greer, C.W. Changes in microbial community composition and function during a polyaromatic hydrocarbon phytoremediation field trial. *Appl. Environ. Microbiol.* **2003**, *69*, 483–489. [CrossRef]
46. Macek, T.; Macková, M.; Káš, J. Exploitation of plants for the removal of organics in environmental remediation. *Biotechnol. Adv.* **2000**, *18*, 23–34. [CrossRef]
47. Kuiper, I.; Lagendijk, E.L.; Bloemberg, G.V.; Lugtenberg, B.J.J. Rhizoremediation: A Beneficial Plant-Microbe Interaction. *Mol. Plant-Microbe Interact.* **2004**, *17*, 6–15. [CrossRef] [PubMed]
48. Lugtenberg, B.J.; Dekkers, L.; Bloemberg, G.V. Molecular determinants of rhizosphere colonization by Pseudomonas. *Ann. Rev. Phytopathol.* **2001**, *39*, 461–490. [CrossRef] [PubMed]

49. Whipps, J.M. Carbon economy. In *The Rhizosphere*; Lynch, J.M., Ed.; Wiley: New York, NY, USA, 1990; pp. 59–97.
50. Ahemad, M.; Kibret, M. Mechanisms and applications of plant growth promoting rhizobacteria: Current perspective. *J. King Saud Univ. Sci.* **2014**, *26*, 1–20. [CrossRef]
51. Puga-Freitas, R.; Blouin, M. A review of the effects of soil organisms on plant hormone signalling pathways. *Environ. Exp. Bot.* **2015**, *114*, 104–116. [CrossRef]
52. Firdaus-e-Bareen; Nazir, A. Metal decontamination of tannery solid waste using Tagetes patula in association with saprobic and mycorrhizal fungi. *Environmentalist* **2010**, *30*, 45–53. [CrossRef]
53. Camp, P.J.; Huber, S.C.; Burke, J.J.; Moreland, D.E. Biochemical Changes that Occur during Senescence of Wheat Leaves. *Plant Physiol.* **1982**, *70*, 1641–1646. [CrossRef] [PubMed]
54. Romanova, A.K.; Semenova, G.A.; Ignat'ev, A.R.; Novichkova, N.S.; Fomina, I.R. Biochemistry and cell ultrastructure changes during senescence of Beta vulgaris L. leaf. *Protoplasma* **2016**, *253*, 719–727. [CrossRef]
55. Girondé, A.; Poret, M.; Etienne, P.; Trouverie, J.; Bouchereau, A.; Le Cahérec, F.; Leport, L.; Orsel, M.; Niogret, M.-F.; Deleu, C.; et al. A profiling approach of the natural variability of foliar N remobilization at the rosette stage gives clues to understand the limiting processes involved in the low N use efficiency of winter oilseed rape. *J. Exp. Bot.* **2015**, *66*, 2461–2473. [CrossRef]
56. Nath, K.; Phee, B.-K.; Jeong, S.; Lee, S.Y.; Tateno, Y.; Allakhverdiev, S.I.; Lee, C.-H.; Gil Nam, H. Age-dependent changes in the functions and compositions of photosynthetic complexes in the thylakoid membranes of Arabidopsis thaliana. *Photosynth. Res.* **2013**, *117*, 547–556. [CrossRef]

© 2019 by the authors. Licensee MDPI, Basel, Switzerland. This article is an open access article distributed under the terms and conditions of the Creative Commons Attribution (CC BY) license (http://creativecommons.org/licenses/by/4.0/).

Article

Microcosm Experiment to Assess the Capacity of a Poplar Clone to Grow in a PCB-Contaminated Soil

Isabel Nogues [1], Paola Grenni [2,*], Martina Di Lenola [2], Laura Passatore [1], Ettore Guerriero [3], Paolo Benedetti [3], Angelo Massacci [1], Jasmin Rauseo [2] and Anna Barra Caracciolo [2]

1. National Research Council, Institute on Terrestrial Ecosystems (IRET-CNR), Area della Ricerca RM1, 00015 Moterotondo, Rome, Italy; isabel.nogues@cnr.it (I.N.); laura.passatore@iret.cnr.it (L.P.); angelo.massacci@ibaf.cnr.it (A.M.)
2. National Research Council, Water Research Institute (IRSA-CNR), Area della Ricerca RM1, 00015 Moterotondo, Rome, Italy; dilenola@irsa.cnr.it (M.D.L.); rauseo@irsa.cnr.it (J.R.); barracaracciolo@irsa.cnr.it (A.B.C.)
3. National Research Council, Institute of Atmospheric Pollution Research (IIA-CNR), Area della Ricerca RM1, 00015 Moterotondo, Rome, Italy; guerriero@iia.cnr.it (E.G.); p.benedetti@iia.cnr.it (P.B.)
* Correspondence: grenni@irsa.cnr.it

Received: 12 July 2019; Accepted: 22 October 2019; Published: 25 October 2019

Abstract: Polychlorinated byphenyls (PCBs) are a class of Persistent Organic Pollutants extremely hard to remove from soil. The use of plants to promote the degradation of PCBs, thanks to synergic interactions between roots and the natural soil microorganisms in the rhizosphere, has been proved to constitute an effective and environmentally friendly remediation technique. Preliminary microcosm experiments were conducted in a greenhouse for 12 months to evaluate the capacity of the Monviso hybrid poplar clone, a model plant for phytoremediation, to grow in a low quality and PCB-contaminated soil in order to assess if this clone could be subsequently used in a field experiment. For this purpose, three different soil conditions (Microbiologically Active, Pre-sterilized and Hypoxic soils) were set up in order to assess the capacity of this clone to grow in the polluted soil in these different conditions and support the soil microbial community activity. The growth and physiology (chlorophyll content, chlorophyll fluorescence, ascorbate, phenolic compounds and flavonoid contents) of the poplar were determined. Moreover, chemical analyses were performed to assess the concentrations of PCB indicators in soil and plant roots. Finally, the microbial community was evaluated in terms of total abundance and activity under the different experimental conditions. Results showed that the poplar clone was able to grow efficiently in the contaminated soil and to promote microbial transformations of PCBs. Plants grown in the hypoxic condition promoted the formation of a higher number of higher-chlorinated PCBs and accumulated lower PCBs in their roots. However, plants in this condition showed a higher stress level than the other microcosms, producing higher amounts of phenolic, flavonoid and ascorbate contents, as a defence mechanism.

Keywords: natural-based remediation strategies; Monviso clone; plant physiology; antioxidant defence; soil microbial communities

---

1. Introduction

Polychlorinated byphenyls (PCBs) are a class of Persistent Organic Pollutants (POPs) differing in the number of chlorine atoms attached to their biphenyl rings. Their characteristics (high molecular stability, low solubility in water and high tendency to adsorb to the particulate phase) make PCBs particularly hard to eliminate from different matrices like soils and sediments. Owing to their widespread use in industry in the past and their persistence, and although they have been banned in several countries since 1979, there is still much environmental contamination. Their elimination from contaminated areas is therefore a challenge [1].

Natural restoration strategies are preferred since they use existing flows of energy and matter, take advantage of local solutions and follow seasonal and climatic changes in ecosystems [2,3]. Among natural-based remediation strategies, the use of plants to promote PCB degradation in the rhizosphere (plant assisted bioremediation, [4]) can be an effective, cost-competitive and environmentally friendly alternative to the most traditional remediation techniques [5].

Despite the high chemical stability and low bioavailability of PCBs, they can potentially undergo biological degradation. The latter involves bacteria, fungi and plants, and can occur differently in aerobic and anaerobic conditions. Anaerobic reductive dechlorination happens when PCBs serve as electron acceptors, thus being turned into less chlorinated congeners, and aerobic transformation involves the lower-halogenated congeners (<5 Cl) and leads to the breakdown of the biphenyl structure [6]. Numerous studies have in this context shown an increase in degradation of PCBs, involving mostly the low-halogenated congeners in vegetated soil as compared with non-vegetated soil [1,7–9]; this is the so-called plant-assisted bioremediation [4]. In the rhizosphere, the plant-microorganism association can increase the degradation of PCBs due to synergic exchanges between the natural soil microbial community and roots [9–12]. In fact, plant roots provide a large surface on which microbial cells can increase in number and be helped to spread through the soil. Some secondary plant compounds exudated by roots can have several functions [13], including acting as growth substrates and/or chemical signals, helping the bacterial enzymes involved in the degradation of PCBs [7,14,15] and promoting the growth of PCB-degrading bacteria [16,17]. Roots can also favour degradation of these contaminants by increasing the permeability of soil and oxygen transfer [18].

Poplar is a model plant for phytoremediation. In fact, it has a fast growth rate and a root system that is able to grow in a wide area, and it is capable of growing in nutrient-poor soil and resisting high concentrations of metal in soil [19,20]. It is well known that plant root exudates facilitate soil microbe activity by providing carbon and nitrogen sources and promoting the growth of PCB-degrading bacteria [16,17]. Various studies involving PCBs and poplar have been undertaken in microcosm studies [11,21–23] and very few have been performed as field studies of historically contaminated soils with PCBs. In the studies, little attention has been paid to poplar growth, physiology and biochemistry during the phytoremediation process. The selection of plant species for remediation purposes has to take into consideration not only the success of previous studies about the same clone, but also the site-specific conditions, which can influence the effectiveness of the strategy. Plant species can adapt to a specific environment and/or respond to some soil threats by using different strategies, including ecophysiological, structural and biochemical responses Moreover, some soil threats such as contamination, nutrient deficiencies, flooding and warming may alter plant morphology, physiology and biochemistry [24–26]. For instance, it is known that hypoxic conditions may lead to a decrease in photosynthetic performance [27–30] and that the extent of this decrease depends on a species' tolerance to soil hypoxia. Trees with a high level of tolerance can maintain photosynthetic rates at relatively high levels. On the other hand, the CO_2 assimilation rates of less tolerant species are strongly reduced [30–32].

On the other hand, though each of the various stress conditions raises different physiological and biochemical plant responses such as stomatal closure (drought) [33], photo-inhibition (high light) [34] or induction of ethanolic fermentation (hypoxia) [35], all of them can lead to an accumulation of Reactive Oxygen Species (ROS) [36]. ROS act normally as signalling molecules [37,38] involved in growth regulation, development and responses to environmental stimuli. Indeed, ROS signalling can lead to plant adaptation to stress through the activation of acclimation pathways [39–42]. However, ROS can also damage cellular components when they overwhelm antioxidant defence mechanisms [43]. The redox equilibrium and its capacity to scavenge ROS thus have a key role in the normal development of a plant and for perception, signalling and acclimation to stress [42]. To maintain balanced ROS levels under stress, a common response of plants is the activation of the enzymatic and non-enzymatic antioxidant system. Non-enzymatic antioxidants comprise ascorbic acid, reduced glutathione, alpha-tocopherol, carotenoids, phenolic compounds and flavonoids, a particular group of phenolic compounds widely distributed in plants [44,45].

This study aimed at evaluating the capacity of the Monviso poplar hybrid clone (*Populus generosa* A. Henry × *P. nigra* L.) [46] to grow in a low-quality soil sampled from a site where different kinds of waste, including dielectric fluids (containing PCBs), were present. Three different soil conditions (microbiologically active, pre-sterilized and hypoxic) were set up in a greenhouse experiment in order to assess the poplar capacity to sustain degradative microbial activity under these soil conditions. The results of this experiment were useful for the subsequent application of a phytoremediation strategy using the Monviso clone in an area chronically contaminated by PCBs.

2. Material and Methods

2.1. Soil Collection from the Historically Contaminated Area and Characterization

The soil was sampled from an area close to the city of Taranto (southern Italy). The sampling site was used for several decades (about 40 years) as an improper dump for dielectric fluids (oil containing polychlorinated biphenyls) and different kind of waste. The latter have accumulated above the original limestone soil with the result that the soil consisted of inhomogeneous materials and was unsuitable for plant growth. A previous analysis by the local environmental agency found a heterogeneous contamination by PCBs and their concentrations exceeded the national legal limits (60 ng/g) for garden, parks and residential areas in numerous soil samples [47].

Equal aliquots of surface soil (0–20 cm) were collected from three different contaminated places in order to obtain a composite sample. After the removal of stones and other residues, the soil samples were air-dried (room temperature) and sieved (2 mm). The soil was classified under the USDA soil classification system as a sandy loam (sand 58%, silt 27%, clay 15%). It had a mildly alkaline pH (about 8), with a total organic carbon content of 14.94 g/Kg, and total nitrogen content of 0.2 g/Kg.

2.2. Microcosm Experimental Design

Aliquots of the composite soil were used to fill 16 microcosms (pots, 3 L capacity). The experimental set comprised three different conditions:

- Microbiologically active soil (MA): Historically polluted soil where a poplar cutting was planted.
- Pre-sterilized soil (Pre-sterilized): Historically polluted soil previously sterilized by autoclaving it (at 121 °C, 20 min), where a poplar cutting was subsequently planted;
- Microbiologically active soil under hypoxic conditions (Hypoxic): Historically polluted soil where a poplar cutting was planted; then each pot was submerged in water for all the experimental period. This treatment was intended to limit the oxygen concentration in the soil in order to reproduce a hypoxic environment for promoting the transformation of higher-chlorinated PCBs.

Un-planted soil microcosms were used as controls (Control).

Each condition was performed in four replicates. Poplars were planted as 20 cm long unrooted cuttings of the Monviso clone (*Populus generosa* A. Henry × *P. nigra* L.), supplied by Alasia Franco Vivai (Savigliano, CN, Italy). All microcosms were maintained in a greenhouse under natural light and at an environmental temperature for more than 12 months (364 days).

The MA, Pre-sterilized and Control microcosms were regularly watered, and the soil water content was maintained at approximately 65% of its field capacity throughout the experiment.

2.3. Sampling of Soil and Plant for Various Analysis

The microcosms were sampled at 6 and 12 months after the experimental setup. At each sampling time, two replicates were sacrificed for each condition (MA, Pre-sterilized and Hypoxic).

From each microcosm, the soil sampled was homogenized and divided into two portions. One was immediately used for the microbial analysis (total microbial abundance and dehydrogenase activity), and the other one was stored at −20 °C for the subsequent PCB analysis.

Roots were also sampled and lyophilized for PCB analysis at 6 and 12 months. A preliminary step was performed by washing the roots to eliminate soil particles attached to them. Firstly, each root was manually shaken down for 10 min, washed (0.9% NaCl) and finally rinsed quickly under running water in a sieve [48].

Plant biomass (roots, leaves, branches) was also assessed in the various conditions at 6 and 12 months. Finally, total phenolic compounds, flavonoids and ascorbate were analysed in the leaves (when present), stems and roots of each poplar tree at the end of the experiment.

Each chemical, biochemical or microbiological analysis was performed in at least three replicates from the same microcosm. Each datum presented is the average of six values.

2.3.1. PCB Markers in Soil and Roots

The PCB congeners analysed were those of the PCB markers (28, 52, 101, 153, 138 and 180). The latter are commonly analysed in environmental studies because they are the most frequently found and are PCB pattern indicators in various sample types [49]. Each congener was named in accordance with the IUPAC numbering system. The PCB analysis was performed using the EPA method 1668 [50].

The PCB extraction from soil and roots was achieved using an ASE 200 (accelerated solvent extraction), as reported in Technical Note 210 (Thermo Fisher Scientific, MA USA). This technique makes it possible to remove interferences due to the sample matrix using adsorbents, joining extraction and purification in a single stage. The extracts were analysed with the Finnigan TRACE GC ultra-chromatograph (Thermo Fisher Scientific, MA USA), coupled with a mass spectrometer in accordance with Muir and Sverko [51]. The detection limit for each PCB congener analysed was 0.5 µg/Kg dry soil. The quantification of the individual PCB congeners was performed with a $^{13}C_{12}$ internal standard multi-point calibration using six calibration standard solutions (P48-M, Wellington Laboratories) in three replicates, from 0.1 pg/µL to 5 ng/µL. The compounds were quantified using the ratio of the analyte and internal standard response (peak area). The instrument limit of quantification (LOQ, the concentration at which quantitative results can be reported with a high degree of confidence), was determined with an approach based on parameters from the analytical curve [52]. LOQ values for each indicator PCB congener are reported in Table 1.

Table 1. Limits of quantification for PCB marker congeners.

PCB Congener (IUPAC Number)	LOQ (pg)
28 (7012-37-5)	43.73
52 (35693-99-3)	25.46
101 (37680-73-2)	11.00
138 (35065-28-2)	29.65
153 (35065-27-1)	26.59
180 (35065-29-3)	27.97

2.3.2. Microbial Abundance and Dehydrogenase Activity

Soil samples from the various microcosms were analysed in order to assess the abundance and activity of the microbial community, evaluating their changing over time in the presence of poplar. Microbial abundances (No. cells/g soil) were determined by the epifluorescence direct count method, with DAPI (4′,6-diamidino-2-phenylindole) as the fluorescent dye [53]. For each analysis, 1 g of soil was put in a test tube with a filter-sterilized fixing solution, as previously described [54]. To detach microbial cells from soil particles, the test tube was shaken for 15 min (400 rpm), and the suspension was then left for 24 h. An aliquot of supernatant (100 µL) was put in contact (20–30 min) with a DAPI solution. The supernatant was then filtered through a 0.2 µm Nuclepore Polycarbonate Black Membrane Filter

(Whatman, Maidston, UK) which was subsequently mounted on a glass slides, and the microbial cells were counted with a Leica epifluorescence microscope (DM 4000B, Leica Microsystems GmbH, Wetzlar, Germany).

Soil dehydrogenase activity was used as a microbiological indicator for the overall activity of the microbial community and how it was influenced by the presence of poplar. Soil dehydrogenase activity was determined using the reduction of 2,3,5-triphenyltetrazolium chloride (TTC) solution to triphenylformazan (TPF), measured in two replicates. 6 g of soil were collected and analysed as reported in Grenni et al. [55]. Soil dehydrogenase activity was expressed as µg TPF/g dry soil.

2.4. Analysis for Growth Monitoring, Plant Physiology and Plant Antioxidants

2.4.1. Growth Monitoring Measurements

During the vegetative growth, plant biomass was recorded and reported on a dry weight basis. Plants were carefully removed from the pots and washed with distilled water to remove any particles attached. Plant organs (roots, shoots, leaves) were then separated and dried at 60 °C in an air-forced oven and 72 h later their dry weights were determined [56]. The root biomass and the branch biomass of each plant were measured at 6 months (autumn) and at 12 months (spring) after planting. Leaf biomass was measured only at 12 months, due to a lack of leaves on the trees in late autumn (six months).

2.4.2. Plant Physiology Measurements

The physiological status of the poplar plants was assessed through their leaf chlorophyll content and fluorescence measurements. A Minolta chlorophyll meter (SPAD) was used to estimate the leaf chlorophyll content as previously described [57]. The following equation was used to convert the SPAD measurement into chlorophyll content.

$$Chlorophyll\ content\ \left[\frac{\mu g}{cm^{-2}}\right] = \frac{99 \times SPAD}{144 - SPAD}$$

On the same leaves chosen for SPAD readings, the chlorophyll fluorescence transient (OJIP transients) was measured using a plant efficiency analyser (PEA, Hansatech Instruments Ltd., King's Lynn, UK) as reported in Pietrini et al. [58].

The chlorophyll content and fluorescence were measured on five leaves per poplar all over the tree at 4 and 12 months, when there were leaves.

2.4.3. Plant Antioxidants

Phenolics, flavonoids and the ascorbate content were used as indicators of plant antioxidant status for evaluating any stress caused by treatments. Total phenolic compounds, flavonoids and ascorbate were analysed (three replicates) in leaves, stems and roots for each poplar tree at six months and the end of the experiment.

The extraction of total phenolic compounds and flavonoids was performed from 200 mg of plant material with 80% methanol (1.5 mL) for 3 min in an ultrasonic bath. The extraction was repeated twice. The amount of extracted total phenolic compounds was determined with the Folin–Ciocalteu reagent [59]. Each analysis was performed in duplicate for each extract. The gallic acid was used as the standard and the total phenolic compounds were expressed as mg of gallic acid equivalents (GAE) per g of fresh weight.

Total flavonoid content was measured using the aluminium chloride method described by Chang et al. [60]. The absorbance was read at a 415 nm wavelength. Analysis was done in triplicate for each extract. Standard solutions of quercetin were used to obtain a standard curve. The total flavonoid content was reported as mg of total quercetin equivalents per g of fresh weight.

Ascorbate was extracted from the plant tissue (about 100 mg fresh weight) in 1.5 mL 3% perchloric acid, and the mixture was centrifuged (5000 rpm, for 20 min) at 4 °C. The reduced ascorbate (ASC) and oxidized ascorbate (DHA) measurements are based on the reduction of Fe^{3+} to Fe^{2+} by ascorbic acid in acidic solution. Fe^{2+} forms complexes with bipyridyl that absorb at 525 nm. Sample pre-incubation with dithiothreitol (DTT) reduced DHA to ASC. The excess DTT was removed with N-ethylmaleimide, and the total ascorbate was determined. The amount of DHA was calculated by subtracting ASC from total ascorbate. The contents were calculated using a standard curve [61]. The ascorbate ratio was then calculated as the proportion between reduced ascorbate and total ascorbate and expressed as [reduced-/total-ascorbate].

2.5. Statistical Analysis

Analysis of variance (one-way analysis of variance) was used to assess the significant differences among treatments in PCB concentration, dehydrogenase activity, total microbial abundance in soil samples and antioxidant content in root, leaf and branch samples. The PC Program used was SIGMASTAT 3.1 software (Systat Software Inc., Point Richmond, CA, USA). The significance level of 0.05 was utilized to indicate whether the treatments were significantly different from each other and from the control. A multiple comparison procedure (Dunn's method) was used in order to isolate the group or groups that differed from the others. Finally, the Post-hoc test was performed on plant antioxidant results.

3. Results

3.1. Soil PCB Concentration

At the start of the experiment (experimental set-up), the sum of the PCB markers analysed (PCB 28, 52, 101, 153, 138, 180) was 47.6 ± 2.5 µg/Kg.

Figure 1 shows photos of the Monviso clone at three and six months. The concentrations of PCB markers in the various soil microcosms at 6 months and compared to the control soil are reported in Figure 2A. A general decrease in PCB concentration, with the exception of PCB 180, was observed in all the plant-treated microcosms. This reduction was significant ($p < 0.05$) for PCBs 101 (2,2′,4,5,5′-Pentachlorobiphenyl), 138 (2,2′,3,4,4′,5′-Hexachlorobiphenyl) and 153 (2,2′,4,4′,5,5′-Hexachlorobiphenyl) in the microbiologically active conditions (MA), with decrease percentages ranging from 20 to 64%.

Figure 1. Photos of the Monviso clone plants. (**A**): Poplars at three months. (**B,C**): Pictures of the roots at six months.

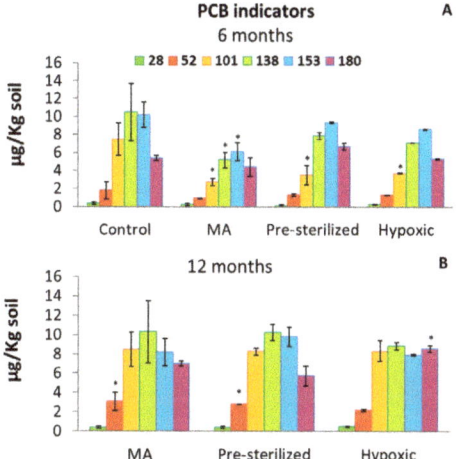

Figure 2. PCB markers (28, 52, 101, 138, 153 and 180) in the various soil microcosms at 6 months (**A**) and at 12 months (**B**). MA: Microbiologically active soil; pre-sterilized: Soil previously sterilized by autoclaving it; hypoxic: Microbiologically active soil microcosms immerged in water; control: Microbiologically active soil, un-planted. Significant differences ($p < 0.01$) among the different sampling times are indicated with an asterisk.

The overall concentrations of PCBs analysed at 12 months increased in all conditions as compared to 6 months, with differences in some congeners (Figure 2B). The low-chlorinated PCB 52 was higher in the MA and Pre-sterilized and the high-chlorinated congener 180 in the hypoxic microcosms ($p < 0.05$).

3.2. PCB Concentrations in Roots

The PCB markers detected in the roots of the planted poplars (MA, Pre-sterilized and Hypoxic) at 6 and 12 months are shown in Figure 3. At month 6, PCB concentrations in roots ranged from 1.31 µg/Kg (PCB 28, Hypoxic) to 89.5 µg/Kg (PCB 153, MA), (Figure 3A). The average values of PCBs in the MA microcosms were significantly higher ($p < 0.05$) than in the other conditions.

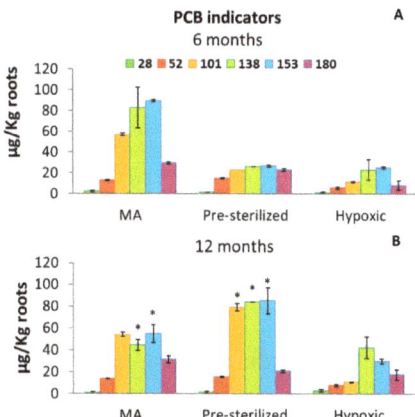

Figure 3. PCB markers in plant roots at 6 months (**A**) and at 12 months (**B**). MA: Microbiologically active soil; hypoxic: Microbiologically active soil under hypoxic conditions; pre-sterilized: Soil previously sterilized by autoclaving it. Significant differences ($p < 0.01$) between 6 and 12 months for the same PCB congener are indicated with an asterisk (*).

At 12 months (Figure 3B), PCBs were found in poplar roots even in the Pre-sterilized condition. In the Hypoxic microcosms, PCBs remained relatively low, with the highest values for the higher-chlorinated congeners (PCB 138, 153 and 180).

3.3. Microbiological Analysis

A significant increase ($p < 0.01$) in microbial abundance (No. cells/g soil) was observed from day 0 to 6 months in the plant presence (MA, Pre-sterilized and Hypoxic soils); at the end of the experiment, although the average values in all planted microcosms were higher than in the control ones, the number of cells decreased (Figure 4).

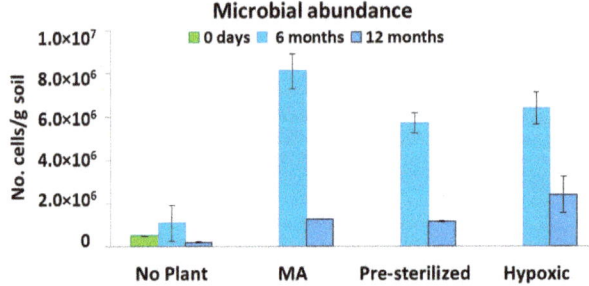

Figure 4. Microbial abundance (No. cells/g soil) in the various conditions. Control: Microbiologically active soil, un-planted; MA: Microbiologically active soil; pre-sterilized: Soil previously sterilized by autoclaving it; hypoxic: Microbiologically active soil under hypoxic conditions.

The average values of dehydrogenase activity were higher ($p < 0.01$) in the planted soils than in the Control at 6 months; the highest value of DHA was observed in the Pre-sterilized soil (Figure 5). At the end of the experiment, the lowest value was found in the hypoxic condition.

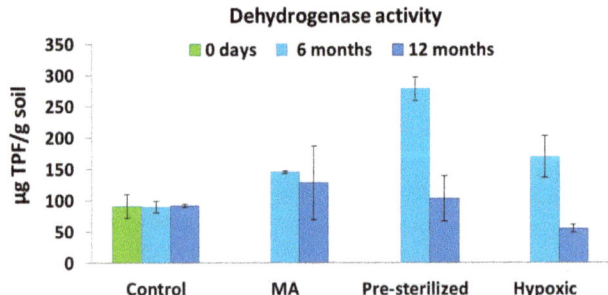

Figure 5. Dehydrogenase activity (μg TPF/g soil) in the various conditions in the three sample times. Control: Microbiologically active soil, un-planted; MA: Microbiologically active soil; pre-sterilized: Soil previously sterilized by autoclaving it; hypoxic: Microbiologically active soil under hypoxic conditions.

3.4. Plant Growth and Physiology

The plant biomass (root, leaves and branches) in the various conditions (MA, Hypoxic, Pre-sterilized) at 6 and 12 months is reported in Table 2.

At 6 months, the lowest root and branch biomass was observed in the MA condition; the differences in root and branch biomass were significant ($p < 0.05$) between the MA condition and the Pre-sterilized one. The branch biomass was significantly lower ($p < 0.05$) in plants from MA than in those from the hypoxic condition. However, these differences were never observed in the subsequent sampling (12 months).

Table 2. Plant biomass (roots, leaves, branches, ±standard deviations) in the various conditions at 6 and 12 months. MA: Microbiologically active soil; pre-sterilized: Soil previously sterilized by autoclaving it; hypoxic: Microbiologically active soil under hypoxic conditions. Letters (a, b) indicate significant differences ($p < 0.05$).

Conditions	Roots (g)	Leaves (g)	Branches (g)
	6 months		
MA	1.31 a ± 1.11	—	5.42 a ± 1.95
Pre-sterilized	4.55 b ± 1.30	—	11.98 b ± 1.84
Hypoxic	3.35 ab ± 2.46	—	11.00 b ± 0.27
	12 months		
MA	5.14 a ± 0.03	5.77 a ± 2.26	8.35 a ± 0.03
Pre-sterilized	7.64 b ± 2.06	8.28 a ± 1.25	8.02 a ± 0.37
Hypoxic	5.40 ab ± 1.34	7.53 a ± 1.00	10.34 a ± 1.61

Maximum quantum yield of PSII (Fv/Fm) and leaf chlorophyll content were measured at 4 and 12 months (Table 3). Fluorescence analysis showed Fv/Fm indices ranging between 0.76 and 0.82. The highest values were observed in the Hypoxic and Pre-sterilized conditions although with no significant differences among treatments and sampling times (Table 3).

Table 3. Maximum quantum yield of PSII (Fv/Fm) and leaf chlorophyll content (±s.d.). MA: Microbiologically active soil; hypoxic: Microbiologically active soil under hypoxic condition; pre-sterilized: Soil previously sterilized by autoclaving it. Letters (a, b) indicate significant differences ($p < 0.05$).

Treatment	Fv/Fm	Chlorophyll ($\mu g/cm^2$)
	4 months	
MA	0.76 a ± 0.11	37.40 a ± 5.76
Hypoxic	0.80 a ± 0.04	28.93 b ± 2.98
Pre-sterilized	0.79 a ± 0.08	27.97 b ± 3.96
	12 months	
MA	0.80 a ± 0.06	31.88 a ± 2.41
Hypoxic	0.82 a ± 0.05	32.90 a ± 4.70
Pre-sterilized	0.82 a ± 0.11	31.59 a ± 5.68

The average values of leaf chlorophyll ranged between 27.97 and 37.40 $\mu g/cm^2$ (Table 3). The highest value ($p \leq 0.001$) was found for the MA condition four months after planting. However, at 12 months after planting, no significant differences were found for chlorophyll contents among the different treatments.

3.5. Plant Antioxidants

In general, the phenolic compounds, flavonoids and total ascorbate were significantly higher in leaves than in roots and shoots ($p < 0.001$) (Figure 6A–D).

Overall values for plant antioxidants were higher in the Hypoxic condition than in the MA and Pre-sterilized ones and these values were significantly higher ($p < 0.05$) at 12 months. It is also noticeable that whereas at 6 months the phenolics and flavonoids values were significantly ($p < 0.05$) lower in plants from the Pre-sterilized treatment than in plants from the MA and Hypoxic treatments, at 12 months these differences were never observed.

In leaves at 12 months the phenolic compound content (Figure 6A) was about 1.3-fold higher in the Hypoxic than in the MA and Pre-sterilized conditions. Leaf flavonoid content (Figure 6B) was also 1.8–2.0-fold higher in the Hypoxic than in the MA and pre-sterilized microcosms, respectively. Moreover, the total ascorbate (Figure 6C) was higher at 12 months in leaves and shoots, with significantly higher values for leaves compared to other plant parts.

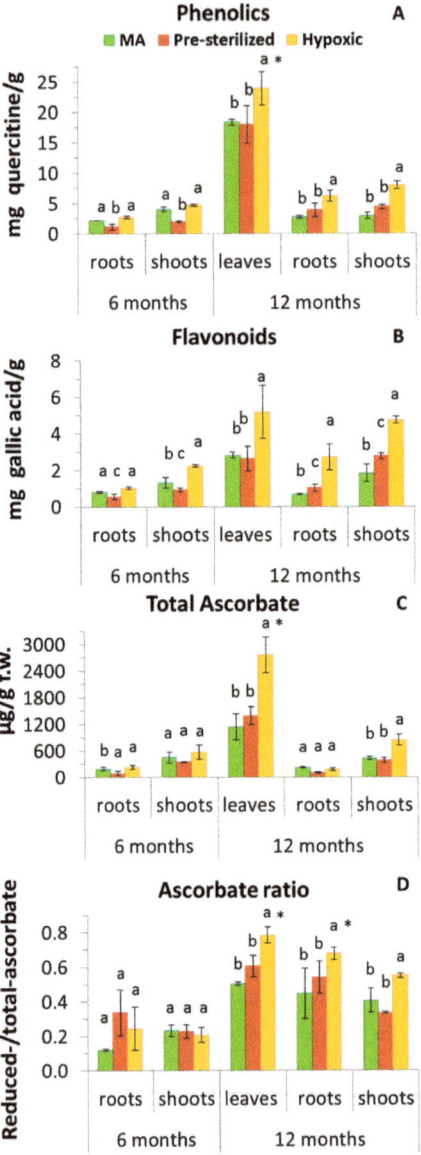

Figure 6. Plant antioxidants found in roots, shoots and leaves at 6 months and at 12 months. (**A**) Phenolics (mg gallic acid/g); (**B**) Flavonoids (mg quercetin/g biomass); (**C**) Total Ascorbate (µg/g fresh weight); (**D**) Ascorbate ratio expressed as [reduced-/total-ascorbate]; MA: Microbiologically active soil; hypoxic: Microbiologically active soil under hypoxic condition; pre-sterilized: Soil previously sterilized by autoclaving it. Letters (a, b) indicate significant differences among treatments ($p < 0.05$). Asterisks (*) indicate significant differences among plant parts ($p < 0.001$).

Finally, the ascorbate ratio (Figure 6D) was generally higher in plants grown under hypoxic soil conditions than in any other treatment in all plant materials (Figure 6D) and these differences were statistically significant for leaves and shoots ($p < 0.05$) at 12 months.

4. Discussion

Poplar trees have been found to be tolerant to several contaminants [62,63] and thanks to their high transpiration, deep root system and fast growth rate, are considered suitable for phytoremediation of both organic and inorganic contaminants [21,64]. For this reason, we tested the capacity of the Monviso clone to grow in a chronically contaminated soil, analysing its physiological response, and to stimulate microbial activity, which can improve PCB transformations.

In our experiment, only after 6 months did the plant promote a general decrease in the soil in concentrations of the PCBs analysed, and this was particularly evident in the microbiologically active condition (MA), (Figure 2A). A synergic effect between plant roots and the natural microbial community of soil was also found. In fact, an increase in the cell number and activity was observed (Figures 4 and 5). Positive effects of the plant and rhizosphere soil microbial community in improving the quality of a PCB contaminated soil were found in a recent work using the Alfalfa plant [48]. In the rhizosphere, plant metabolites can act as chemical signals for inducing the microbial enzymes involved in PCB degradation [65–67]. In return, PCB-degrading bacteria can produce plant growth stimulators [68,69]. At 12 months a general increase in PCB markers was observed in all conditions (Figure 2B). This result could be presumably ascribed to the microbial-root association effect of stimulating the desorption of some non-analysed PCBs, which were attached to the historically contaminated soil, favouring their transformation to the six PCB markers analysed in this work (i.e., 28, 52, 101, 138, 153 and 180), as also found by other authors [6,11,21,70]. Another possibility could be the transformation of higher chlorinated PCBs to the PCB markers analysed, as also found in other studies [1,9]

The quantity of PCB markers detected in the roots (Figure 3) was ascribable to the wide growth of the root system (Figure 1, Table 2), which filled the pots and acts as a strong bioconcentrator. Plant exudates can act as surfactants, forming hydrophilic complexes with PCBs and making possible their transportation into roots [71]. PCBs are not expected to enter the transpiration stream due to their high hydrophobicity and other studies have shown that they were not found in the upper part of the plant [72].

It is interesting to note that when the highest PCB concentrations were found in roots (MA at 6 months, Figure 3A), the root and branch biomasses were significantly the lowest (Table 2).

Overall, the results suggest that, in association with the natural microbial community of soil (MA), this poplar clone was well adapted to the PCBs and the microbial populations responded differently at 6 and 12 months. In fact, they initially (6 months) increased in number with the decrease of the six PCB markers; however, they maintained their general activity, but diminished in number when the PCBs increased again (12 months).

The pre-sterilized soil condition was set-up to evaluate how the natural microbial community of this historically contaminated soil had an influence on the plant physiology of this clone. In this condition the poplar cuttings were planted in a soil where the microbial community was not initially present (Figure 4). However, the microbial populations which survived the sterilization treatment (i.e., those able to produce resistant cysts) and derived from the environment developed abundantly, as expected in a non-colonized substratum (Figures 4 and 6) and found in other studies using pre-sterilised soils [55,73,74]. The overall Pre-sterilized results showed that, although more active than in the other conditions, the microbial community colonizing the rhizosphere in the first 6 months (Figure 6) was initially able to stimulate more plant growth than PCB transformation. However, in the subsequent sampling (12 months) the overall increase in PCB markers and the microbial abundance, activity and plant growth and physiology values were comparable to MA, suggesting that the interactions established over time between bacterial populations and the rhizosphere were similar in both treatments. In fact, the roots were able to concentrate PCB markers in the rhizosphere in an amount comparable to those observed in MA at 6 months.

The hypoxic condition was set up for assessing if some PCB congener degradation could be favoured in an anaerobic situation. In this case, each poplar cutting was planted in a soil immersed in water for all the experimental period. The analysis of the six PCB markers at month 6 showed that their overall decrease was lower than in the MA condition; moreover, at month 12 higher amounts (e.g., of the high-chlorinated 180) were found than in the MA and Pre-sterilized conditions. The hypoxic environmental condition therefore presumably promoted PCB transformations that were more chlorinated (>7 Cl) than the 180 congener and were transformed to some less-chlorinated PCBs, including the 180 one, as also found by other authors [75,76]. In this condition, a marked difference was also found in the PCB markers detected in the roots (less than the other conditions) and in the antioxidants produced by the plant, which were significantly higher than in the other conditions, showing that in this case the plant was more stressed.

The different experimental conditions (natural soil, Pre-sterilized, Hypoxic condition) influenced the development of different microbial populations in the rhizosphere and these in turn affected the plant growth and stress response. For example, in the MA plants as compared to the others a less plant growth occurred. Chekol et al. [8] affirmed that different effects of PCBs on plant biomass can be related to a different sensitivity to PCBs of the root-microorganisms system established in the rhizosphere. For example, there are microorganisms that promote plant growth through the action of 1-aminocyclopropane-1-carboxylate (ACC) deaminase. This enzyme can cleave the plant ethylene precursor ACC, and thereby lower the ethylene levels of stressed plants [77]. However, at 12 months, the upper part of plant biomass under the three different treatments was similar, despite the higher PCB accumulation found in roots in the MA and Pre-sterilized soils. This result confirms the adaptation of both rhizosphere microorganisms and this poplar clone to a PCB presence. Other studies considering other plant species have shown no significant differences in plant biomass between plants grown in a control soil and in PCB contaminated soil [8,78].

The photosynthetic performance of the poplar plants was measured in terms of leaf chlorophyll content and chlorophyll fluorescence measurements. Photosynthesis is a pivotal process for plant growth and biomass production, and it has thus often been used as a stress bioindicator [79]. Moreover, it has been shown that, the photosynthetic apparatus is sensitive to different soil conditions, such as hypoxia, soil drought and high soil temperature. For instance, hypoxia can alter the maximum quantum yield of PSII [80], as well as chlorophyll synthesis [81]. Chlorophyll (Chl) content and Chl fluorescence values for the MA treatment turned out to be similar to other values reported for Monviso poplar plants [82], indicating that PCBs did not alter these parameters (and did not cause any severe stress to the poplar plants). However, both the Hypoxic and Pre-sterilized conditions induced a reduction of Chl content in poplar plants 4 months after planting, though leaf chlorosis did not lead to a decrease in maximum quantum yield of PSII photochemistry (Fv/Fm) in hypoxic plants.

The overall results show that, although the soil pre-sterilization and the hypoxic conditions did not promote a general decrease in the soil of the PCBs analysed and instead promoted a general increase in plant oxidants (in particular for the hypoxic one), these conditions did not negatively influence the plant biomass or the plant physiology measurements, highlighting a good adaptability by this clone. The clone tested was therefore able to grow in a low-quality soil, contaminated by PCBs, and to promote the transformation of these contaminants in the presence of the natural soil microbial community. The sterilization and the hypoxic condition acted as selective forces on the soil microbial community and for this reason they were less efficient in PCB transformation.

This paper reports for the first-time data regarding the poplar physiology in a hypoxic condition for its possible application for bioremediation purposes. Watering the soil can be an occasional strategy to improve the anaerobic degradation of PCBs. We cannot exclude a subsequent long-term decrease in PCBs in this historically contaminated soil, as reported in recent review papers [1,9]. However, knowledge of plant-microbe interactions in PCB degradation is far from complete and further studies are necessary to better investigate the chemical dialogue between plant and microbes in the rhizosphere. In any case, thanks to this preliminary greenhouse experiment, a phyto-assisted bioremediation strategy with the Monviso clone has been subsequently performed in the field and is currently in progress.

5. Conclusions

The Monviso clone was able to grow in the low-quality soil studied and to significantly improve microbial activity and promoted an overall PCB transformation (an overall decrease in PCB marker concentration at 6 months and increase at 12 months).

Further research is needed to better understand the metabolic degradation pathways of PCBs in the rhizosphere. In particular, the analysis of a higher number of important congeners, the molecular feedback mechanisms that cause the transformation of PCBs in the rhizosphere and the specific bacterial populations involved, is currently in progress.

This preliminary research was very useful in encouraging us to plant the Monviso poplar cuttings in a field where the soil quality was very low and made it possible to have some information on the physiology of this clone under different conditions. Thanks to this experiment, a phyto-assisted bioremediation strategy using this clone Monviso has been subsequently performed at this experimental site and is currently in progress.

Author Contributions: Conceptualisation, A.M., A.B.C. and P.G.; methodology, I.N., P.G., M.D.L., L.P., E.G., P.B., A.M., J.R. and A.B.C.; formal analysis, I.N., P.G., M.D.L., L.P., E.G., P.B., J.R. and A.B.C.; data curation, I.N., P.G., L.P., E.G., P.B., and A.B.C.; writing—original draft preparation, I.N., P.G., and A.B.C.; writing—review and editing, I.N., P.G., L.P., E.G., J.R. and A.B.C.; project administration, A.M.

Funding: This research was partially funded by CISA S.p.A. (Massafra, Italy) grant number 0005159/2012.

Acknowledgments: The authors thank Valeria Ancona (IRSA-CNR) for soil sampling and Francesca Falconi (IRSA-CNR) for her help in microbiological analyses.

Conflicts of Interest: The authors declare no conflict of interest.

References

1. Terzaghi, E.; Zanardini, E.; Morosini, C.; Raspa, G.; Borin, S.; Mapelli, F.; Vergani, L.; Di Guardo, A. Rhizoremediation half-lives of PCBs: Role of congener composition, organic carbon forms, bioavailability, microbial activity, plant species and soil conditions, on the prediction of fate and persistence in soil. *Sci. Total Environ.* **2018**, *612*, 544–560. [CrossRef] [PubMed]
2. Meli, P.; Rey Benayas, J.M.; Balvanera, P.; Martínez Ramos, M. Restoration enhances wetland biodiversity and ecosystem service supply, but results are context-dependent: a meta-analysis. *PLoS ONE* **2014**, *9*, e93507. [CrossRef] [PubMed]
3. Keesstra, S.; Nunes, J.; Novara, A.; Finger, D.; Avelar, D.; Kalantari, Z.; Cerdà, A. The superior effect of nature based solutions in land management for enhancing ecosystem services. *Sci. Total Environ.* **2018**, *610*, 997–1009. [CrossRef] [PubMed]
4. Wenzel, W.W. Rhizosphere processes and management in plant-assisted bioremediation (phytoremediation) of soils. *Plant Soil* **2009**, *321*, 385–408. [CrossRef]
5. Mercado-Blanco, J.; Abrantes, I.; Barra Caracciolo, A.; Bevivino, A.; Ciancio, A.; Grenni, P.; Hrynkiewicz, K.; Kredics, L.; Proença, D.N. Belowground Microbiota and the Health of Tree Crops. *Front. Microbiol.* **2018**, *9*, 1006. [CrossRef]
6. Field, J.A.; Sierra-Alvarez, R. Microbial transformation and degradation of polychlorinated biphenyls. *Environ. Pollut.* **2008**, *155*, 1–12. [CrossRef]
7. Singer, A.C.; Crowley, D.E.; Thompson, I.P. Secondary plant metabolites in phytoremediation and biotransformation. *Trends Biotechnol.* **2003**, *21*, 123–130. [CrossRef]
8. Chekol, T.; Vough, L.R.; Chaney, R.L. Phytoremediation of polychlorinated biphenyl-contaminated soils: The rhizosphere effect. *Environ. Int.* **2004**, *30*, 799–804. [CrossRef]
9. Vergani, L.; Mapelli, F.; Zanardini, E.; Terzaghi, E.; Di Guardo, A.; Morosini, C.; Raspa, G.; Borin, S. Phyto-rhizoremediation of polychlorinated biphenyl contaminated soils: An outlook on plant-microbe beneficial interactions. *Sci. Total Environ.* **2017**, *575*, 1395–1406. [CrossRef]
10. Sharma, J.K.; Gautam, R.K.; Nanekar, S.V.; Weber, R.; Singh, B.K.; Singh, S.K.; Juwarkar, A.A. Advances and perspective in bioremediation of polychlorinated biphenyl-contaminated soils. *Environ. Sci. Pollut. Res.* **2018**, *25*, 16355–16375. [CrossRef]

11. Meggo, R.E.; Schnoor, J.L. Rhizospere redox cycling and implications for rhizosphere biotransformation of selected polychlorinated biphenyl (PCB) congeners. *Ecol. Eng.* **2013**, *57*, 285–292. [CrossRef] [PubMed]
12. Germaine, K.J.; McGuinness, M.; Dowling, D.N. Improving Phytoremediation through Plant-Associated Bacteria. In *Molecular Microbial Ecology of the Rhizosphere*; John Wiley & Sons, Inc.: Hoboken, NJ, USA, 2013; pp. 961–973.
13. Olanrewaju, O.S.; Glick, B.R.; Babalola, O.O. Mechanisms of action of plant growth promoting bacteria. *World J. Microbiol. Biotechnol.* **2017**, *33*, 197. [CrossRef] [PubMed]
14. Uhlik, O.; Musilova, L.; Ridl, J.; Hroudova, M.; Vlcek, C.; Koubek, J.; Holeckova, M.; Mackova, M.; Macek, T. Plant secondary metabolite-induced shifts in bacterial community structure and degradative ability in contaminated soil. *Appl. Microbiol. Biotechnol.* **2013**, *97*, 9245–9256. [CrossRef] [PubMed]
15. Slater, H.; Gouin, T.; Leigh, M.B. Assessing the potential for rhizoremediation of PCB contaminated soils in northern regions using native tree species. *Chemosphere* **2011**, *84*, 199–206. [CrossRef]
16. Segura, A.; Rodríguez-Conde, S.; Ramos, C.; Ramos, J.L. Bacterial responses and interactions with plants during rhizoremediation. *Microb. Biotechnol.* **2009**, *2*, 452–464. [CrossRef]
17. Macek, T.; Macková, M.; Káš, J. Exploitation of plants for the removal of organics in environmental remediation. *Biotechnol. Adv.* **2000**, *18*, 23–34. [CrossRef]
18. Gomes, H.I.; Dias-Ferreira, C.; Ribeiro, A.B. Overview of in situ and ex situ remediation technologies for PCB-contaminated soils and sediments and obstacles for full-scale application. *Sci. Total Environ.* **2013**, *445*, 237–260. [CrossRef]
19. Di Baccio, D.; Tognetti, R.; Sebastiani, L.; Vitagliano, C. Responses of Populus deltoides x Populus nigra (Populus x euramericana) clone I-214 to high zinc concentrations. *New Phytol.* **2003**, *159*, 443–452. [CrossRef]
20. Sebastiani, L.; Scebba, F.; Tognetti, R. Heavy metal accumulation and growth responses in poplar clones Eridano (Populus deltoides × maximowiczii) and I-214 (P. × euramericana) exposed to industrial waste. *Environ. Exp. Bot.* **2004**, *52*, 79–88. [CrossRef]
21. Meggo, R.E.; Schnoor, J.L.; Hu, D. Dechlorination of PCBs in the rhizosphere of switchgrass and poplar. *Environ. Pollut.* **2013**, *178*, 312–321. [CrossRef]
22. Liu, J.; Schnoor, J.L. Uptake and translocation of lesser-chlorinated polychlorinated biphenyls (PCBs) in whole hybrid poplar plants after hydroponic exposure. *Chemosphere* **2008**, *73*, 1608–1616. [CrossRef] [PubMed]
23. Zhai, G.; Hu, D.; Lehmler, H.-J.; Schnoor, J.L. Enantioselective biotransformation of chiral PCBs in whole poplar plants. *Environ. Sci. Technol.* **2011**, *45*, 2308–2316. [CrossRef] [PubMed]
24. Hu, L.; Robert, C.A.M.; Cadot, S.; Zhang, X.; Ye, M.; Li, B.; Manzo, D.; Chervet, N.; Steinger, T.; van der Heijden, M.G.A.; et al. Root exudate metabolites drive plant-soil feedbacks on growth and defense by shaping the rhizosphere microbiota. *Nat. Commun.* **2018**, *9*, 2738. [CrossRef] [PubMed]
25. Chibuike, G.U.; Obiora, S.C. Heavy Metal Polluted Soils: Effect on plants and bioremediation methods. *Appl. Environ. Soil Sci.* **2014**, *2014*, 1–12. [CrossRef]
26. Gargallo-Garriga, A.; Wright, S.J.; Sardans, J.; Pérez-Trujillo, M.; Oravec, M.; Večeřová, K.; Urban, O.; Fernández-Martínez, M.; Parella, T.; Peñuelas, J. Long-term fertilization determines different metabolomic profiles and responses in saplings of three rainforest tree species with different adult canopy position. *PLoS ONE* **2017**, *12*, e0177030. [CrossRef] [PubMed]
27. Ojeda, M.; Schaffer, B.; Davies, F.S. Flooding, root temperature, physiology and growth of two Annona species. *Tree Physiol.* **2004**, *24*, 1019–1025. [CrossRef] [PubMed]
28. Fernandez, M.D. Changes in photosynthesis and fluorescence in response to flooding in emerged and submerged leaves of Pouteria orinocoensis. *Photosynthetica* **2006**, *44*, 32–38. [CrossRef]
29. Kreuzwieser, J.; Rennenberg, H. Molecular and physiological responses of trees to waterlogging stress. *Plant. Cell Environ.* **2014**, *37*, 2245–2259. [CrossRef]
30. Ferner, E.; Rennenberg, H.; Kreuzwieser, J. Effect of flooding on C metabolism of flood-tolerant (*Quercus robur*) and non-tolerant (*Fagus sylvatica*) tree species. *Tree Physiol.* **2012**, *32*, 135–145. [CrossRef]
31. Jaeger, C.; Gessler, A.; Biller, S.; Rennenberg, H.; Kreuzwieser, J. Differences in C metabolism of ash species and provenances as a consequence of root oxygen deprivation by waterlogging. *J. Exp. Bot.* **2009**, *60*, 4335–4345. [CrossRef]
32. Parent, C.; Crèvecoeur, M.; Capelli, N.; Dat, J.F. Contrasting growth and adaptive responses of two oak species to flooding stress: Role of non-symbiotic haemoglobin. *Plant. Cell Environ.* **2011**, *34*, 1113–1126. [CrossRef] [PubMed]

33. Agurla, S.; Gahir, S.; Munemasa, S.; Murata, Y.; Raghavendra, A.S. Mechanism of stomatal closure in plants exposed to drought and cold stress. In *Survival Strategies in Extreme Cold and Desiccation*; Springer: Singapore, 2018; pp. 215–232.
34. Wang, F.; Wu, N.; Zhang, L.; Ahammed, G.J.; Chen, X.; Xiang, X.; Zhou, J.; Xia, X.; Shi, K.; Yu, J.; et al. Light signaling-dependent regulation of photoinhibition and photoprotection in Tomato. *Plant Physiol.* **2018**, *176*, 1311–1326. [CrossRef] [PubMed]
35. Banti, V.; Giuntoli, B.; Gonzali, S.; Loreti, E.; Magneschi, L.; Novi, G.; Paparelli, E.; Parlanti, S.; Pucciariello, C.; Santaniello, A.; et al. Low oxygen response mechanisms in green organisms. *Int. J. Mol. Sci.* **2013**, *14*, 4734–4761. [CrossRef] [PubMed]
36. Sharma, P.; Jha, A.B.; Dubey, R.S.; Pessarakli, M. Reactive oxygen species, oxidative damage, and antioxidative defense mechanism in plants under stressful conditions. *J. Bot.* **2012**, *2012*, 1–26. [CrossRef]
37. Mittler, R.; Vanderauwera, S.; Suzuki, N.; Miller, G.; Tognetti, V.B.; Vandepoele, K.; Gollery, M.; Shulaev, V.; Van Breusegem, F. ROS signaling: The new wave? *Trends Plant Sci.* **2011**, *16*, 300–309. [CrossRef] [PubMed]
38. Baxter, A.; Mittler, R.; Suzuki, N. ROS as key players in plant stress signalling. *J. Exp. Bot.* **2014**, *65*, 1229–1240. [CrossRef] [PubMed]
39. Mittler, R. ROS Are Good. *Trends Plant Sci.* **2017**, *22*, 11–19. [CrossRef]
40. Dietz, K.-J. Efficient high light acclimation involves rapid processes at multiple mechanistic levels. *J. Exp. Bot.* **2015**, *66*, 2401–2414. [CrossRef]
41. Mignolet-Spruyt, L.; Xu, E.; Idänheimo, N.; Hoeberichts, F.A.; Mühlenbock, P.; Brosché, M.; Van Breusegem, F.; Kangasjärvi, J. Spreading the news: Subcellular and organellar reactive oxygen species production and signalling. *J. Exp. Bot.* **2016**, *67*, 3831–3844. [CrossRef]
42. Choudhury, F.K.; Rivero, R.M.; Blumwald, E.; Mittler, R. Reactive oxygen species, abiotic stress and stress combination. *Plant J.* **2017**, *90*, 856–867. [CrossRef]
43. Mitra, T.; Singha, B.; Bar, N.; Das, S.K. Removal of Pb(II) ions from aqueous solution using water hyacinth root by fixed-bed column and ANN modeling. *J. Hazard. Mater.* **2014**, *273*, 94–103. [CrossRef] [PubMed]
44. Miller, G.; Suzuki, N.; Ciftci-Yilmaz, S.; Mittler, R. Reactive oxygen species homeostasis and signalling during drought and salinity stresses. *Plant. Cell Environ.* **2010**, *33*, 453–467. [CrossRef] [PubMed]
45. Das, K.; Roychoudhury, A. Reactive oxygen species (ROS) and response of antioxidants as ROS-scavengers during environmental stress in plants. *Front. Environ. Sci.* **2014**, *2*, 53. [CrossRef]
46. Bianconi, D.; De Paolis, M.R.; Agnello, M.C.; Lippi, D.; Pietrini, F.; Zacchini, M.; Polcaro, C.; Donati, E.; Paris, P.; Spina, S.; et al. Field-scale rhyzoremediation of a contaminated soil with hexachlorocyclohexane (HCH) isomers: The potential of poplars for environmental restoration and economical sustainability. In *Handbook of Phytoremediation*; Nova Science Publishers: New York, NY, USA, 2011; pp. 783–794. ISBN 978-161728753-4.
47. Italian Ministry of the Environment Legislative Decree no. 152, 2006, Rules in environmental field. *Ital. Off. J.* **2006**, *88*, 1–425.
48. Di Lenola, M.; Barra Caracciolo, A.; Grenni, P.; Ancona, V.; Rauseo, J.; Laudicina, V.A.; Uricchio, V.F.; Massacci, A. Effects of apirolio addition and Alfalfa and compost treatments on the natural microbial community of a historically PCB-contaminated soil. *Water Air Soil Pollut.* **2018**, *229*, 143. [CrossRef]
49. EFSA. Opinion of the scientific panel on contaminants in the food chain on a request from the commission related to the presence of non dioxin-like polychlorinated biphenyls (PCB) in feed and food. *EFSA J.* **2005**, *284*, 1–137.
50. US-EPA. *Method 1668, Revision A Chlorinated Biphenyl Congeners in Water, Soil, Sediment, Biosolids, and Tissue by HRGC/HRMS*; U.S. Environmental Protection Agency, Office of Water, Office of Science and Technology, Engineering and Analysis Division (4303T), 1200 Pennsylvania Avenue, NW: Washington, DC, USA, 2003; p. 20460.
51. Muir, D.; Sverko, E. Analytical methods for PCBs and organochlorine pesticides in environmental monitoring and surveillance: A critical appraisal. *Anal. Bioanal. Chem.* **2006**, *386*, 769–789. [CrossRef]
52. Ribani, M.; Collins, C.H.; Bottoli, C.B.G. Validation of chromatographic methods: Evaluation of detection and quantification limits in the determination of impurities in omeprazole. *J. Chromatogr. A* **2007**, *1156*, 201–205. [CrossRef]
53. Barra Caracciolo, A.; Grenni, P.; Cupo, C.; Rossetti, S. In situ analysis of native microbial communities in complex samples with high particulate loads. *FEMS Microbiol. Lett.* **2005**, *253*, 55–58. [CrossRef]

54. Barra Caracciolo, A.; Giuliano, G.; Grenni, P.; Guzzella, L.; Pozzoni, F.; Bottoni, P.; Fava, L.; Crobe, A.; Orrù, M.; Funari, E. Degradation and leaching of the herbicides metolachlor and diuron: A case study in an area of Northern Italy. *Environ. Pollut.* **2005**, *134*, 525–534. [CrossRef]
55. Grenni, P.; Barra Caracciolo, A.; Rodríguez-Cruz, M.S.; Sánchez-Martín, M.J. Changes in the microbial activity in a soil amended with oak and pine residues and treated with linuron herbicide. *Appl. Soil Ecol.* **2009**, *41*, 2–7. [CrossRef]
56. Barra Caracciolo, A.; Bustamante, M.A.; Nogues, I.; Di Lenola, M.; Luprano, M.L.; Grenni, P. Changes in microbial community structure and functioning of a semiarid soil due to the use of anaerobic digestate derived composts and rosemary plants. *Geoderma* **2015**, *245–246*, 89–97. [CrossRef]
57. Pietrini, F.; Di Baccio, D.; Iori, V.; Veliksar, S.; Lemanova, N.; Juškaitė, L.; Maruška, A.; Zacchini, M. Investigation on metal tolerance and phytoremoval activity in the poplar hybrid clone "Monviso" under Cu-spiked water: Potential use for wastewater treatment. *Sci. Total Environ.* **2017**, *592*, 412–418. [CrossRef] [PubMed]
58. Pietrini, F.; Passatore, L.; Patti, V.; Francocci, F.; Giovannozzi, A.; Zacchini, M. Morpho-Physiological and Metal Accumulation Responses of Hemp Plants (*Cannabis Sativa* L.) Grown on Soil from an Agro-Industrial Contaminated Area. *Water* **2019**, *11*, 808. [CrossRef]
59. Ugulin, T.; Bakonyi, T.; Berčič, R.; Urbanek Krajnc, A. Variations in leaf total protein, phenolic and thiol contents amongst old varieties of mulberry from the Gorizia region. *Agricultura* **2015**, *12*, 41–47. [CrossRef]
60. Chang, C.C.; Yang, M.H.; Wen, H.M.; Chern, J.C. Estimation of total flavonoid content in propolis by two complementary colometric methods. *J. Food Drug Anal.* **2002**, *10*, 178–182.
61. Hernandez, M.; Fernandez-Garcia, N.; Diaz-Vivancos, P.; Olmos, E. A different role for hydrogen peroxide and the antioxidative system under short and long salt stress in *Brassica oleracea* roots. *J. Exp. Bot.* **2010**, *61*, 521–535. [CrossRef]
62. Pietrini, F.; Zacchini, M.; Iori, V.; Pietrosanti, L.; Bianconi, D.; Massacci, A. Screening of poplar clones for cadmium phytoremediation using photosynthesis, biomass and cadmium content analyses. *Int. J. Phytoremediation* **2009**, *12*, 105–120. [CrossRef]
63. Marmiroli, M.; Pietrini, F.; Maestri, E.; Zacchini, M.; Marmiroli, N.; Massacci, A. Growth, physiological and molecular traits in Salicaceae trees investigated for phytoremediation of heavy metals and organics. *Tree Physiol.* **2011**, *31*, 1319–1334. [CrossRef]
64. Pajević, S.; Borišev, M.; Nikolić, N.; Arsenov, D.D.; Orlović, S.; Župunski, M. phytoextraction of heavy metals by fast-growing trees: A review. In *Phytoremediation*; Springer International Publishing: Cham, Vietnam, 2016; pp. 29–64. ISBN 9783319401485.
65. Sylvestre, M. Prospects for using combined engineered bacterial enzymes and plant systems to rhizoremediate polychlorinated biphenyls. *Environ. Microbiol.* **2013**, *15*, 907–915. [CrossRef]
66. Meggo, R.E.; Schnoor, J.L. Cleaning polychlorinated biphenyl (PCB) contaminated garden soil by phytoremediation. *Environ. Sci.* **2013**, *1*, 33–52. [CrossRef]
67. Qin, H.; Brookes, P.C.; Xu, J. *Cucurbita* spp. and *Cucumis sativus* enhance the dissipation of polychlorinated biphenyl congeners by stimulating soil microbial community development. *Environ. Pollut.* **2014**, *184*, 306–312. [CrossRef] [PubMed]
68. Doty, S.L. Enhancing phytoremediation through the use of transgenics and endophytes. *New Phytol.* **2008**, *179*, 318–333. [CrossRef] [PubMed]
69. Lugtenberg, B.; Kamilova, F. Plant-growth-promoting rhizobacteria. *Annu. Rev. Microbiol.* **2009**, *63*, 541–556. [CrossRef] [PubMed]
70. Ancona, V.; Barra Caracciolo, A.; Grenni, P.; Di Lenola, M.; Campanale, C.; Calabrese, A.; Uricchio, V.F.; Mascolo, G.; Massacci, A. Plant-assisted bioremediation of a historically PCB and heavy metal-contaminated area in Southern Italy. *N. Biotechnol.* **2017**, *38*, 65–73. [CrossRef]
71. Campanella, B.F.; Bock, C.; Schröder, P. Phytoremediation to increase the degradation of PCBs and PCDD/Fs. *Environ. Sci. Pollut. Res.* **2002**, *9*, 73–85. [CrossRef]
72. Whitfieldaslund, M.; Zeeb, B.; Rutter, A.; Reimer, K. In situ phytoextraction of polychlorinated biphenyl—(PCB)contaminated soil. *Sci. Total Environ.* **2007**, *374*, 1–12. [CrossRef]
73. Barra Caracciolo, A.; Ademollo, N.; Cardoni, M.; Di Giulio, A.; Grenni, P.; Pescatore, T.; Rauseo, J.; Patrolecco, L. Assessment of biodegradation of the anionic surfactant sodium lauryl ether sulphate used in two foaming agents for mechanized tunnelling excavation. *J. Hazard. Mater.* **2019**, *365*, 538–545. [CrossRef]

74. Barra Caracciolo, A.; Grenni, P.; Ciccoli, R.; Di Landa, G.; Cremisini, C. Simazine biodegradation in soil: Analysis of bacterial community structure byin situ hybridization. *Pest Manag. Sci.* **2005**, *61*, 863–869. [CrossRef]
75. Fagervold, S.K.; May, H.D.; Sowers, K.R. Microbial reductive dechlorination of Aroclor 1260 in Baltimore harbor sediment microcosms is catalyzed by three phylotypes within the phylum Chloroflexi. *Appl. Environ. Microbiol.* **2007**, *73*, 3009–3018. [CrossRef]
76. Imamoglu, I.; Christensen, E.R. PCB sources, transformations, and contributions in recent Fox River, Wisconsin sediments determined from receptor modeling. *Water Res.* **2002**, *36*, 3449–3462. [CrossRef]
77. Glick, B.R. Modulation of plant ethylene levels by the bacterial enzyme ACC deaminase. *FEMS Microbiol. Lett.* **2005**, *251*, 1–7. [CrossRef] [PubMed]
78. Low, J.E.; Whitfield Åslund, M.L.; Rutter, A.; Zeeb, B.A. Effect of plant age on PCB accumulation by *Cucurbita pepo* ssp. *pepo*. *J. Environ. Qual.* **2010**, *39*, 245. [CrossRef] [PubMed]
79. Sitko, K.; Rusinowski, S.; Kalaji, H.M.; Szopiński, M.; Małkowski, E. photosynthetic efficiency as bioindicator of environmental pressure in *A. halleri*. *Plant Physiol.* **2017**, *175*, 290–302. [CrossRef] [PubMed]
80. Rzepka, A.; Krupa, J.; Lesak, I. Effect of hypoxia on photosynthetic activity and antioxidative response in gametophores of *Mnium undulatum*. *Acta Physiol. Plant.* **2005**, *27*, 205–212. [CrossRef]
81. Horchani, F.; Aloui, A.; Brouquisse, R.; Aschi-Smiti, S. Physiological responses of tomato plants (*Solanum lycopersicum*) as affected by root hypoxia. *J. Agron. Crop Sci.* **2008**, *194*, 297–303. [CrossRef]
82. Iori, V.; Pietrini, F.; Bianconi, D.; Mughini, G.; Massacci, A.; Zacchini, M. Analysis of biometric, physiological, and biochemical traits to evaluate the cadmium phytoremediation ability of eucalypt plants under hydroponics. *iForest-Biogeosciences For.* **2017**, *10*, 416–421. [CrossRef]

© 2019 by the authors. Licensee MDPI, Basel, Switzerland. This article is an open access article distributed under the terms and conditions of the Creative Commons Attribution (CC BY) license (http://creativecommons.org/licenses/by/4.0/).

Communication

Design of a Smart System for Rapid Bacterial Test

Rajshree Patil [1], Saurabh Levin [2], Samuel Rajkumar [2] and Tahmina Ajmal [3],*

[1] Institute of Chemical Technology (ICT), Matunga, Mumbai 400019, India; rajshree@ffem.io
[2] Foundation for Environmental Monitoring, Shivajinagar, Bangalore 560001, India; saurabh@ffem.io (S.L.); srajkumar@ffem.io (S.R.)
[3] School of Computer Science and Technology, University of Bedfordshire, Luton LU1 3J, UK
* Correspondence: Tahmina.Ajmal@beds.ac.uk

Received: 18 November 2019; Accepted: 15 December 2019; Published: 19 December 2019

Abstract: In this article, we present our initial findings to support the design of an advanced field test to detect bacterial contamination in water samples. The system combines the use of image processing and neural networks to detect an early presence of bacterial activity. We present here a proof of concept with some tests results. Our initial findings are very promising and indicate detection of viable bacterial cells within a period of 2 h. To the authors' knowledge this is the first attempt to quantify viable bacterial cells in a water sample using cell splitting. We also present a detailed design of the complete system that uses the time lapse images from a microscope to complete the design of a neural network based smart system.

Keywords: microscope; bacterial contamination; water contamination; artificial intelligence

1. Introduction

Microbial contamination of drinking water is a daunting challenge that has severe impacts on human health [1]. According to the World Health Organization (WHO) statistics in 2017, 785 million people lack even a basic drinking water service, including 144 million people who are dependent on surface water. In India alone, annually around 37.7 million people are affected by waterborne diseases which result in 1.5 million child deaths due to diarrhea [2–4]. For the past few decades, across the world various organizations—governments, non-governmental organizations (NGOs), scientists—are actively working to understand the challenges and provide a sustainable solution for supply water fit for human consumption [5–7]. One of the United Nations' Sustainable Development Goals is "By 2030, achieve universal and equitable access to safe and affordable drinking water for all" [8]. The first step in these efforts is to get an accurate assessment of the water quality for the presence of bacterial contamination. Bacteria like *Escherichia coli* (*E. coli*), which has been universally accepted as an indicator of fecal contamination, forms a useful indicator of microbial water contamination [9,10]. The basic concept of using coliforms as indicator bacteria is due to its abundance in the feces of humans and other warm-blooded animals. If fecal pollution has entered drinking water, it is likely that these bacteria will be present even after significant dilution. Thus, the presence of pathogens is determined with indirect evidence by testing for indicator organisms such as coliform bacteria which is from the same source of pathogenic bacteria. Moreover, monitoring these indicator bacteria is safe, easy, and economical as compared to other pathogens [11].

However, the tests for detecting the presence of coliforms, like any other bacterial contaminant, takes 24 to 48 h for confirmation and requires a degree of technical expertise [12]. This duration is long for any remedial action to have an effect and this delay results in a rapid spread of disease. Moreover, these methods require transportation of samples to laboratories, which can be challenging in remote locations [13]. Thus, there is an urgent need for a rapid method for on-site detection and to simultaneously communicate contamination in drinking water. Most of the methods [14,15] rely on allowing the growth of bacteria in selective growth mediators in the presence of a specific substrate which atleast requires 24 h; then they quantify the resulting growth by an indication of gradients of turbidity or color in a liquid media which could be detected visually, or by using a photometer or camera. Alternatively, these can also be quantified as visible detectable colonies on a solid media, giving estimates of viable bacteria in the sample. These standard methods are bounded by some theoretical limitations in terms of the time needed to detect the results. These methods [16] can sense cell density only when the population reaches 10^6 to 10^7 CFU/mL (colony forming unit) per milliliter. The molecular biology methods, such as PCR [17,18], are exorbitantly expensive, require trained personal, and still need at least 4–6 h for data interpretation. Direct observation of microbial contaminants using microscopy, which is one of the oldest methods of detecting bacterial cells at the single cell level, can provide a solution. The bacterial identification method uses differential staining procedure for classifying bacteria in different groups (e.g., gram positive or gram negative, cocci or rods, etc.). However, despite its tremendous potential of direct observation of bacterial cells under the microscope, the method is not used extensively for monitoring indicator bacterial pathogens such as coliform in water bodies for two reasons: it requires a trained eye and it does not discriminate between living and dead cells.

Similarly, application of high-throughput microscopy such as atomic force and optical microscopy are commonly used method for monitoring bacterial division at the single cell level. This requires immobilizing cells via some means of physical attachment such as agarose gel or within polydimethylsiloxane (PDMS)-based microfluidic channels [19,20]. Physical attachment is required to confine the cells movement so these can be observed under the microscope. PDMS is usually used to develop microfluidic devices using microfabrication techniques. However, PDMS microfluidic devices are complicated to fabricate and expensive. Use of agarose gel is the inexpensive way to immobilize cells where the cells are trapped between a thin layer of agarose and a transparent glass slide or coverslip. The function of agarose is similar to that of agar powder most commonly used in microbiological growth media as a solidifying agent where growth media can also act as a source of nutrients for the immobilized cells. However, such a system gets dehydrated rapidly due to the evaporation of moisture and thus becomes disturbed in a short span of time.

The other major disadvantage of the simple microscopic method is that the morphological features of bacteria observed under the microscope may not be sufficient for specific identification of bacteria. However, with advances in fluorescence microscopy and by using different vital dyes, such as propidium iodide, it is possible to differentiate between viable and non-viable bacterial cells [21,22]. These dyes are also instrumental in eliminating interference of any physical contamination. Epifluorescence microscopy can be used to rapidly estimate total cell numbers in water samples by using dyes, such as 4,6-diamidino-2-phenylindole dihydrochloride, acridine orange, and SYBR Green I (SG), that bind to nucleic acids which differentiate between live and dead bacteria. In these techniques, specificity of bacterial detection can be achieved by using specific probes such as 16 s, rRNA probe, antibodies or aptamers specific for target bacterial population [23,24]. However, the regents and the cost of the instrument involved are very high [25].

Another limitation of the microscopy-based method is the low sample volume that can be used for analysis (at microliter level), hence this doesnot meet the WHO recommended limit of detection (i.e., 1 CFU/100 mL of sample) [26]. Concentration of bacteria from a large volume of water using membrane filtration and back elution in a small volume of water is one of the possible approaches to enhance the detection limit of microscopic observation [27].

The common factor between all bacterial measurement approaches is detection of viable bacterial cells under favorable growth conditions. In this work, we propose a method which still uses the traditional method of filtration to condense the amount of fluid and then specific growth media is used for bacteria cells to multiply. The proposed smart system uses a microscope lens together with a camera to observe their multiplication. The novelty of the idea is the capability to capture the splitting events (binary fission) of bacteria which usually occur in 30–60-min intervals if the cells are alive and surrounded by favorable conditions of growth (e.g., nutrient and temperature). Thus, here detection is carried out at a single cell level, which could occur within 1–2 h unlike in conventional methods that detect the bacterial population after a lengthy incubation period of 24–72 h. The concept is to immobilize concentrated bacterial cells in a small volume of media that are then continuously observed and recorded. Since bacteria are immobilized in a very small area the probability of observing bacteria under the field of view of microscope increases many folds as compared to the traditional turbidimetric method of sensing bacteria. The turbidimetric method requires that the bacterial growth reach at least 10^6–10^7 cells/mL to visibly detect their presence, whereas a bacterial growth of less than 100 CFU/mL can be observed under the microscope. Since this method is based on capturing the splitting event of bacteria, the dead cells present under the microscopic view are naturally eliminated in the detection process. Our initial findings indicate that an initial bacterial concentration of 1–10 CFU/mL can be detected within a 1–2 h window. In this study, we used *Escherichia coli* (*E. coli*) as a means of interest in water contamination monitoring. However, the methods used can be applied to other bacteria also with suitable changes.

2. Materials and Methods

Live-Cell Imaging of Bacteria

Escherichia coli NCIM 2277 obtained from National Culture Collection of Industrial Microorganisms (NCIM), Pune, India was used in this study. The stock suspension of bacteria was prepared by growing the bacterial strains in A1 agar (HiMedia® Laboratories, Mumbai, India) at 37 °C for 24 h. The grown cells were washed off using normal saline (0.85% sodium chloride in distilled water) and pelleted out by centrifugation at 6000 rpm for 10 min. The pellet washing procedure was repeated twice by centrifugation for 5 min using normal saline. The cell density of the culture was adjusted to obtain a final cell concentration in the range of 10^1–10^2 CFU/mL. During the test diluted stock cell suspension was used for preparing test water contaminated with 10 CFU/mL of *E. coli*. The experimental procedure involves filtering 100 mL of artificially contaminated water sample through a sterile 0.45μm membrane filter (Millipore, Bangalore, India). After filtration the bacterial cells trapped on the membrane filter were eluted back in 1 mL of sterile coliform specific growth media (i.e., Rapid HiColiform broth) (HiMedia® Laboratories, Mumbai, India). The presence of sodium lauryl sulphate makes the medium selective for coliform by inhibiting accompanying microflora, especially the gram-positive organisms. The trapped cells were then transferred in an Eppendorf tube of 2 mL capacity. Thus, now the concentration of bacteria per mL of sample increases to 100 fold to that of the original concentration. To 0.5 mL of this suspension, 0.5 mL of coliform specific agar media (containing 1% of agar) is added and mixed thoroughly. Before the mixture gets solidified it was loaded in-between the coverslip place on the graded area of the counting chamber so as to form a thin film of solid layer of immobilized bacterial cell below the coverslip. To prevent dehydration of this film a liquid growth medium was added to the side trough. The open end of the trough at four corners was sealed using araldite to prevent leakage of the liquid media before loading the slide. A heating pad was attached to the slide to create a warm environment which is required for optimal growth of bacteria. The heating pad consists of a heating element (resistive heating pad by SparkFun Electronics, Bangalore, India® was used here). A microcontroller-based thermostat was placed in a small plastic box to maintain the optimum multiplication for the target bacteria, typically this is either 37 °C or 44 °C. The system was powered by a 9V standard battery that supplies 9V/1Amp to maintain the required temperature. The temperature can be set to 37 °C and itvaries ±2 around the target value due to periodic on/off of the heating pad by the microcontroller. The temperature was controlled by the thermistor reading and three light-emitting diodes (LEDs) (red, blue, and yellow) are placed to give a visual indication of temperature. When the set-up starts, all the three LEDs, were in the on position; later on when the system stabilizes only one LED was on depending on temperature. A red LED indicates that temperature is higher than set, blue LED indicates when temperature is lower than set value, and yellow LED indicates normal functioning. The slide was fixed under the microscope (Labomed Lx 300i) and observed under 40× objective lens viewed using the camera (MICAPS, FERLAF, FO50). The bacterial multiplication was continuously recorded by using the auto capture tool at regular intervals of time. A schematic diagram of the microscope setup is shown below in Figure 1 together with two photographs of the experimental setup.

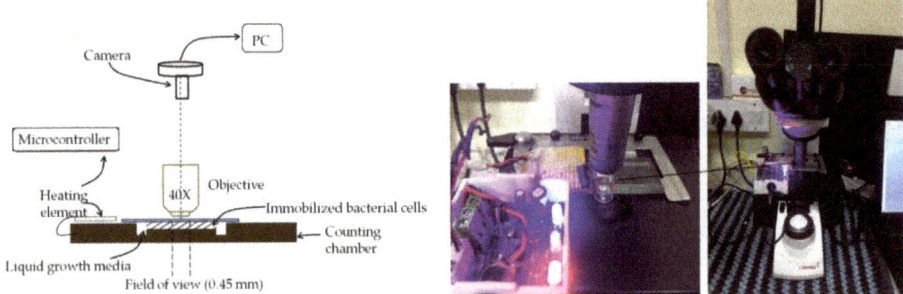

Figure 1. Experimental setup for observing bacterial multiplication.

The bacterial multiplication can be evaluated by observing any change in bacterial cell numbers or absolute change in elongation or movement of individual bacterial cells or their chain relative to a fixed reference point. The grids present on the counting chamber were used as a reference point because they remained stable on the final images.

3. Results and Discussion

Our aim is to develop a simple and rapid method for detection of bacteria in water samples by combining a traditional culture-based detection method and simple bright field microscopy. The idea is to reduce the detection time by using microscope so that an observation can be made at the single cell level. Immobilization of bacteria in the specific growth media on the surface of a specialized slide captured the growth pattern of single bacteria under the microscope. Figure 2 shows multiplication events over a period of 4 h with an initial concentration of bacterial cells in a water sample of 10 CFU/mL. During the experiment the immobilized agar layer did not show signs of desiccation and was thin enough to facilitate the exchange of air required for *E. coli*.

As seen in the Figure 2, in the initial image at 0:00 (H:min) no specific form of bacterial shape can be identified and only a few dark dots can be seen distributed on the surface. After 20 min one of the dark spots indicated by an arrow in the image started elongating and acquiring a distinct shape. In-between 40–50 min a recognizable shape of bacteria appears in the image. The 1:00 h image shows initiation of the splitting event and at 1:20 h, a complete splitting of bacteria in two distinct halves can be seen. This splitting time is considered as an evidence of the presence of viable bacteria in the water sample. As seen in Figure 2, all the images after 1:20 clearly show the multiplication of bacteria within an aggregate and the size of the lumps start increasing. At the end of 4:00 h the form is clearly identifiable and has a distinct morphology and clear movement. The time required to clearly distinguish the bacteria shape from other spots present in the initial image is 1–2 h and can be considered as detection time for this method. As mentioned in Section 2, the specificity of coliform detection is achieved in this experiment by using a commercial coliform specific growth media similar to that used in most of the traditional commercial coliform detection methods.

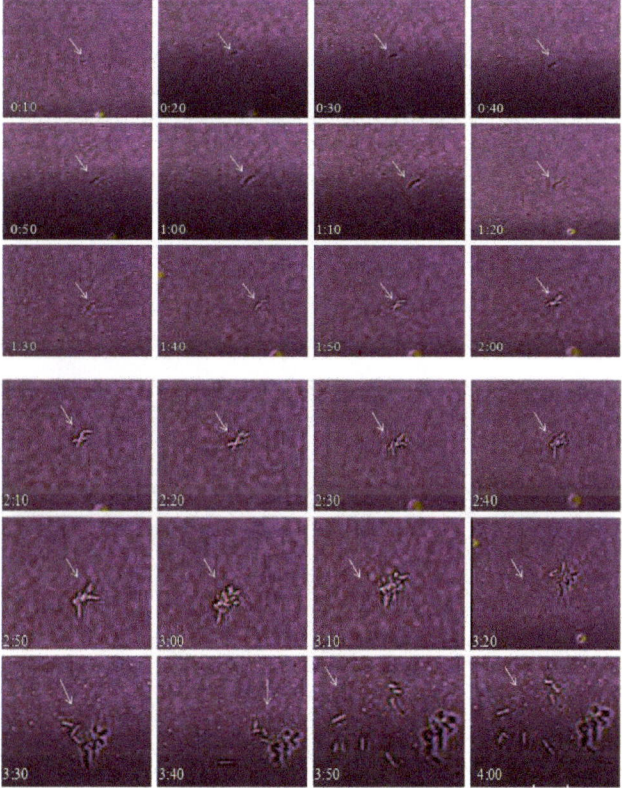

Figure 2. Bacterial multiplication observed over a period of 4 h at an interval of 10 min (H:min).

In this paper we aim to demonstrate the design of a test for coliforms in drinking water that can be completed in one hour. The device will need to be designed to be portable with the ability to transmit results to an online database, or to a user's phone. These additions and the proposed design of the whole system are discussed in some detail in the next sections. The concept rests on being able to detect bacterial growth quickly and correlating the amount of growth to the quantity of coliforms present in a sample. The work presented in the paper is to prove the concept using this methodology. The data shown in Figure 2 is used as evidence for the proof of concept on which we present our future design. The future work will be on developing a user-friendly, rugged prototype that can be used in the field. The subsequent section describes this proposed design in detail.

4. Configuration of the Proposed Device

The description given in the previous section concerns the preliminary investigation that was made to precede with the development of the design as shown in subsequent section. The device would comprise of an incubation chamber with a sample holder, a disposable sample tray, a microscope camera, and a microcontroller with software to track the bacterial growth at a single cell level in water (or food sample).

4.1. Test Chamber

The proposed design of the test chamber is given in Figure 3. It has a central, optically transparent area for viewing the splitting event of bacteria through the microscopic lens and recording by the camera module. The surface of this is treated with 3-aminopropyltriethoxysilane (APTES) to localized (immobilized) bacterial cells. This process has been developed inhouse. The test chamber is provided with an outer trough for holding selective growth media for target contaminant which provides a continuous supply of growth media to the cell localized at the central transparent area. This arrangement also prevents dehydration of the bacterial cell localized in the transparent area. During the test the pre-enriched sample processed in a custom-designed cartridge, as described in the subsequent section (Figure 4), is loaded in the test chamber through an input port. The test chamber is then placed in a sample holder of the device. The application of software will be developed to analyze at least 10 splitting events before predicting evidence of bacterial growth.

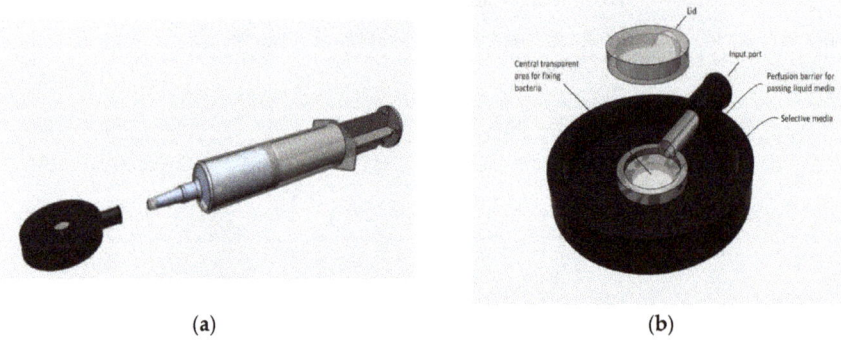

Figure 3. Internal design of proposed test chamber (**a**) Sample injection port (**b**) internal design of the test chamber.

The sample holder containing the pre-processed sample is then placed in an incubation chamber which maintains the temperature using a simple combination of a heating coil and a thermistor

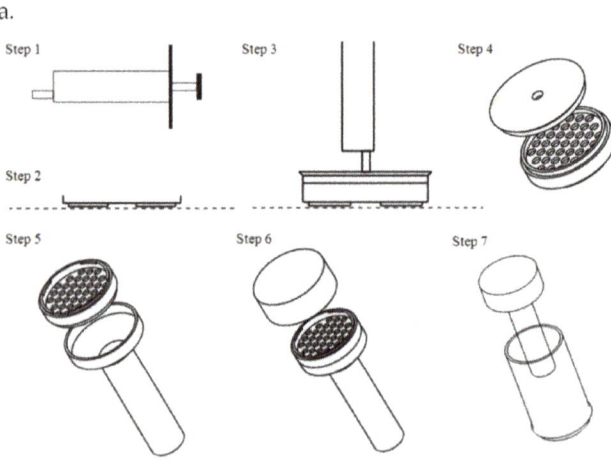

Figure 4. *Cont.*

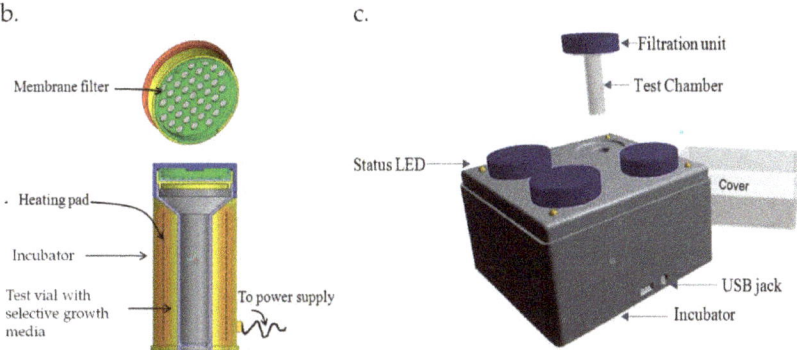

Figure 4. (a) Sample pre-enrichment test procedure: step 1: Collect 100 mL of water sample in a sterile disposable 2: set the disposable filter down on a flat surface; step 3: dispense the contents of the syringe into the filter allowing the water to drain out from underneath; step underneath; step 4: remove the syringe fixture part; step 5: unscrew the lid and place the disposable filter into the test chamber; step 6: close the lid back tightly; step 7: place the test chamber inside the single unit or multiple unit incubator. (b) Single unit incubator for incubating a single sample and (c) multiple unit incubator for incubating multiple samples. LED: light-emitting diode; USB.

4.2. Sample Pre-Enrichment

Figure 4 illustrates the eventual test procedure to be performed while testing field samples. The steps involve passing 100 mL of water sample, and filtering through 0.45 µm one-time use filter under pressure through a disposable syringe. The filter with the trapped bacteria would be enclosed in a disposable test chamber filled with a coliform selective growth medium. Finally, the test chamber can be placed in the portable incubator, which maintains a steady 37°C incubation temperature a for testing a single sample (Figure 4b) and for testing multiple samples (Figure 4c).

4.3. The Complete System

The complete system is illustrated below in Figure 5, indicating the various components necessary for optimum bacterial growth, its capture, analysis, and transmission. The proposed prototype is 10 cm in height and 4 cm in width. The camera module, placed at the bottom, takes a number of images of the sample through the lens attachment at regular intervals. These images are stored in the microcontroller for analysis by making use of artificial intelligence tools—image analysis and neural networks. These give a result within 1 h from the start.

The test results can then be sent to the user's phone via Bluetooth or USB. A phone app could be provided to communicate this data to a central database allowing for other actions such as sending alerts, real-time mapping, and analysis. Moreover, considering the cost of lenses and electronic components used in the system, the initial cost of the device could be in the range of 140–280 United States Dollars (USD), whereas the consumable cost per test would be 0.7 USD, which is comparatively minimal when compared to different low-cost kits available in the market for bacterial contaminants which are in the range of 0.5 to 21 USD [11,28].

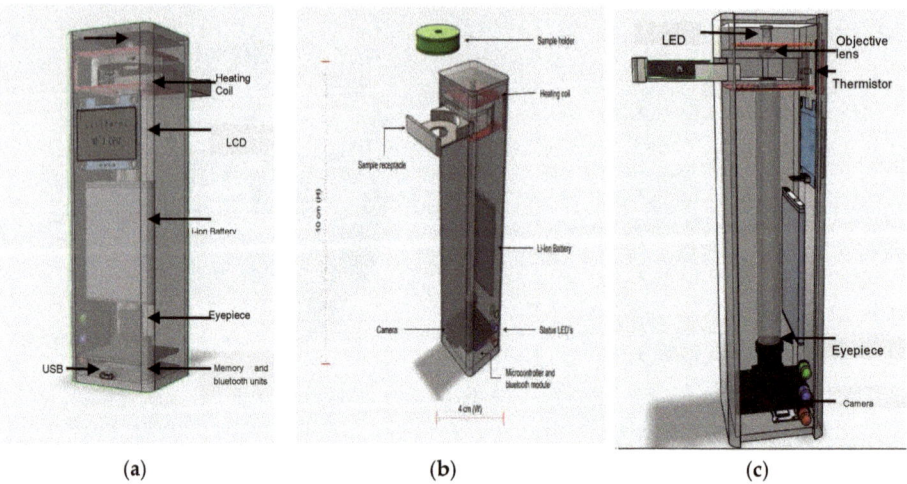

Figure 5. Design of proposed smart device. LCD: (**a**) Outer body of the device showing display, (**b**) test chamber holder within the device, (**c**) internal components of the device.

5. The Decision System

A BactImAS (Bacterial Image Analysis Software) is an open source software that can be used for cell segmentation and tracking, as well as storage, analysis, and visualization of acquired timelapse data. The detailed procedure and application of software is described by Mekterović et al, [29].

Developing an algorithm based on a convolution neural network to interpret the presence and absence of bacteria based on a splitting event will be the next step. The ANN (artificial neural network) system architecture shown in Figure 6 is suitable for this application. The software would be in the microcontroller system and would need to be trained using training images first. The system would analyze at least 10 splitting events before making the decision of bacterial growth.

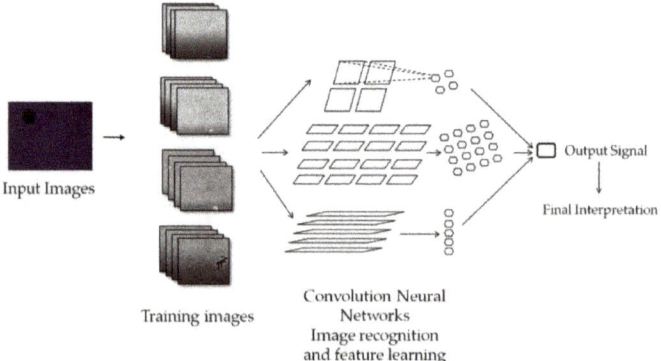

Figure 6. Neural network for tracking bacterial multiplication.

The images for training neural network were collected using time lapse microscopy at an interval of 10 min. Figure 7 shows two sets of these images after a duration of 1 h each. Figure 7a,b shows positive growth and Figure 7c,d shows negative growth. The images processed by threshold application are used to count the increase in bacterial growth over time. A simple count of black pixels, after applying image thresholding would increase over time if growth of bacteria takes place.

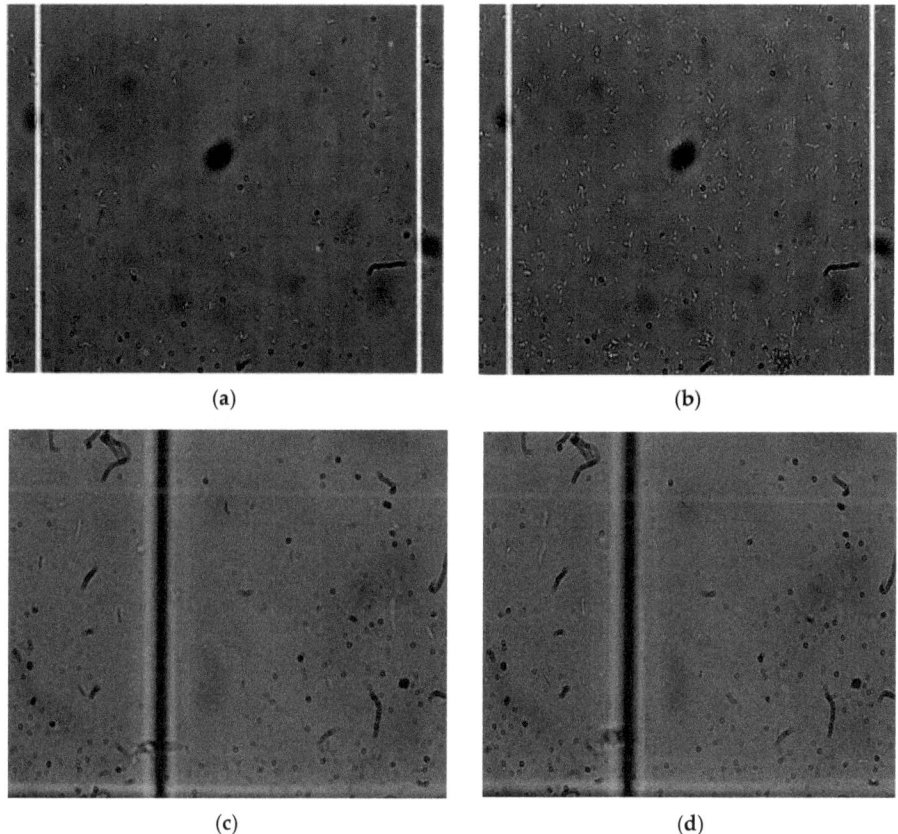

Figure 7. Training images: cell growth pattern of *E. coli*. (**a**) Positive images at t = 00:00 and (**b**) t = 01:00 h Negative training image at (**c**) 00:00 and (**d**) 01:00. Magnification 450×.

6. Conclusions

This paper describes the concept and initial findings for the design of a smart system for early detection of bacteria in a water sample. The system is based on the application of microscope image processing and neural network for rapid detection of bacterial contaminants in a water sample. If successfully productized, the device will offer a number of advantages over bacterial testing kits currently available in the market. Results are obtained much faster (1–2 h) than the time required for traditional tests (24–48 h) and can be automatically geotagged and uploaded online. This makes it much easier to survey and monitor water points. Setup and per test costs are much lower than what is required for a fully functional lab and thus easy to use for in-field situations. This provides an easy way to monitor water sources in rural areas for fecal contamination, and appropriate measures can then be taken by the government or NGOs. It can greatly reduce the incidents and spread of waterborne diseases as it would allow the source of contamination to be determined much quicker than by traditional means.

Such a device could make it an invaluable tool during disaster relief efforts, when determining the location of clean water is pertinent. Similarly, using the device, water infrastructure providers can ensure that the outputs of their filtration systems are attaining the prescribed standards. The device could be equally useful in detecting pathogens in food and beverages, thus preventing incidences of food poisoning. It could also be suitable for testing the sterility of food items in food processing industries. Moreover, applying such rapid methods would help companies to have better and faster control of raw materials and final products resulting in savings in time and costs. Such rapid methods can also offer a better reactivity throughout the manufacturing process.

The investigation on bacterial cell biology is important for a number of fields, including genetic engineering and biotechnology, synthetic biology, infectious disease research, antibacterial development, molecular and systems biology, and clinical screening. Current methods in microbiology address bacteria at a population level, with measurements averaged over millions or billions of individual cells and thus difficult to monitor changes at the individual cell level. However, the device proposed can do bacterial analysis at the single cell level and thus will provide more detailed information such as cell variation, dynamic response profiles, spatial information, and morphological features of bacteria in different environments and thus will be highly beneficial to academic and research institutes.

We present here proof of concept; however, more research is still required to make such an innovation acceptable in the market for field testing. To take it further, the next steps in this direction for the research team include testing the concept with real water samples to test the efficiency of the specificity of the system and validate the process as well as device prototyping and testing the data analytics system.

Author Contributions: R.P. conceived of the presented idea, design and performed experiment, studied different aspects of project, wrote the manuscript with support of T.A., S.L. designed the device and performed all electronic work required during the experiment. S.R. developed the theory of machine learning for bacteria detection, supervised the findings of this work. T.A. writted the manuscript, verified the analytical methods, technical inputs. All authors discussed the results and contributed to the final manuscript. All authors have read and agreed to the published version of the manuscript.

Funding: This research received no external funding.

Conflicts of Interest: The authors declare no conflict of interest.

References

1. Ashbolt, N.J. Microbial contamination of drinking water and human health from community water systems. *Curr. Environ. Health Rep.* **2015**, *2*, 95–106. [CrossRef] [PubMed]
2. Parsai, A.; Rokade, V. Water Quality Monitoring Infrastructure for Tackling Water-Borne Diseasesin the State of Madhya Pradesh, India, and Its Implication on the Sustainable Development Goals (SDGs). In *The Relevance of Hygiene to Health in Developing Countries*; Potgieter, N., Hoffman, A.T., Eds.; IntechOpen: London, UK, 2019. [CrossRef]
3. Holm, R.; Singini, W.; Gwayi, S. Comparative evaluation of the cost of water in northern Malawi: From rural water wells to science education. *Appl. Econ.* **2016**, *48*, 4573–4583. [CrossRef]
4. Tortajada, C.; Biswas, A.K. Achieving universal access to clean water and sanitation in an era of water *scarcity*: Strengthening contributions from academia. *Curr. Opin. Environ. Sustain.* **2018**, *34*, 21–25. [CrossRef]
5. Kuylenstierna, J.L.; Björklund, G.; Najlis, P. Sustainable water future with global implications: everyone's responsibility. In *Natural Resources Forum*; Blackwell Publishing Ltd.: Oxford, UK, 1997; Volume 21, pp. 181–190.
6. Batchelor, C.H.; Rama Mohan Rao, M.S.; Manohar Rao, S. Watershed development: A solution to water shortages in semi-arid India or part of the problem? *Land Use Water Resour. Res.* **2003**, *3*, 1–10. [CrossRef]
7. Hanchett, S.; Akhter, S.; Khan, M.H.; Mezulianik, S.; Blagbrough, V. Water, sanitation and hygiene in Bangladeshi slums: An evaluation of the WaterAid–Bangladesh urban programme. *Environ. Urban.* **2003**, *15*, 43–56.
8. Osborn, D.; Cutter, A.; Ullah, F.; Universal Sustainable Development Goals. Understanding the Transformational Challenge for Developed Countries. Report of a Study by Stakeholder Forum. May 2015. Available online: https://sustainabledevelopment.un.org/content/documents/1684SF_-_SDG_Universality_Report_-_May_2015.pdf (accessed on 10 December 2019).
9. Martin, N.H.; Trmčić, A.; Hsieh, T.H.; Boor, K.J.; Wiedmann, M. The evolving role of coliforms as indicators of unhygienic processing conditions in dairy foods. *Front. Microbiol.* **2016**, *7*, 1–8. [CrossRef]

10. Odonkor, S.T.; Ampofo, J.K. Escherichia coli as an indicator of bacteriological quality of water: An overview. *Microbiol. Res.* **2013**, *4*. [CrossRef]
11. Tallon, P.; Magajna, B.; Lofranco, C.; Leung, K.T. Microbial indicators of faecal contamination in water: A current perspective. *Water Air Soil Pollut.* **2005**, *166*, 139–166. [CrossRef]
12. Bain, R.B.; Elliott, M.; Matthews, R.; Mcmahan, L.; Tung, R.; Chuang, P.; Gundry, S. A summary catalogue of microbial drinking water tests for low and medium resource settings. *Int. J. Environ. Res. Pub. Health* **2012**, *9*, 1609–1625. [CrossRef]
13. Gunda, N.S.K.; Naicker, S.; Shinde, S.; Kimbahune, S.; Shrivastava, S.; Mitra, S. Mobile water kit (MWK): A smartphone compatible low-cost water monitoring system for rapid detection of total coliform and *E. coli*. *Anal. Methods* **2014**, *6*, 6236–6246. [CrossRef]
14. Rompré, A.; Servais, P.; Baudart, J.; De-Roubin, M.R.; Laurent, P. Detection and enumeration of coliforms in drinking water: Current methods and emerging approaches. *J. Microbiol. Methods* **2002**, *49*, 31–54. [CrossRef]
15. Warren, L.S.; Benoit, R.E.; Jessee, J.A. Rapid enumeration of Fecal Coliforms in water by a colorimetric beta-galactosidase assay. *Appl. Environ. Microbiol.* **1978**, *35*, 136–141. [PubMed]
16. Lewis, C.L.; Craig, C.C.; Senecal, A.G. Mass and density measurements of live and dead Gram-negative and Gram-positive bacterial populations. *Appl. Environ. Microbiol.* **2014**, *80*, 3622–3631. [CrossRef] [PubMed]
17. Law, J.W.; Ab Mutalib, N.S.; Chan, K.G.; Lee, L.H. Rapid methods for the detection of foodborne bacterial pathogens: Principles, applications, advantages and limitations. *Front. Microbiol.* **2015**, *5*, 770. [CrossRef] [PubMed]
18. Molina, F.; López-Acedo, E.; Tabla, R.; Roa, I.; Gómez, A.; Rebollo, J.E. Improved detection of Escherichia coli and coliform bacteria by multiplex PCR. *BMC Biotechnol.* **2015**, *15*, 48. [CrossRef]
19. Priest, D.G.; Tanaka, N.; Tanaka, Y.; Taniguchi, Y. Micro-patterned agarose gel devices for single-cell high-throughput microscopy of *E. coli* cells. *Sci. Rep.* **2017**, *7*, 17750. [CrossRef]
20. Joyce, G.; Robertson, B.D.; Williams, K.J. A modified agar pad method for mycobacterial live-cell imaging. *BMC Res. Notes* **2011**, *4*, 73. [CrossRef]
21. Fernández-Miranda, E.; Majada, J.; Casares, A. Efficacy of propidium iodide and FUN-1 stains for assessing viability in basidiospores of Rhizopogonroseolus. *Mycologia* **2017**, *109*, 350–358. [CrossRef]
22. Gao, P.; Sun, L.; Li, Y.; Zou, X.; Wu, X.; Ling, Y.; Luan, C.; Chen, H. Vital staining of bacteria by sunset yellow pigment. *Pol. J. Microbiol.* **2017**, *66*, 113–117. [CrossRef]
23. Haines, A.M.; Tobe, S.S.; Kobus, H.; Linacre, A. Finding DNA: Using fluorescent in situ detection. *Forensic Sci. Int. Genet. Suppl. Ser.* **2015**, *5*, e501–e502. [CrossRef]
24. Haines, A.M.; Tobe, S.S.; Kobus, H.; Linacre, A. Duration of in situ fluorescent signals within hairs follikes. *Forensic Sci. Int. Genet. Suppl. Ser.* **2015**, *5*, 175–176. [CrossRef]
25. Mishra, M.; Chauhan, P. Applications of Microscopy in Bacteriology. *Microsc. Res.* **2016**, *4*, 1–9. [CrossRef]
26. World Health Organization (WHO). *Guidelines for Drinking-Water Quality*; World Health Organization: Geneva, Switzerland, 2011.
27. Pettipher, G.L.; Mansell, R.; McKinnon, C.H.; Cousins, C.M. Rapid membrane filtration-epifluorescent microscopy technique for direct enumeration of bacteria in raw milk. *Appl. Environ. Microbiol.* **1980**, *39*, 423–429. [PubMed]
28. Patil, R.; Levin, S.; Halery, N.; Gupta, I.; Rajkumar, S. A smartphone-based early alert system for screening of coliform contamination in drinking water. *J. Microbiol. Biotechnol. Food Sci.* **2019**, 539–547. [CrossRef]
29. Mekterović, I.; Mekterović, D. BactImAS: A platform for processing and analysis of bacterial time-lapse microscopy movies. *BMC Bioinform.* **2014**, *15*, 251. [CrossRef]

© 2019 by the authors. Licensee MDPI, Basel, Switzerland. This article is an open access article distributed under the terms and conditions of the Creative Commons Attribution (CC BY) license (http://creativecommons.org/licenses/by/4.0/).

MDPI
St. Alban-Anlage 66
4052 Basel
Switzerland
Tel. +41 61 683 77 34
Fax +41 61 302 89 18
www.mdpi.com

Water Editorial Office
E-mail: water@mdpi.com
www.mdpi.com/journal/water

www.ingramcontent.com/pod-product-compliance
Lightning Source LLC
LaVergne TN
LVHW071944080526
838202LV00064B/6672